T0074197

Lecture Notes in Physics

Volume 943

The Lecture Notes in Physics

The series Lecture Notes in Physics (LNP), founded in 1969, reports new developments in physics research and teaching-quickly and informally, but with a high quality and the explicit aim to summarize and communicate current knowledge in an accessible way. Books published in this series are conceived as bridging material between advanced graduate textbooks and the forefront of research and to serve three purposes:

- to be a compact and modern up-to-date source of reference on a well-defined topic
- to serve as an accessible introduction to the field to postgraduate students and nonspecialist researchers from related areas
- to be a source of advanced teaching material for specialized seminars, courses and schools

Both monographs and multi-author volumes will be considered for publication. Edited volumes should, however, consist of a very limited number of contributions only. Proceedings will not be considered for LNP.

Volumes published in LNP are disseminated both in print and in electronic formats, the electronic archive being available at springerlink.com. The series content is indexed, abstracted and referenced by many abstracting and information services, bibliographic networks, subscription agencies, library networks, and consortia.

Proposals should be sent to a member of the Editorial Board, or directly to the managing editor at Springer:

Christian Caron
Springer Heidelberg
Physics Editorial Department I
Tiergartenstrasse 17
69121 Heidelberg/Germany
christian.caron@springer.com

More information about this series at http://www.springer.com/series/5304

Raúl Sánchez • David Newman

A Primer on Complex Systems

With Applications to Astrophysical and Laboratory Plasmas

 Springer

Raúl Sánchez
Departamento de Física, Grupo de Fusion
Universidad Carlos III de Madrid
Madrid, Spain

David Newman
Physics Department
University of Alaska-Fairbanks
Fairbanks, Alaska
USA

ISSN 0075-8450   ISSN 1616-6361   (electronic)
Lecture Notes in Physics
ISBN 978-94-024-1227-7   ISBN 978-94-024-1229-1   (eBook)
https://doi.org/10.1007/978-94-024-1229-1

Library of Congress Control Number: 2017964210

Printed on acid-free paper

This Springer imprint is published by Springer Nature
The registered company is Springer Science+Business Media B.V.
The registered company address is: Van Godewijckstraat 30, 3311 GX Dordrecht, The Netherlands

For the young person in all of us
who thinks all that matters is already known.
For the older person in some of us
who realizes that we hardly know anything
completely.

Preface

When Maria Bellantone from Springer first proposed that we wrote this book in the hot Spanish summer of 2011, we asked ourselves whether there was room for yet another book on complex systems. The truth is that there are quite a few good books out there that discuss complex systems and complexity. There are even some which do so in the context of plasma science, although mostly focusing on astrophysical plasmas. So why should we write another book? What needs do still exist out there that have not been properly addressed and that might justify writing, printing and selling another book on the subject? Why would any student or researcher feel a need to buy this book and to use it?

Our take on the subject of complex systems is that of researchers whose primary interest is not to understand complexity as an abstract concept but to identify the mechanisms within a particular physical system that make its behaviour complex. A first step is to prove that complex dynamics are actually at work. In some cases, this can be shown rather easily. Sometimes, even a simple inspection of the system is sufficient. That is certainly the case of swarming behaviour, or of the evolutionary processes that biological systems undergo, to name just a few. In the study of plasma systems, where both of us have carried out the majority of our work for the last three decades, things are rather subtle and much more complicated. For starters, experimental data is often not easily accessible. And when it is, it does not often correspond to *what one would have liked to measure* in the first place. Instead, it is *what is possible to measure* in the experiment, usually in the form of time series that require a significant amount of interpretation to make any sense of.

Neither of us considers himself an expert on stochastic theory, on statistics, or on the theory of complexity. We are plasma physicists by trade, with a large history in the study of turbulence in magnetically confined fusion plasmas, such as those inside a tokamak or a stellarator. Our profile is one of scientists that feel as comfortable when handling mathematical equations as when running numerical simulations in a supercomputer or when analyzing experimental data provided to us by friends and colleagues. As such, we need a clear—simple, but not dumbed down—description of the main ideas behind complex dynamics and succinct prescriptions of the best methods available to detect them and to characterize them. There are possibly better books out there that describe in painful detail the mathematical properties of fractional Brownian motion, the Langevin equation or the rigorous mathematical theory of fractional derivatives. Our approach in this book is not the one of any of those, very respectable, texts. We intend instead to cover the bare essentials on these topics, in a mathematically sound but not excessively rigorous way, and to blend them together with sufficient basic concepts of plasma physics as to form a mix that could be useful for plasma scientists of many flavours: from hardcore fusion scientists to space physicists and beyond, from pure theoreticians to computer scientists, from phenomenological modellers to pure experimentalists, and from aspiring graduate students to established experts. This is the way in which we would like that this book excels. We expect it to offer a useful experience that, in the sense previously described, cannot be found elsewhere.

We would like to conclude by saying that we are already too old to claim that no errors, misconceptions or mistakes have made it into our book. The only thing we can swear by is that, from the bottom of our hearts, we have worked hard to make this book the best possible and that we have not intentionally obscured or embellished any part of it, or selectively left out any unpleasant details for fear of our own contributions being looked at in a less favourable light. On the contrary, we have purposely tried to make it as clear, direct and easy to use as possible. Our hope is that our readers will benefit from it and consider it a worthwhile purchase. It is now up to them to judge our work and decide whether we have achieved successfully what we set ourselves to do, or whether we have failed some place along the way.

Leganés, Madrid, Spain   Raúl Sánchez
Fairbanks, AK, USA   David Newman
August 2017

Acknowledgements

This book has taken way longer to be completed that we ever thought possible. Initially, it was going to be a 2-year-long project that then became a 3-year-long one, then a 4-year-long one, a 5-year-long one and ended up taking 6 full years to complete. This continuous chain of unintended delays does not happen without causing all kinds of despair, suffering, back pain, teeth-grinding and general nagging that has had an impact not only on the both of us but on all the people around us, co-workers, friends and family as well. Therefore, it is to each and every one of them that we would like to extend our very special thanks, since without them this project would never have seen the light at the end of a very long tunnel.

First, we would like to thank our editor, Maria Bellantone, and the rest of the people at Springer. Maria's patience with us has been enormous since she first approached us in the summer of 2011, and her kind words of advice and support have been unending. It is certain that, just by ourselves, we would not have been able to stay with this (or any other) project for this long. So kudos to her and her team.

Secondly, we would like to acknowledge the constant support of the two academic institutions where we are lucky to teach and carry out our research: the Universidad Carlos III de Madrid in Spain and the University of Alaska Fairbanks, in the United States. Their Departments of Physics provide a suitable environment to carry out projects like this one. The constant interactions with students, researchers and colleagues that they make possible have certainly enriched our personal and professional lives as much as the outcome of this particular adventure.

Next, we would like to thank our many students, collaborators and colleagues in Spain, the United States and all around the world that have helped us shape the book by means of many suggestions and criticisms, by providing examples and corrections, by reading parts of the document and by pointing errors to be fixed and holes to be filled in some of its contents.

Special thanks go to our past and present PhD students, Jorge Alcusón, Isabel Fernandez, José Angel Mier, Douglas Ogata, Soma Panta, Hugo Rodríguez, Debasmita Samaddar, Luis Vela, Toon Weyens and Ryan Woodard. Although many of them started their PhD after this project was born, most become doctors well before it was finished due to the long process that giving birth to this book finally became. We also want to send many special thanks to our day-to-day research collaborators, in particular Andrés Bustos and José Miguel Reynolds, who read considerable parts of the manuscript, provided data and figures, detected countless mistakes and helped us make it much more readable.

We would also like to send our warmest thanks to our mentors, Benjamin Carreras, Luis García and Paul Terry, with whom we started to work in this "complex field" and from whom we have learned almost all that we know today, or, at least, the things among what we think we know that are actually correct.

We are also greatly indebted to the many wonderful colleagues with whom we have collaborated in the past, or remain doing so, in the field of plasma physics and in complex systems. Among them, we would like to mention people like Lee Berry, Iván Calvo, Luis Chacón, Viktor Decyk, Diego del-Castillo-Negrete, Guilhem Dif-Pradalier, Ian Dobson, Ivo Furno, Xavier Garbet, Joachim Geiger, Tobias Görler, Carlos Hidalgo, Guido Huijsmans, Steve Hirshman, Juan Antonio Jiménez, Jean-Noel Leboeuf, Alberto Loarte, Vickie Lynch, José Ramón Martín, Piero Martin, Yanick Sarazin, Gianluca Spizzo, Don Spong, David Terranova, Víctor Tribaldos, Boudewijn van Milligen, Andrew Ware, Nicholas Watkins and Pavlos Xanthopoulos. We also want to acknowledge the insight we gained by interacting with some of the best scientists that work on complexity in the context of solar and magnetospheric science and thank two international workshops organized by Markus Aschwanden in Bern during the autumns of 2012 and 2013, under the auspices of the International Space Science Institute.

It is time to continue now with the real deal, our families and closest friends. We will always be in debt with them. This is a debt that we will not be able to ever pay back, since they are the ones that had to cope with us while sinking in the deepest holes of despair and agony while producing this manuscript: Ramón, Antonio, Víctor, Susana, Paco, José Angel, Juan and José Luis in Madrid and Renate and Brian in Fairbanks. Countless coffees and meals, hours and hours of conversations, of kind smiles and of pats on the back. We could not have done it without them.

Our brothers and sisters, their spouses and children have always been there for us. Dara and Scott, their daughters Leah and Tessa and their son Ari Jose. Ruben and Sonia and their sons Iván and David. Nine-year-old Iván, in particular, helped his desperate uncle to cope with frustration by playing countless hours of PlayStation 4 videogames. His efforts are really appreciated. Our parents, Luis and Carmina and Ted and Sally. They were all greatly proud of their sons, and this work is a tribute to all their efforts towards making us good scientists and, more importantly, better people. Luis regretfully died in January 2017, a few months before this project was complete. He could not see it completed and published. This book is specially dedicated to his memory.

Last, but not least, our deepest gratitude goes to our respective spouses, Estefanía and Uma, and to Tilahun, that although he is the son of just one of us, who has been spoilt by the other one so much and for so long, that it actually makes no difference. We could try to enumerate the many things they have done for us during these years, but we would surely fail. We could praise the great work that Estefanía, a very talented designer, did in preparing many of the figures that are included in the book, or the many hours of conversation with Uma, a world-famous atmospheric physicist, regarding the best way to explain certain concepts or how to approach better our target audience. Instead, we prefer to praise their unending love and support, their smiles and their self-control to avoid punishing us for our unforgivable lack of attention while being busy trying to move this project forward. We will probably not have time to pay them back in our lifetime, but we will do our best to do so.

We have tried to remember and thank all the people that have contributed, in some way or another, to make this project a reality. Regretfully, we have certainly failed to include all of them in these paragraphs. We apologize to each and every one of the ones that do not appear here in advance. They should remember, before getting mad at us, that the fact that you were not mentioned does not mean that your help was not greatly appreciated.

Leganés, Madrid, Spain   Raúl Sánchez
Fairbanks, AK, USA   David Newman
August 2017

Contents

Part II   Complex Dynamics in Magnetized Plasmas

Acronyms

Lists of abbreviations used in the book:

$Ac(\tau)$	Autocorrelation function
ACS	Adaptive complex system
$Ad(\tau)$	Autodifference function
BBF	Bursty bulk flow (in the magnetotail)
BC	Box-counting (fractal dimension)
BTW	Bak, Tang and Wiesenfeld (sandpile)
CBC	Constant bin content (method)
CBS	Constant bin size (method)
$cdf(x)$	Cumulative distribution function
CLT	Central limit theorem
CME	Coronal mass ejection (in the Sun)
$Co(\tau)$	Covariance function
CTRW	Continuous time random walk
DFA	Detrended fluctuation analysis
DTEM	Dissipative trapped electron mode (in tokamaks)
$E(x)$	Exponential (probability density function)
$e(x)$	Stretched exponential (probability density function)
ELM	Edge localized mode (in tokamaks)
ETB	Edge transport barrier (in tokamaks)
fBm	Fractional Brownian motion
fGn	Fractional Gaussian noise
fLe	Fractional Langevin equation
fLm	Fractional Lévy motion
fLn	Fractional Lévy noise
$Fr(x)$	Frechet (probability density function)
fTe	Fractional transport equation
$Gu(x)$	Gumbel (probability density function)
ICF	Inertial confinement fusion
IMF	Interplanetary magnetic field

ITB	Internal transport barrier (in tokamaks)
ITG	Ion temperature gradient mode (in tokamaks)
KH	Kelvin-Helmholtz instability (in fluids and plasmas)
L(x)	Lévy (probability density function)
lhs	Left-hand side
LogCau(x)	Log-Cauchy (probability density function)
LogN(x)	Log-normal (probability density function)
LogSt(x)	Log-stable (probability density function)
Lp(x)	Laplace (probability density function)
mBm	Multifractional Brownian motion
MCF	Magnetic confinement fusion
mLm	Multifractional Lévy motion
N(x)	Gaussian (probability density function)
ODE	Ordinary differential equation
PDE	Partial differential equation
pdf	Probability density function
PIC	Particle-in-cell (numerical method)
Po(k,t)	Poisson (discrete distribution)
PP	Predator-prey (model)
rhs	Right-hand side
RW	Random walk
SF	Survival function (method)
sf(x)	Survival function
SOC	Self-organized criticality
SW	Solar wind
We(x)	Weibull (probability density function)

Part I
Characterization of Complex Systems

In the first part of this book, the most important properties and features usually found in complex systems are introduced, discussed and illustrated. The style attempts to be direct, clear and precise, purposely avoiding any excessive mathematical rigour. It is our intention to help readers to develop the sometimes elusive intuition needed to distinguish actual complex behaviour from a merely complicated one. In parallel, the first part of the book also provides a collection of popular analysis tools, together with relatively easy-to-follow instructions to use them, with which complex properties can be identified and quantified in real data. The running sandpile provides a safety line that interconnects all the chapters of the first part of the book, serving both as a paradigm for complex behaviour and as a testbed where many of the analysis tools are illustrated. In addition, a list of problems is proposed at the end of each chapter on which readers may try to practice, challenge and hone their new skills, if they so desire.

Chapter 1
Primer on Complex Systems

1.1 Introduction

Complexity has become one of the buzzwords of modern science. It is almost impossible to browse through a recent issue of any scientific journal without running into terms such as *complexity, complex behaviour, complex dynamics* or *complex systems* mentioned in one way or another. But what do most scientists actually mean when they use these terms? What does it take for a system to become complex? Why is it important to know if it is complex or not?

1.1.1 What Is a Complex System?

Most scientists that work on complex systems will state, if asked, their own idea of what a complex system is. Most probably, their answers will be different, at least in the specific details. However, most of them would probably agree that certain features are more frequent than others when complex behaviour is observed. It is the collection of all of these features that provide, in our opinion, the closest thing we have to a "definition" of what a complex system is.

We will try to illustrate the situation by means of an example. Everyone is familiar with biological **swarms**. Roughly speaking, **swarming** *is a behaviour exhibited by many animal species, by which individuals tend to move together in large aggregations, but as if governed by a single mind instead of by their own wills.* Swarming behaviour is exhibited, among others, by many insects, birds, fish and mammals. Let's consider, for instance, a bee hive [1]. If taken alone, individual bees move around minding their own business. Within the hive, however, individuals interact with their closest neighbours and these neighbours with their respective own and, quite magically, swarming behaviour sets on. The colony appears to move as a single entity and responds to any external stimulus extremely quickly. Why do

© Springer Science+Business Media B.V. 2018
R. Sánchez, D. Newman, *A Primer on Complex Systems*,
Lecture Notes in Physics 943, https://doi.org/10.1007/978-94-024-1229-1_1

Fig. 1.1 Examples of swarming behaviour. Starting from the left, upper corner and going around in a clockwise sense: bees swarm for protection and effective foraging; birds form flocks while migrating; jellyfish gather in big schools to survive large predators such as fish and seabirds; sheep form herds for warmth and protection [Credits: all free public domain images from Pixabay.com]

swarms happen? Probably, because the laws of natural evolution have dictated that swarms are more efficient than isolated individuals in order to provide protection, to fight off predators, to migrate over long distances or to optimize foraging. It is for similar reasons that mammals form herds, fish form schools and birds form flocks (see Fig. 1.1).

Swarms provide a good example of complex behaviour, and many of their properties are indeed shared by most complex systems. Consider, for instance, the fact that the features that define each of their **many individual constituents** (i.e., each bee) *become unimportant* to explain the dynamics of the system (the swarm) while acting as a whole. It is only the **type of interactions** that exist among the individual elements that matters. The interaction could be as simple as following the motion of your closest neighbour or the chemical trace it leaves behind! But these local interactions *permit the emergence of intricate, unexpected behaviours and patterns in an unguided, self-organized way, when the system has to react to external stimuli.* In fact, **self-organization** and **emergence** are, in our opinion, the two most defining properties of all complex systems.

It is apparently simple interactions among many individual constituents that drive the emergence of complex behaviours in many other systems. Some of them, known as **adaptive complex systems** (ACS), are even capable of tuning the strength and type of these interactions to better adapt to the external environment. Although

Fig. 1.2 Examples of complex systems. Starting from the left, upper corner and going around in a clockwise sense: weather systems and hurricanes, natural evolution and extinctions, forest fires, the stock market and economic crises, the spreading of wars, electric power grids, and (center) traffic in big cities [Credits: all free public domain images from Pixabay.com]

ACS are often found in biological contexts, they can also be found elsewhere. For instance, in technological applications, as it is the case of smart power grids, or in social systems, such as the global stock market (see Fig. 1.2).

Ultimately, there is always an **external force** (or principle) that drives complexity in a system. This force can vary greatly from one to another. It usually has to do with how the system interacts with the external environment to optimize some goal. In the case of swarming, it is probably the desire to better survive as a species. In other cases, it might be the need to dissipate, transport or distribute energy effectively, or the desire to make the largest possible profit at the lowest cost.

1.1.2 Examples of Complex Systems

Complex systems are plentiful in nature. As mentioned previously, they usually share some general properties. To start with, they are **open systems**, in the sense that they interact with their surroundings. They are composed of **many individual parts** that interact among themselves in such a way (nonlinearly) that multiple **feedbacks and feedback loops** are established. These feedbacks are the transmission chain through which a change in some part of the system conditions the evolution of other parts. As a result of these interactions, **emergence** of unexpected behaviour takes place, almost always through **self-organization**. Complex systems often exhibit **long-term memory**, in the sense that the past history of the system conditions its future behaviour. Another typical feature is the particular significance of **extreme events** in their dynamics. These events could be described as coherent, often unpredictable, phenomena that suddenly affect a very large part (if not all) of the system.

The **world stock market** provides another good example of a complex system [2]. It is an open institution where human and legal beings can buy or sell shares, and interact among them by exchanging currency. Of course, **greed** is the main driving force behind the stock market. Money is to be made by buying stocks when prices are low and selling them when they are high. The price of the shares is driven by their demand (the more demanded they are, the higher their price) and availability (the more available they become, the lower their price), which respectively provide examples of a positive and a negative feedback. Many other feedbacks exist in the world stock market. Some come from the fact that it is *coupled to other complex systems*. Thus, global geopolitics, technological progress, peer-pressure and corruption affect the way in which the world stock market evolves. As a result, the value of shares changes with time in a very complex, seemingly unpredictable, manner. But this evolution is not random. It is well documented that share prices exhibit a very long memory, with actual prices being affected by the previous history of the share value, as well as by the performance of other shares in the same business sector, and even by the behaviour of the global market as a whole. It is also particularly interesting (and painful!) to observe that the stock market can suddenly undergo large **market crashes** in which the value of a large fraction of all shares suddenly decreases very quickly.[1] These crashes are not random, but caused by the intricate interaction of the many events that take place in the system over long periods of time, thus making them very difficult to predict. The significance and the way in which these extreme events appear are both good illustrations of the kind of self-organized, emergent behaviour than can appear in a complex system.

Other important complex systems of particular interest to us are **the human brain** [3, 4], **human languages** [4, 5], **biological systems** [6, 7], **electrical power grids** [8] or **weather systems** [9]. As previously advertised, all of them are open systems formed by many elements that interact mostly locally and that are subject to multiple feedbacks and feedback loops. It is easy to identify emergent behaviours that appear via self-organization in all of them. In the case of the brain, the combination of the interaction between millions of neurons under the influence of external stimuli creates a complex system whose most remarkable emergent behaviour is the **ability of thinking**. In the case of languages, it is the need to communicate among humans and the interactions between those speaking the same language and the influence of other populations speaking other languages that drive the **creation and evolution of languages** over time. For biological systems, the

---

[1]At the same time, it is during these large crashes that new fortunes are more easily made!

interaction among species as well as among members of each species, driven by the need to feed and reproduce and the desire to survive, are behind the process of **natural evolution**. Extreme events happen, in this case, in the form of **mass extinctions**. And, of course, one should always consider the greatest emergent event of all: the appearance of **life** itself [7]. In the case of electrical power grids, the complicated networks that connect generators, transformers and many other elements to best provide our homes with energy are driven by the combination of the energy production units, the systems that store it and transport it, and its final users, both domestic and industrial. Emergent catastrophic events appear here in the form of **blackouts** that, in some cases, can leave large fractions of a whole continent without access to energy [8]. Finally, in weather systems, the turbulent interaction between many packets of atmospheric air, affected by the changes induced in their state by the heat coming from the Sun, the mass leaving in the form of rain or incoming from the evaporating ocean water, and the interactions with masses of land and ice, condition the temporal evolution of both short-term weather and long-term climate. Any emergent behaviours here? One could quickly mention **tornadoes** or **hurricanes**, for instance, at the level of the short-term weather. Or, at a much longer scale, the onset of **glaciation periods**.

Many other examples of complex systems can be easily found in the literature, in almost every field of science. A few more that we find particularly engaging are, for instance, the dynamics of **earthquakes** [10], the behaviour of **forest fires** [11], the workings of **the human body** and its **immune system** [3], the dynamics of **telecommunication grids** [12], the **internet** [13], the **history of wars** [14], the dynamics **manufacturing processes and logistics** [15], the spreading of **political ideas** or **infectious diseases**, or the behaviour of **traffic in large cities** [16].

1.1.3 Complex Systems in Plasma Science

Plasmas, that will be the focus of the second part of this book, provide another playground where complex behaviours can often emerge (see Fig. 1.3). If one keeps in mind the discussion we just had, it is easy to understand the reasons. Loosely speaking, plasmas are composed of charged particles (ions and electrons) that interact among themselves in such a way that they are able to maintain (a close to) overall spatial charge neutrality. These charges, in their motion, modify any electric and magnetic fields that might confine them as well as those through which they interact among themselves. The situation is thus ripe for the establishment of feedbacks and feedback loops of all signs. Not surprisingly, emergence and self-organization abound in plasma systems. We will enumerate a few examples now, many of which will be revisited in later chapters.

Starting at some of the largest scales in our universe, one could first consider the case of **galaxies** [17], in which a partially ionized background plasma and billions of stars interchange mass and energy in the presence of multiple feedbacks that couple all scales, from the smallest ones (related to the interactions between

Fig. 1.3 Examples of plasma complex systems. Starting from the left, upper corner and going around in a clockwise sense: the formation of stars within dust clouds, the flaring Sun, galaxy formation and its organization within clusters, the Earth's magnetosphere and its auroras, and a tokamak reactor (in this case, the JET tokamak in the UK) used in magnetic confinement fusion [Credits: nebula (public domain image from *Pixabay.com*); flaring Sun (© ESA/NASA - SOHO/LASCO); Pinwheel galaxy (©ESA/NASA - Hubble); JET tokamak (© EFDA-JET, Eurofusion.org); auroral view from space (© ESA/NASA - ISSS; photo taken by Scott Kelly)]

stars) to the largest ones (intragalactic distances within galaxy clusters). The local availability of mass, the evolution of its metallicity, the local interactions mediated by gravitational and tidal forces and the many heating and cooling processes at work (star explosions, nuclear fusion, radiation, plasma compression, accretion, etc.) are some of the main players. It is apparent that galaxies eventually self-organize themselves internally, often producing beautiful spiral patterns that are still not well understood. They also order themselves within very large clusters that may confine many millions of them. Complex behaviour has often been invoked to explain the resulting spatial distribution of galaxies inside these clusters, that exhibits an apparent fractal structure [18], since fractality is a well known trademark of complexity (see Chap. 2). In fact, observations like these have even lead to the recent proposal of fractal theories of the cosmos, although it must be acknowledged that these ideas still have a minority support among cosmologists [19].

It has also been suggested that complex dynamics could be at play in the **accretion disks** that form around black holes and some massive stars, as they engulf mass and energy from neighbouring objects such as other stars, interplanetary nebulae and others [20]. Depending on the mass of the central object, the disk material might be in a state that would range from partially to completely ionized. Magnetic fields are also thought to be present, mostly in a direction perpendicular to the plane of the disk. The reason why complex behaviour has been suggested as relevant in this context is the well documented observation of very long **power-law decaying tails** in the statistics of the energy of the X-ray bursts that accretion disks produce, another typical signature of complex dynamics (see Chap. 3).

Individual **stars** also seem to exhibit behaviours characteristic of a complex system [21]. These balls of very hot plasma burn huge amounts of hydrogen in order to keep the thermal pressure that prevents their gravitational collapse. The energy and heat produced is transferred to the surface of the star via several mechanisms that include radiation, convection and conduction. Some of this energy eventually leaves the star in the form of **flare** and **coronal mass ejection** (CME) explosive events, in which local reconnection of the surface magnetic field appears to take place [22]. Similarly to the case of accretion disks, the statistics of the energy released in flare events consistently exhibit a **power-law decay** for several decades [23, 24]. In addition, stars have long puzzled us with their ability to self-generate and sustain their own magnetic field [22, 25], as do also galaxies and some planets. The **stellar dynamo** is another example of emergent behaviour. We will discuss solar flares and CMEs in more detail in Chap. 7.

Complex behaviours involving plasmas are also found at planetary scales. For instance, in **planetary magnetospheres**. The magnetosphere is the region around a planet where the motion of charged particles coming from outer space is modified by the planetary magnetic field. In the case of our Earth, the majority of these particles come from the **solar wind**, that deforms its dipole-like magnetic field by forming a long magnetotail that extends away from the Sun for very long distances [26, 27]. The magnetosphere can be disturbed by any solar flare and CME that leaves the Sun in the direction of the Earth, since they can lead to violent reconnections of large regions of the magnetotail. These events, known as **geomagnetic storms**, can affect communications on the Earth's surface but are also responsible for the beautiful aurora seen over the Earth's poles. In addition to storms, smaller disturbances of the magnetotail, known as **geomagnetic substorms**, do also happen in the absence of large flares or CMEs. Substorms are apparently driven by the constant, intermittent forcing provided by the solar wind. Although they are rather unpredictable, substorms are far from random. It has been known for some time that their statistics also exhibit self-similarity (as does the aurora's intensity!) and that their temporal indices exhibit long-term correlations and other fractal characteristics, which points to some kind of complex dynamics [28]. Geomagnetic storm and substorms will be discussed in more detail in Chap. 8.

Last, but not least, we finally arrive to **laboratory plasmas**, such as those confined in a tokamak in order to produce fusion energy. Tokamak plasmas also appear to exhibit rather interesting emergent behaviours in regimes particularly relevant for reactor operation. **Tokamaks** are toroidal magnetic configurations that confine a fully ionized plasma by means of a poloidal magnetic field, generated by the current flowing in the plasma in the toroidal direction, and a toroidal magnetic field which is generated by external coils [29]. For some time now, it has been

known that tokamak plasmas exhibit so-called **canonical profiles** for sufficiently
high power. That is, the radial profiles of plasma temperature and pressure seem
rather insensitive to the strength or location of the external heating. This regime of
confinement is known as the L-mode. While in this state, the plasma energy leaks
out of the toroidal trap at a rate which is much larger than what is predicted from the
expected collisions among plasma particles. This excess has been long attributed
to the action of **plasma turbulence**. Even more interestingly, the scaling of the
energy leaking rate with the tokamak radius is rather odd, being much worse than
what should be expected if energy was being transported out of the device by some
turbulence-enhanced eddy diffusivity. Some authors have attributed this behaviour
to the rather singular features that transport seem to be endowed with when the local
plasma profiles sit very close to the threshold values that determine the onset of local
turbulence [30, 31]. In this so-called **near-marginal** regime, turbulent dynamics
do certainly appear complex, as will be argued in Chap. 6. Luckily for the future
of fusion energy production, tokamak plasmas seem to spontaneously transit into
a better confinement regime as the external heating is further increased. In this
enhanced confinement regime (known as H-mode), the plasma edge is affected by
a high poloidal and toroidal rotating motion (known as a **zonal flow**) over a narrow
region known as the **edge transport barrier** [32, 33]. This self-organization of the
tokamak plasma constitutes another example of emergent behaviour that will be
discussed in Chap. 9.

1.1.4 Complexity Science

We conclude this introductory section by making some general remarks about this
relatively new branch of science that investigates the properties of complex systems.
That is, of systems that, due to the way in which their many parts are interconnected
and how they interact with their external environment, experience the **emergence** of
unexpected collective behaviours and features in an unguided, **self-organized** way.

The first important point to make is that, from a *formal perspective*, complexity
science is rather different from traditional theories such as Classical Mechanics,
Electromagnetism or Quantum Mechanics. Any of these classical theories is based
on a finite number of unprovable axioms, that are either inferred from observation (*inductive approach*) or taken for granted, pending later confirmation from
experimentation of the outcomes of the theory (*deductive approach*). Theorems
and predictions are then derived rigorously from the axioms, sometimes with the
aid of careful approximations to enable further analytical progress. Finally, these
predictions are contrasted against the real world, process through which the validity
of the axioms—and of the approximations that made the derivation possible—is
confirmed.

Complexity Science, on the other hand, does not work in this way. It does
not follow a systematic path starting from a few basic axioms. Instead, it is built
organically, typically starting with the careful observation of the overall behaviour

of the systems under study. In a sense, one could even say that most approaches to complexity[2] can be categorized within at least one of the following strategies [34]:

1. *the determination of whether a system is complex or not*;
2. *the quantification of how complex the system is*;
3. *the understanding of how the system became complex*;
4. *the investigation of the consequences of being complex.*

The first approach is mostly descriptive. To determine whether a system is complex one basically looks for self-organized emergent behaviours. In some cases this is quite evident, as in the case of swarms. In others, the emergent feature may be more subtle or the system of interest may be not easily accessible (say, stars or galaxies). Then, one often looks for as many of the characteristic features of complex systems as possible. Many tools are available for this task: *correlation analysis*, *Fourier analysis*, *statistical analysis*, *time series analysis* and many more. We will have an opportunity to discuss many of these tools soon.

The quantification of the degree of complexity of a complex system is however, more subjective and controversial, since there is not such a thing as a unique definition of a measure of complexity. Each author uses its favourite one, of which many examples exist: *entropies* of different kinds, *fractal dimensions*, *fractional exponents*, *scale-invariance measures*, *information theory*, etc.

Finding out how and why the system became complex, on the other hand, is a completely dynamical task. One tries to discover why complex behaviour appears by investigating the interactions between the parts that form the system and from the knowledge of the physical, biological or social mechanisms at work. Concepts such as *self-organization*, *criticality*, *feedback loops*, *free energies*, *thresholds* and so on are often called upon.

Finally, the investigation of the consequences of being complex is the most predictive of all the aforementioned approaches, and possibly the more important one for engineering and practical applications. How it is carried out, depends on the system under study and on which our predictive needs are. These predictions, however, are often quite different from what more conventional theories usually provide.[3] Most of the time, they come in the form of statistical laws and trends. Predictions can also be made about what qualitative changes in dynamics could be expected if the system is manipulated in certain ways, which could have important practical applications. In particular, rather trustful predictions are often possible about the emergent, large-scale, ordered features that appear in a complex system, even if the detailed evolution of its state is out of reach. That is the reason why it is possible to predict, with some degree of confidence, what the climate will be

---

[2]Many would even go further and claim that there are as many approaches as there are investigators working in this context!

[3]In fact, some authors feel that the complex approach is useless to predict the future. We feel that this criticism is however rather unfair since, even if the level of quantitative predictability associated to classical theories is well beyond the capabilities of complexity theory, a certain level of predictability does indeed exist.

like decades from now, even if the local weather cannot be predicted beyond a few days. Or why the changes in the overall mortality in city traffic resulting from some change in transportation regulation can be estimated, even if the fate of a particular driver is completely unpredictable.

To sum up, it should be apparent to the reader that complexity science is an attempt at enabling a level of understanding of the dynamics of a system much deeper than what could be obtained by simply considering the detailed evolution of the governing dynamical equations. It does it in a somewhat unorthodox way, that often makes it difficult to derive specific quantitative predictions, but that permits a more intimate grasp of what is actually happening. In addition, it also provides tools and methods to look for and to establish the existence of certain features and behaviours, typical of complex systems. These tools can be of great value in practical applications. A favourite of ours is their use in **verification and validation** exercises of available numerical models for physical phenomena. Complexity science can help to carry out these tests at a dynamical level that is out of reach of more traditional theories. Take the case, for instance, of any of the huge numerical codes that try to simulate the Earth's climate by solving hundreds of coupled ordinary differential equations that theoretically describe the joint dynamics of the atmosphere and the ocean, as well as of ice and land masses. In order to fit these calculations within the supercomputers available, severe simplifications must be done in the models. There is a risk, however, that these simplifications may inadvertently restrict the kind of dynamics that the simulation is able to capture, restricting the validity (and usefulness!) of its predictions. The good news is that, at least in principle, the same tools that complex scientists use to detect complex features in real Earth climate data could be used to check whether similar behaviour is captured or not in the simulations. In this way, complexity science can provide additional tests of confidence on the numerical models and help to identify their shortcomings. Similar statements could be made about many other fields beyond climate science, such as fusion or space plasma simulations.

1.2   Key Concepts in the Study of Complex Systems

There are several concepts that appear recurrently when discussing complex systems. We have chosen to categorize them within three large classes: (1) **defining properties** that characterize a system as being complex; (2) **basic ingredients** that are *often* present in systems that exhibit complex dynamics, and (3) **emergent features** that *may* be exhibited by a complex system.

1.2.1 Defining Characteristics

By defining characteristics of complexity we mean those fundamental properties that a complex system should *always* exhibit. They are, in our opinion and that of many others, **self-organization** and **emergence**.

1.2.1.1 Self-organization

A system is said to self-organize when it *spontaneously rearranges itself (spatially, temporally or spatio-temporally) in a purposeful manner in order to better accomplish some goal*. In physical systems, this goal often is the efficient dissipation of the incoming energy, given the constraints imposed by the laws of Physics or by its boundary conditions [35]. Many other goals are also possible, however, particularly for non-physical systems. For instance, in biological systems the final goal is often survival. In economical systems, it is to maximize profits. In political systems, to maximize power or control.

A natural consequence of self-organization is that the *internal order of the system increases*. For that reason, it has sometimes been argued that self-organization somewhat contradicts the mandate of the second law of Thermodynamics, at least in the sense that it opposes the tendency towards maximizing disorder that is linked to an ever increasing entropy. Since such a violation is impossible, any rise of complexity must always be accompanied by the simultaneous exportation of disorder to the surroundings. This is the reason why complex systems are usually **open systems driven** from the outside. Furthermore, it is the external drive that is ultimately responsible for keeping the system at its complex state, in which energy is dissipated most effectively given the constraints of the system. If the drive is removed, complexity eventually disappears [36].

1.2.1.2 Emergence

Emergence is the process by which *novel and a priori unexpected patterns, entities, properties or behaviours arise in a system, usually at a level higher than the one at which the interactions between its constituents take place*. Emergence takes place only if the nature of the interactions allows the establishment of **feedbacks and feedback loops** at these higher levels, through which parts of the system react to changes in others. These actions and reactions ultimately drive self-organization. Since the number of possible interactions that can be established in a system grows quickly with the number of its constituents, complex systems are usually **composed of many elements**. It must be noted, however, that it is not just the sheer number of interactions that matters, but their ability to constitute feedback loops. Or in other words, not every system with many constituents behaves as a complex system. On the other hand, the formation of a tepid network of feedback loops is what makes prediction of the system behaviour extremely difficult, if not virtually impossible.

1.2.2 Basic Ingredients

Basic ingredients of complexity are those elements that are often present in systems that end up exhibiting some kind of complex dynamics. The reader should be aware that not every element discussed next will be present in every system. However, the detection of their presence should always ring a loud bell in our minds to warn about the possibility of complex dynamics being at work.

1.2.2.1 Openness

Most complex systems are *open and driven in the sense that they can interact with their surroundings*. In physical systems, this usually means that the system can interchange mass, momentum or energy with the outer world. In other cases, the interaction may take the form of an interchange of currency, the influence of external stimuli, the availability of food, etc. The need for openness is ultimately dictated by the second principle of Thermodynamics, as we discussed previously.

1.2.2.2 Non-determinism

Most complex systems are *non-deterministic in the sense that their future evolution is not uniquely determined by their previous state, but also by the action of its surroundings, which is often unknown*.[4] In spite of their non-deterministic nature, it is interesting to note that many of the emergent features that will be discussed later do admit a certain level of prediction.

1.2.2.3 Dynamical Feedbacks

In the theory of dynamical systems, a **feedback** is an action by which *the output of one process serves as input to another*. A **feedback loop** consists of any number

---

[4]It is interesting to note that non-determinism is often invoked to differentiate complex systems from chaotic ones. Indeed, ***chaotic systems*** *are usually deterministic, closed and low-dimensional*, while ***complex systems*** *are often non-deterministic, open and high-dimensional*. This distinction also relies on the fact that chaotic systems are usually simplified mathematical models, whilst complex systems are often real systems. It is however worth pointing out that when we *simulate* complex systems in a computer, nondeterminism disappears since computers are perfectly deterministic. However, we can still observe many emergent features. This is the case, for instance, of any numerical simulation of a fully-developed turbulent system with a large number of degrees of freedom. In the computer, these simulations are driven by a deterministic source (even pseudo-random numbers are deterministic!). However, they still behave like a complex system in many ways. One is the development of a scale-free inertial range; another, the exhibition of long-term correlations and scale-free statistics. This fact suggests that openness and high-dimensionality are probably stronger requirements than non-determinism in order to achieve complex behaviour!

of processes successively connected by feedbacks, with the output of the last one serving as input of the first process of the chain. A loop can consist of just one element subject to its **self-interaction**. Or of two elements subject to their **mutual interaction**. Or it could involve many elements, each one successively affecting the next in line via a **directed, one-way interaction**.

It is traditional to distinguish between **positive** and **negative** feedbacks. A feedback is **positive** if the two processes that are connected tend to change *in phase*. For instance, the price of shares in the stock market and their demand are in phase, since a higher [lower] demand drives a higher [lower] price. Feedback is **negative**, however, if both processes tend to be *out of phase*. That is, one is suppressed [increased] if the other grows [decreases]. Going back to the stock market example, the price of shares and their availability are connected through a negative feedback. The more [less] shares are available, the less [more] will buyers be willing to pay for them.

A positive feedback promotes growth and thus, change. In complex systems, it is through positive feedbacks that small local variations can end up growing into the much larger structures or patterns that we describe as emergent. In contrast, *negative feedbacks introduce constrains that limit or suppress growth or change*. Therefore, negative feedbacks can, if dominant, suppress the appearance of emergent features in complex systems. At the same time, negative feedbacks also enable the maintenance, over long periods of time, of any emergent property that might have appeared during a period of dominance of positive feedbacks.

It is also traditional to introduce the concept of **positive and negative feedback loops**. Loops may connect any arbitrary number of processes but, for simplicity, we discuss feedback loops that connect just two processes. In the **two-process positive loop**, the two feedbacks that form it must have the same sign. That is, both will be either positive or negative. This means, that the growth [decrease] of the first process would drive the growth [decrease] of the second, which would then also drive the growth [decrease] of the first one. Clearly, positive loops, if unchecked, lead to uncontrolled growth or decrease, and eventually to a catastrophe. For example, in the case of the stock market, a large and sudden decrease in the price of share may decrease its demand, if buyers perceive it as an unsafe investment, which will further decrease the value of the share. If unchecked, the share will soon turn worthless. When this happens on a massive scale, a global economic crisis follows! **Two-process negative loops**, on the other hand, are formed by two feedbacks with different signs. For example, a sudden increase in the demand of a particular share will increase its price (positive feedback), but the larger price will ultimately diminish its demand (negative feedback) as less and less buyers can afford to buy them. Clearly, when more than two processes are involved in a loop, negative feedback loops are easier to establish than positive ones, since a positive loop always requires all feedbacks to be of the same sign.

1.2.2.4 High-Dimensionality

Complex systems have a very large number of dimensions. These dimensions refer however to *the number of components of the system*, not to the dimensionality of the space in which the system lives. This is easier to visualise if one thinks, for instance, of the numerical representation of a weather system, whose elements are the many (say N) parcels of air that fill the atmosphere and whose velocity, pressure or temperature we would like to track over time. The position and velocity of each of these parcels in the three-dimensional space the atmosphere occupies is given by three numbers each, and thus $6N$ numbers are required to establish the state of the atmosphere at any given time. The number of dimensions (or **degrees of freedom**) of the **phase space** of the problem is thus said to be $6N$ and is usually very large for complex systems.[5] This need stems from the fact that the number of possible interactions and feedback loops grows very quickly with the number of elements.[6] Since the fraction of feedback loops that are positive diminishes quickly when the number of interconnected processes increases, the probability of extreme events, although still significant when compared to non-complex systems, remains relatively small. Otherwise economic crisis would happen every other month, instead of being separated by decades.

1.2.2.5 Nonlinearity

The term "nonlinear" often appears in relation to complexity. However, being nonlinear is a mathematical, not a physical property.[7] Thus, the use of the term in this context has to do with the fact that the dynamical equations that describe most complex systems, either ordinary (ODEs) or partial differential equations (PDEs), are almost inevitably nonlinear. The reason is that *feedback loops involving two or more elements of a system often enter its evolution equations through non-linear terms*.

To illustrate the relation between feedbacks and nonlinearities, we will use a famous example, the *predator-prey (PP) model*.[8] The version of the model

---

[5]The set of partial differential equations that describe the weather system have, in fact, an infinite number of dimensions since $N \to \infty$ as the limit of zero parcel size is taken.

[6]In contrast, chaotic systems are usually low-dimensional systems defined by a small number of ordinary differential equations, usually $N < 10$.

[7]Nonlinear means that the *superposition principle*, that states that the any linear combination of solutions of a problem is again a solution of the same problem, is no longer valid.

[8]The reader should be aware that this predator-prey model is low-dimensional, closed and deterministic, and does not exhibit complex dynamics. In fact, the version described here is not even chaotic since less than two interacting populations are considered. We feel, however, that thanks to its simplicity, this model works great to illustrate the connection between nonlinearities and feedbacks and feedback loops. It is also great to introduce the concept of threshold, that will be of great importance later on.

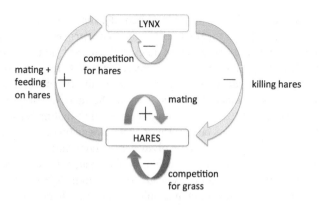

Fig. 1.4 Feedbacks in the PP model. Signs give feedback polarity (positive or negative)

considered here has only one prey and one predator, say hares and lynx, which makes it very easy to analyze. Hares feed primarily on grass. We will assume that their numbers grow because of mating and that they diminish only because of the action of predators (i.e., lynx) or because of the competition for food (i.e., grass) with other hares. The number of lynx, on the other hand, also grows through mating but always in the presence of food (i.e., hares), being diminished by the competition for food (i.e., hares) with other lynx. By naming the two populations n_h (hares) and n_l (lynx), a possible mathematical model to study their interplay in these conditions is provided by the following pair of ODEs:

$$\frac{dn_h}{dt} = \alpha n_h - \beta n_h n_l - \epsilon n_h^2 \tag{1.1}$$

$$\frac{dn_l}{dt} = \delta n_l n_h - \gamma n_l \tag{1.2}$$

where all parameters are assumed constant and positive.

The feedbacks present in the equations are schematically presented in Fig. 1.4. There are five different feedbacks, each of them associated to one of the terms appearing on the right hand side (rhs) of the PP equations. Let's consider the first nonlinear term on the right hand side of the first equation: $-\beta n_h n_l$. It represents the decrease in the hare population due to their killing by lynx. *The nonlinearity appears because the killing process naturally requires the simultaneous presence of both lynx and hares!* The negative sign then tells us that the feedback, that connects the hare population to the *product of lynx and hare populations*, is negative. The larger this product is, the more the hare population will decrease. Why? Because more hares imply an easier kill for lynx, but more lynx also imply more hare killing. Thus, the relevant quantity is the product of the two. A second feedback is provided by the nonlinear term in the second equation: $+\delta n_l n_h$. This time the term is positive since it reflects the fact that, the more hares there are, the more lynx can live off them by mating and feeding on them. Again, the nature of the interaction requires that both hares and lynx be present simultaneously, that leads to the nonlinearity. In fact, these two feedbacks together constitute a **two-process negative feedback loop**, by which an increase [decrease] in hares will cause by an increase [decrease] in lynx that will be followed a decrease [increase] in both hares and lynx.

A different type of feedback loop also present in the PP model has to do with **self-interactions**. For instance, the $+\alpha n_h$ term in the first equation represents the growth of the hare population due to mating. Note that mating only requires the presence of hares, and can thus be represented by a **linear term**. Interestingly, there is a second feedback loop that only involves hares, represented by the last term in the first equation: $-\epsilon n_h^2$. This term models the fact that, for a constant supply of food, a overabundance of hares finally leads to starvation. It is however nonlinear because, for a small hare population, it should be unimportant relative to the mating linear term, but it should dominate if the hare population grows very large. The use of the nonlinearity achieves this goal. Finally, there is another self-interaction loop represented by the linear term $-\gamma n_l$ in the second equation. It represents a *negative* feedback loop associated to the effect on the lynx population of the superabundance of lynx.[9]

Once specific values are prescribed for all parameters (i.e., $\alpha, \beta, \gamma, \delta$ and ϵ), the relative importance between all feedbacks is set, and certain outcome will follow. The beauty of the PP model is that the possible outcomes are just a handful. They are illustrated in Fig. 1.5, that shows the phase-space[10] of the model for different parameter choices. The upper, left panel shows the phase-space for $\epsilon = 0$. That is, when there is no limitation to the number of hares that can live on the available amount of grass. In this limit, the model reduces to the famous two-dimensional Lotka-Volterra system [37].[11] Then, hares and lynx numbers vary in periodic fashion (see Problem 1.2), following a closed orbit in phase space around a central point, given by $(n_h, n_l) = (\gamma/\delta, \alpha/\beta)$. Lynx and hares may seem to approach extinction at some moments, but they always manage to rebuild their numbers.

Nothing really interesting happens in the two-dimensional Lotka-Volterra equation when the parameter values are changed while keeping $\epsilon = 0$. It is only the location in phase space of the central point that moves around. The parameter ϵ must

---

[9]The attentive reader might complain that we used a nonlinear term to model a similar process in the case of hares. The explanation is based on the fact that hares can live without lynx, quietly feeding on grass, but lynx cannot live without hares to hunt. Therefore, n_l should naturally go to zero whenever $n_h = 0$. This requires the use of the linear term (see Problem 1.1).

[10]The *phase-space* of the predator-prey model is the two-dimensional space that uses n_l and n_h as coordinates. A point in this phase space represents a possible state of the system, given by a pair of values (n_h, n_l). A trajectory in phase space corresponds to the evolution in time of the hare and lynx populations, starting from given initial conditions, as allowed by the equations of the model.

[11]The two-dimensional Lotka-Volterra equation does not exhibit either chaotic or complex behaviour. However, if one considers N-order Lotka-Volterra equations, $\dot{n}_i = r_i n_i (1 - \sum_{j=1}^{N} \alpha_{ij} n_j)$, chaotic behaviour ensues for $N > 3$, where the system, although deterministic, is no longer integrable [38]. For $N = 3$, limit cycles do appear, to which the trajectories are drawn by the dynamics, but trajectories are not chaotic.

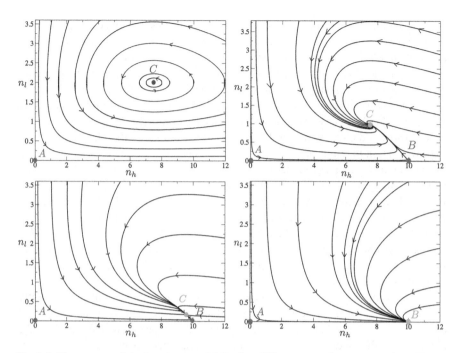

Fig. 1.5 Phase space of the hare-lynx model for four different sets of parameter values: **Upper Left:** $\alpha = 2$, $\beta = 1$, $\delta = 0.2$, $\gamma = 1.5$ and $\epsilon = 0$; **Upper Right:** $\alpha = 2$, $\beta = 0.5$, $\delta = 0.2$, $\gamma = 1.5$ and $\epsilon = 0.2$; **Lower Left:** $\alpha = 2$, $\beta = 0.5$, $\delta = 0.2$, $\gamma = 1.9$ and $\epsilon = 0.2$; **Lower Right:** $\alpha = 2$, $\beta = 0.5$, $\delta = 0.2$, $\gamma = 3$ and $\epsilon = 0.2$. The stable fixed points are shown in green, neutral in blue and unstable, in red

become non-zero for interesting new dynamics to become possible. In that case, the phase space undergoes important changes, as shown in the remaining panels of Fig. 1.5. The best way to illustrate these changes is to calculate the **fixed points**[12] of the dynamical system. Setting $\dot{n}_h = \dot{n}_l = 0$, we find that there are three fixed points (see Problem 1.3):

$$(A)\ n_h = n_l = 0 \tag{1.3}$$

$$(B)\ n_h = \alpha/\epsilon, n_l = 0 \tag{1.4}$$

$$(C)\ n_h = \gamma/\delta, n_f = (\alpha - \epsilon\gamma/\delta)\,\beta^{-1}, \tag{1.5}$$

---

[12]A *fixed point* of a system of ODEs is a point of phase space where the right-hand-side of the system of ODEs vanishes [39]. Since the right-hand-sides give the time derivatives of all relevant dynamical quantities, the evolution of the system remains unchanged if it starts from any fixed point in phase space, or after it reaches any of them over time. Thus, their name. Fixed points maybe **stable**, if the solution tends to come back to the fixed point after being perturbed, **unstable**, if perturbations grow and push the solution away, or **neutral**, if the perturbation neither grows nor is damped (see Appendix 1 in this chapter).

corresponding to the case of (i) extinction of both species (A); (ii) hares growing to the maximum number supported by the grass available in the absence of lynx (B) and (iii) hare and lynx numbers remaining finite, in mutual equilibrium (C). The upper, right panel in Fig. 1.5 shows an example of the phase-space of the predator-prey model for small, but positive ϵ, showing the three fixed points. The reader will note that all phase-space trajectories finally reach point (C) and move away from both (A) and (B), thus escaping partial or total extinction. This happens because (C) is an **stable** fixed point, whilst both (A) and (B) are both **unstable**. In fact, it is easy to prove that, in the limit of $\epsilon \to 0$, (C) becomes neutral and gives the central point of the Lotka-Volterra equation, around which all orbits circulate.[13] The situation however changes if $\epsilon > \epsilon_c := \alpha\delta/\gamma$, since n_l would no longer be positive at the (C) fixed point. This causes point (C) to coalesce with point (B), that becomes the new stable point of the system (see lower, right panel in Fig. 1.5). The result? That the only possible outcome of the evolution is now the extinction of all lynx! What has happened here? That we have encountered a **threshold** (also known as **tipping point**, in the context of biological systems), set by the value ϵ_c, that separates two very different outcomes.

Although the PP model does not exhibit complex behaviour, it illustrates pretty well that any change in the relative importance of feedback loops can lead to a change of the overall behaviour of the system. This type of change will not take place in the PP model as it is advanced in time, since all parameter values are prescribed at the start, and its evolution is thus completely deterministic thereafter. But in a complex system, being open and driven from the outside in an often unpredictable manner, conditions can change at any time in ways that make hidden thresholds be overcome and sudden changes in behaviour take place. The much higher dimensionality of complex systems also implies that many more feedback loops and many more thresholds do exist, and that a much richer variety of dynamical behaviours could be accessed while responding to any changes in the external drive. Some of these new behaviours may lead to the type of emergent features that make systems complex.

1.2.2.6   Thresholds

By threshold we mean *the magnitude or intensity of a certain quantity that must be exceeded for a certain reaction or phenomenon to occur*. The concept of threshold has just been beautifully illustrated within the PP model, in which two very different outcomes were obtained depending on whether ϵ exceeded or not the critical value $\epsilon_c := \alpha\delta/\gamma$. For $\epsilon > \epsilon_c$, lynx became extinct. For $\epsilon < \epsilon_c$, their numbers always reached a finite value $n_f = (\alpha - \epsilon\gamma/\delta)\beta^{-1}$. The parameter ϵ_c is an example of a **global threshold**, that determines the transition between two different types of global behaviour. In high-dimensional complex systems, the much larger number of

---

[13]Note that, in the same limit, (B) and (A) coalesce to $(n_h, n_l) = (0, 0)$, and remain unstable.

feedback loops often present can lead to many different thresholds. Therefore, the number of accessible behaviours becomes also much larger.

Thresholds can also be *local* in the sense that *if a certain quantity exceeds some threshold value at a certain location, some new phenomenon will take place at that particular location*. **Local thresholds** are particularly important in the case of plasmas and fluids in which exceeding a certain local threshold for quantities such as a velocity, temperature or pressure (or, more precisely, their gradients) often leads to the excitation of instabilities at that location. For instance, in fluid turbulence, the local threshold for the excitation of Kelvin-Helmholz (KH) instabilities has to do with the velocity gradient [40]. Or, in the case of interchange instabilities in tokamak plasmas, a local threshold for the pressure gradient value must be overcome [29]. Local thresholds are, of course, not exclusive to turbulent systems. One could consider, for instance, the dynamics of faults in the Earth crust. In them, earthquakes are excited whenever the amount of local stress overcomes what the fault can sustain at each particular location [10].

1.2.3   Main Emergent Features

The term "emergent feature" refers to any of the special patterns, events, properties, phenomena or behaviours that are often observed in complex systems, typically taking place over scales much larger than those characteristic of the constituents of the system. An example is the observation of *fractal* patterns in river beds or in human pulmonary alveoli. Or the formation of banded flow patterns in Jupiter's atmosphere. The observation of emergent features is indicative of the fact that complex dynamics might be at play. It must be kept in mind, however, that the list of features given here is not complete by any means. Other emergent behaviours are possible. On the other hand, it must also be noted that non-complex dynamical explanations could sometimes be more adequate to explain the observation of apparently complex features. Indeed, there are scale-invariant processes which are not due to complex dynamics. One could mention, for instance, the type of *random walk* process that underlies diffusive motion, that is statistically scale-invariant both in space and time [41] but not driven by complex behaviour, in any sense of the word.

1.2.3.1   Scale-Invariance

Scale-invariance means that *the system looks and behaves similarly at every scale*. Or in layman terms: if you could rescale yourself and enter the system at a different scale, everything would look the same as it did before rescaling. Scale-invariance manifests itself in many ways. One is through **fractal spatial patterns**. A spatial pattern is loosely termed *fractal* if it looks *approximately, statistically or exactly the same when examined at different spatial scales* [41, 42]. Fractal patterns are often observed in nature. For example, one could mention the structure of veins in leaves, or the way in which capillaries organize within the pulmonary alveoli in our lungs, or the patterns formed by some rivers as they push water downhill on relatively shallow riverbeds. In any of these examples, as in many others, fractal patterns seem to be the answer found by system to provide an optimal setup to achieve some goal. These goals probably are, for the examples considered, to best interchange gases in our lungs, to efficiently transport the products of photosynthesis to the rest of the tree or to best move large quantities of water through a shallow and rough terrain.

The idea of "invariance upon examination on different scales" is not restricted, in complex systems, to fractal spatial patterns. It is often observed in other forms, such as in **temporal records**. This is the case, for instance, of time traces of the inter-beat interval of the human heart, or of signals of neural activity in the human brain, both of which have a similar appearance when examined over widely varying time scales [43].

Scale-invariance is sometimes also exhibited by the **statistical distributions** associated to the phenomena taking place in the system, that often exhibit power-law decays over a wide range of scales. The reason why power-laws are often linked to scale-invariance is that any function, $f(x)$, that verifies that $f(x) = \lambda^\mu f(\lambda x)$ for some exponent μ is scale-invariant [42]. Clearly, the power law $f(x) = x^{-a}$ is scale-invariant for $\mu = a$. Thus, power-law statistics are often observed in complex systems, that are sometimes said to exhibit *statistical scale-invariance*. For instance, this is the case of the statistical distribution of released energy by earthquakes (basis of the well-known Gutenberg-Richter law [44]), the statistics of the energy released by the Sun in the form of flares [23] or the statistics of casualties in major wars [45], among others.

We will discuss scale-invariance and its properties at length in Chap. 3, including many useful methods to detect it and to quantify scale-invariance exponents. However, it must always be kept in mind that scale-invariance is always limited, in real systems, by **finite-size effects**. That is, scale-invariance only holds within a certain range of scales, usually limited from below by those scales characteristic of the system constituents (their size, lifespan, reaction times, etc.) and from above, by the system largest scales (mainly, the system global size, the amount of time needed to transverse it or its lifespan). The finite range of scales over which scale-invariance is observed in finite systems is often referred to as the **mesorange**, the **self-similar range** or the **scale-invariant range**.

1.2.3.2 Coherent Structures

The term "coherent structure" has been used for many decades in fields related to fluid and magnetic turbulence [46]. It refers to *any large structure created by the system dynamics that remains coherent (i.e., lasts) for time lapses well beyond typical turbulent times*. Coherent structures in turbulent systems may take the form of, among others, long-lived large vortical structures—such as the Jupiter's red spot or the tornadoes and hurricanes that can form in weather systems–, quasi-permanent regions of flow motion sustaining strong shear layers or discontinuities—such as Jupiter's zonal flow patterns, the solar tachocline, propagating fronts associated to combustion or explosions, or the radial transport barriers found in tokamak reactors in certain regimes—, or slow-varying, large-scale magnetic fields—such as those observed stars and galaxies, and planets such as our own Earth.

In complex systems of non-turbulent nature, long-lived structures that keep their coherence for very long periods of time, with sizes/lifespans well beyond those of the system constituents, do also exist. We will abuse language and refer to them also as coherent structures. One could think, for instance, of the several instances of biological swarms previously discussed as coherent structures in biological systems. Or the phenomenon of fashion so prevalent in social systems, by which large fractions of the population follow a particular trend or practice for long periods of time due to peer-influence and information spreading. A similar process takes place in political and economical systems, in which large number of people adhere to specific ideas and opinions that appear, develop and eventually fade away in an apparent, self-organized manner, such as liberalism, marxism, nationalism and many others.

1.2.3.3 Extreme Events

The term "extreme event" refers to *any event that affects a significant fraction (and sometimes all) of the system extension*. In many cases, they have catastrophic consequences—such as a global economic crisis, a massive extinction, a glaciation period, a very large earthquake, blackout or hurricane—, what makes them particularly interest to predict. Extreme events are, however, not exclusive to complex systems, since they may also happen in non-complex systems as well. Their significance is however much larger because statistics in complex systems tend to exhibit power-law decays (instead of exponential, or Gaussian decays). This fact increases the probability of extreme events rather significantly, as will be explained in more depth in Chap. 2. A second distinguishing aspect is the fact that extreme events do not happen randomly in complex systems, being instead dictated by the evolution of the tepid network of mutual interactions between system constituents over very long periods of time (namely, a chain of events that result in an unchecked positive feedback loop coming to dominance). This absence of randomness, that is not exclusive to the excitation of extreme events, is loosely known as **memory** and will be the main focus of Chap. 4.

1.2.3.4 Memory

Complex systems often exhibit "memory", in the sense that the *the future behaviour of the system depends not only on the current system state, but also on its past history under the influence of its environment, for periods of time much longer than any characteristic time that could be associated to the system individual constituents.*

This type of memory is particularly important for the appearance and evolution of any emergent feature in the system. It affects the triggering of extreme events that is often heavily conditioned by the system past history. Let's take, for instance, the triggering of earthquakes at those parts of a fault where the local stress has exceeded a certain threshold value. Most of the time, the fault will be able to release the excess local stress by triggering small earthquakes that move it to other parts of the fault. But eventually, the whole extension of the fault will sustain values of stress so close to the maximum that much larger earthquakes, affecting significant parts of the fault extension, become inevitable. Memory expresses itself here by influencing, for periods well beyond the timescale over which individual earthquakes take place, when future earthquakes of a certain size will take place under the external drive provided by tectonic plate motion [10]. A very similar phenomenology is found in other complex systems, as will be discussed in length in Chap. 4.

1.2.3.5 Criticality

Another behaviour that may emerge in complex systems is "criticality". When this term is used, it is usually meant that *the system displays similar features to those exhibited in critical transitions in equilibrium thermodynamics*. Namely, *scale-invariant fluctuations with divergent correlation scales both in space and time* [47]. In contrast to what happens in equilibrium transitions, criticality emerges here in an open driven system without fine tuning any external parameter. This behaviour is known as **self-organized criticality** (SOC), a concept introduced by P. Bak in the late 1980s [48]. SOC is thought to play an important role in processes as different as the triggering of earthquakes [10], the dynamics of forest fires [11], the dynamics of blackouts in power grids [8] and many others [49]. SOC is also a concept of particular relevance in plasma science, where it might relevant to the understanding of the dynamics of solar flaring [24] (see Chap. 7), of geomagnetic substorms in the Earth's magnetosphere [50] (see Chap. 8) and of near-marginal turbulent transport in tokamaks [30, 31] (see Chap. 6), among others.

1.2.3.6 Fractional Transport

In some cases, transport across a complex system becomes endowed with features that are quite different from the familiar *diffusive transport* in order to make it more efficient. This appears to be the case, for instance, of turbulence when attempting to transport either energy out of a tokamak plasma, at least in certain regimes (see Chap. 6), or angular momentum out of an accretion disk [24].

In order to understand better what these new features are and what they imply, we need to briefly discuss diffusive transport first. The paradigm of diffusive transport was introduced in the 19th century, and assumes that the local flux Γ of some transported quantity, say n, is proportional to its local gradient, ∇n. That is, $\Gamma = -D\nabla n$, expression that is known as **Fick's law** [51]. The proportionality constant, or *diffusivity*, is determinable, under certain hypothesis, from the properties of the microscopic process responsible [52]. For instance, in the simple case of a gas in which molecules transverse on average a mean free path λ before colliding with other molecules with an average frequency v, the diffusivity turns out to be roughly given by $D \simeq \lambda^2 v$. Particle transport throughout the gas can then be well described, at least for distances much larger than λ and times much longer than v^{-1}, by introducing Fick's law in the continuity equation for the gas density,

$$\frac{\partial n}{\partial t} + \nabla \cdot \Gamma = S, \tag{1.6}$$

where S is an arbitrary external source, which yields the usual diffusive equation,

$$\frac{\partial n}{\partial t} = D\nabla^2 n + S. \tag{1.7}$$

A similar reasoning can be made for any other system in which *both a characteristic length and a characteristic timescale for the transport process can be found*. Characteristic transport scales are, however, absent in some complex systems. Let's take, for instance, any of those systems governed by SOC. Criticality implies that fluctuation scales are divergent in both space and time. Therefore, typical scales equivalent to both λ and v^{-1} are nowhere to be found, since they are usually related to (now divergent) moments of the fluctuation statistics. Transport through a SOC system, as will be discussed in the next section, thus becomes rather counter-intuitive, specially for our Fick-trained minds [31]. Similar situations are sometimes also encountered in complex systems that do not exhibit SOC dynamics [53]. For instance, non-diffusive transport often place takes through fractal environments, as it is the case of many porous media [54]. SOC or non-SOC, we will refer to all these instances of non-Fickian transport under the general label of **fractional transport**[14] for reasons that will made clearer in Chap. 5.

1.3 Self-organized Criticality

Self-organized criticality or, more simply, SOC, will play an important role in later chapters. Thus, it is worth to discuss it here in more detail. The concept of SOC was first introduced as an emergent behaviour of certain complex systems in the late 1980s. Originally, the concept and the ingredients needed to make it possible were presented in the context of a cellular automata, known as the Bak-Tang-Wiesenfeld sandpile or, in short, the *BTW sandpile* [48]. However, we will discuss it here using the **running sandpile**, another cellular automata introduced a few years later [55]

---

[14]Various other names such as *non-diffusive transport, anomalous transport, self-similar transport* or *scale-free transport*, are also often used, depending mainly on the specific field and author.

that turns out to be more adequate for some of the plasma applications that will be discussed in the second part of this book (see Chap. 6).

1.3.1   The Running Sandpile

The running sandpile is a one-dimensional cellular automata (see Fig. 1.6). It has L cells containing grains of sand up to a height h_n at each cell n, where $n = 1, \cdots, L$. It is driven by *continuously and randomly* dropping N_b grains of sand, with a frequency p_0, on each cell. As a result, the height of the cells increases over time. The dynamics of the running sandpile are defined by a simple rule: whenever the local slope, defined as $z_n = h_n - h_{n+1}$, exceeds a certain prescribed threshold $Z_c > 0$, N_F grains of sand are transported from the n-th to the $(n + 1)$-th cell. That is, transport is *directed*, taking place always down the gradient. Any grain of sand that reaches the edge (i.e., the L-th cell) is removed from the system. The running sandpile, when advanced according to this rule, eventually reaches a stationary state when the edge losses balance on average the sand dropped throughout the system.

The running sandpile is a paradigmatic example of a complex system of the self-organized critical type. It contains the **main ingredients** needed for SOC dynamics to emerge: *it is open, slowly-driven and contains a local threshold that separates fast local relaxation from periods of local inactivity.* Its more characteristic feature is that, for $N_F > 1$, transport through the system takes place in the form of **avalanches** of sand, with linear extensions ranging anywhere from one cell to the system size. Avalanches happen because, after sand is removed from cell n onto cell $(n + 1)$, the

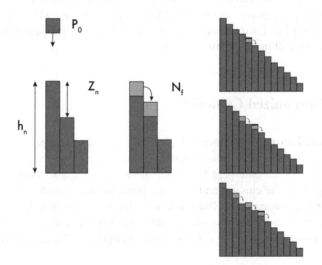

Fig. 1.6 Sketch of the running sandpile. As shown, wherever and whenever the local slope exceeds a critical value Z_c, N_f grains of sand are toppled to the next cell

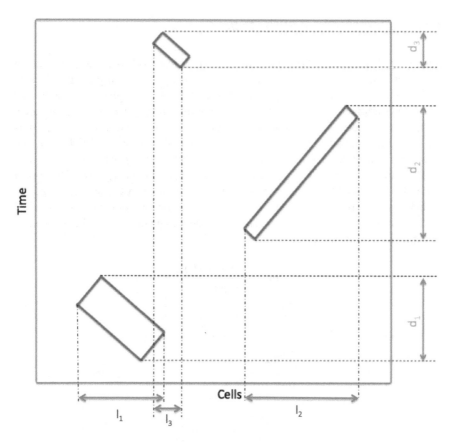

Fig. 1.7 Avalanches in the running sandpile. The horizontal axis corresponds to the cell number; the vertical one, to the iteration number. Three different avalanches are shown in red, containing those cells that have undergone any relaxations. Although sand is always transported down the gradient (i.e, to the right), *holes* might also be transported up the gradient (to the left). This happens if the local removal of N_F grains of sand from cell n causes that cell $n-1$ becomes unstable. The definitions of avalanche linear size (l) and duration (d) are illustrated

latter may become itself unstable if z_{n+1} exceeds Z_c due to the sand just added. If this instability state continues to propagate to more neighbouring cells, an avalanche forms (see Fig. 1.7). The condition $N_F > 1$ is very important for SOC dynamics. It guarantees that the local slope will sit, most of the time, below and not at Z_c. This avoids the possibility of all cells continuously maintaining a slope $Z = Z_c$, which would make every avalanche reach the sandpile edge. Similarly, it is also required that $(p_0 N_b)n < N_F/2$ for all $n \leq L$. Otherwise, the region for $n > n_s$, with $n_s = N_F/2(p_0 N_b)$ will constantly have $Z_n > Z_c$ (i.e., the cell would be overdriven!), and transport would never cease at that location.

1.3.2 Criticality in the SOC State

Interesting things happen once the running sandpile has reached a steady-state under the aforementioned conditions. In particular, the sandpile exhibits properties typical of the critical points at which equilibrium phase transitions happen, despite the fact that the system is not in thermodynamical equilibrium [48]. *Criticality* is apparent in the form of divergent correlations both in time and space. Most interestingly, this state of things is reached without the need of any tuning whatsoever.[15] In fact, that is the main reason for describing this type of criticality as *self-organized*.

The criticality of the SOC state can be made apparent in many ways. The easiest one is by means of the various power laws that abound in the system. Take the probability density function (pdf) of the avalanche linear extensions (see Fig. 1.7 for the definition of linear extension, l). Phenomenologically, it is well described by an expression of the kind (see Fig. 1.8)

$$p(l) \sim \frac{\exp(-l/l_2)}{1 + (l/l_1)^b},$$
(1.8)

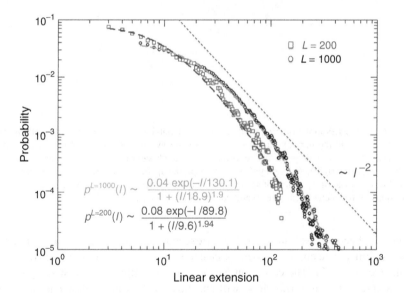

Fig. 1.8 Probability density function of avalanche linear extensions computed for two sandpile sizes, $L = 200$ and $L = 1000$. Both cases are run with $N_F = 30$, $Z_c = 200$, $N_b = 10$ and $p_0 = 10^{-4}$. Best fit to Eq. 1.8 is shown for each case, yielding $b \simeq 1.92 \pm 0.2$. Estimated mesoranges cover the intervals $\sim (10, 90)$ for the $L = 200$ case, and $\sim (20, 130)$ for the $L = 1000$ case

---

[15]This is in contrast to the situation in equilibrium phase transitions, where some physical magnitude, usually temperature, must be finely tuned.

with $b \in [1, 2]$, and with $l_2(L) \gg l_1(L) > 1$. The two scales l_1 and l_2 are associated (but not identical) to the smallest and the largest linear extensions that avalanches can have (the system size and the minimum meaningful length, one cell). But the important point here is that, for avalanches with linear extensions within the so-called *mesorange*, $l_1(L) \ll l \ll l_2(L)$, the pdf behaves as a power-law $p(l) \sim l^{-b}$, that is scale-invariant by definition.[16] As previously mentioned, finite-size effects—due to both the finite sandpile and unit cell sizes, or to the finite duration of the simulation and the fact that the minimum time is one iteration—limit the extension of the mesorange from below and from above.[17]

Criticality in space is associated with the divergent nature of Eq. 1.8 over the mesorange *whenever* $1 < b < 2$. Indeed, if we attempt to compute the average linear extension of avalanches in that case, we would find that,

$$\bar{l}_L \sim \int_1^L l p(l) dl \sim \int_1^L l^{1-b} dl \sim L^{2-b} \longrightarrow \lim_{L \to \infty} \bar{l}_L = \infty. \qquad (1.9)$$

Therefore, the sandpile's SOC steady state lacks a characteristic spatial scale, since the *average distance that sand could travel inside an avalanche* diverges with the system size, L. Or, in other words, such scale is not set by the transport process, since it is only limited by the system size. In contrast, in any diffusive system, the transport spatial scale is independent of (and much smaller than) the system size, as we discussed in the previous section.

Criticality in time, on the other hand, can be detected by looking at the pdf of the lapses of time between successive avalanches, also known as *waiting (or quiet) times*. Waiting-times determine the characteristic timescale in the running sandpile. They do so because, in order to exhibit SOC, the sandpile must not be overdriven, that requires $(N_b p_0) < N_F/2L$. This means that, in practice, the average duration of an avalanche must be much shorter than the typical waiting periods between them.[18] Thus, the waiting-times become the only remaining timescale of interest. Revealing the critical nature of the temporal behaviour of the randomly driven running sandpile is however somewhat subtle.

A direct calculation of the pdf of waiting times between avalanches yields an exponential (or Poisson) distribution. The reason is that avalanches can only start when a drop of sand falls, together with the fact that these grains are dropped

---

[16]Scale-invariance is also apparent in many other quantities of the sandpile. Naturally, the actual value of the exponents and the limits of the mesoscale are different.

[17]In an infinite sandpile (in space and time), the mesoscale would extend over all scales. The sandpile would then behave as a true monofractal. In reality, deviations from scale-invariance should be expected at scales that approach any of the finite boundaries. As a result, he monofractal behaviour is only approximate. In particular, the sandpile size (lifespan) must be significantly larger (longer) than the unit cell (one iteration) in order to display SOC dynamics.

[18]In fact, Bak's original formulation (and most analytical studies, such as those that describe SOC dynamics as an absorbing transition [56]) of SOC assume the limit of zero duration, in which avalanches are relaxed instantaneously once excited.

randomly with probability p_0 [57]. Because of this randomness, the probability of having to wait for an amount of time w for the next avalanche must be independent of how much time we have already been waiting, w'. Mathematically, this condition is expressed as (see Sect. 2.3.4):

$$p(w) = \frac{p(w + w')}{p(w')}, \tag{1.10}$$

that is only satisfied by the exponential distribution, $p(w) = w_0^{-1} \exp(-w/w_0)$, with $w_0 \propto p_0^{-1}$ giving the average waiting time. It would thus seem that a characteristic timescale w_0 does indeed exist, and that all critical temporal behaviour is absent from the system. However, this is not the case.

A random drive is not needed for SOC dynamics to emerge and either non-random drives [58] or even random drives applied at non-random locations [59] could be used without altering the SOC character of the transport dynamics, as long as the system is not overdriven. The only measurable consequence of using non-random drives is that the sandpile waiting-time distribution changes from an exponential to a power-law, while all other avalanche statistics remain unchanged. Interestingly, what is in fact rather insensitive to the details of the drive, is *the pdf of the waiting-times between avalanches sufficiently large so that their linear extensions lie within the mesorange*. Or in other words, when avalanches with size $l < l_1(L)$ are simply ignored. In this case, waiting times are distributed according to an expression like Eq. 1.8, again with $1 < b < 2$ (see Fig. 1.9). The critical nature of the sandpile temporal behaviour becomes now apparent, and is linked to the fact that the *average waiting-time between avalanches within the mesorange again diverges with the system size*. The fact that the waiting-time pdf is no longer exponential also implies that the triggering process of these avalanches ceases to be uncorrelated. Instead, the previous record of waiting times conditions strongly the probability of which the next waiting time will be. We have previously referred to this state of things loosely as *memory*. We discuss it a bit more in what follows.

1.3.3 Memory in the SOC State

How is memory established and maintained in the running sandpile? The relaxation rules are local in time and space, since they require only current information to advance the state. So no help there. Similarly, if the driving of the system is random, it cannot originate there. Only remaining suspect is the *past history of the sandpile height profile* itself. Clearly, avalanches start at cells where the local gradient is below, but very close to the threshold, and will stop at two other cells (one below and one above the original excitation point) where the local profile is well below critical. In the former locations, the addition of a single grain of sand is enough to turn the location unstable; in the latter, the addition of the sand coming from the relaxation of the previous cells is not enough to push the gradient above the critical

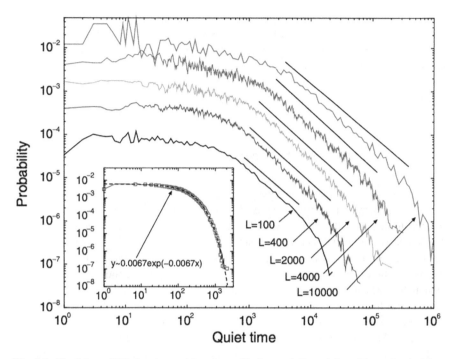

Fig. 1.9 Fig. 2 from [58] showing waiting time pdfs for sandpile activity without thresholding the avalanche duration (inset) and after thresholding. Sandpiles with sizes $L = 100, 400, 2000, 4000$ and 10,000 are used, with duration threshold respectively being $d_t = 30, 75, 200, 400$ and 800. Values are chosen to be well into the mesorange for each sandpile. Reprinted figure with permission. Copyright 2002 by the American Physical Society

value. After the avalanche has ceased, the local gradient has been reduced below marginal at the starting point, and pushed closer to marginal at the two stopping cells that mark the maximum extension of the avalanche. In addition, there is another cell where the local gradient is reduced: the cell two positions after the innermost stopping point [60]. This means that the fact that an avalanche was excited at some starting location increases the probability of another avalanche starting at any of these locations in the future. Hence, as this process happens once and again, the *slope profile becomes "rough" on all scales*. This is how memory is established and stored in the profile, and transmitted through time.

Some astute readers might however object by saying that the distribution of previous positions also determines the future state of a deterministic gas. Therefore, any system should be able to display memory in that sense, right? What is then so special about the running sandpile? The subtlety lies again in the lack of characteristic scales of the SOC state. When finite characteristic scales exist in time and space, memory effectively disappears for times longer and distances larger than those scales. However, spatio-temporal criticality prevents the existence of these characteristic scales in the running sandpile and, therefore, *all the past history and the full spatial extent of the system conditions the future evolution.* This is not the case of the deterministic gas, where the previous history is essentially wiped out after a few collision times. We will have much more to say about memory in the running sandpile (and by extension, in SOC systems) in Chap. 4, where we will also illustrate several methods to detect and quantify the presence and type of memory in a system.

1.3.4 Transport in the SOC State

SOC also has important consequences for the nature of transport through the sandpile. Transport becomes very different from the *diffusive transport* that we briefly discussed at the end of Sect. 1.2. In the SOC steady state, transport events are linked to the avalanche mechanism and to the existence of a local threshold. Spatio-temporal criticality ensures that there is neither a characteristic scale for the avalanche linear size nor a characteristic frequency, since the statistics of both avalanche sizes and waiting-times diverge with the system size. Therefore, the type of transport that emerges exhibits many features that seem nonintuitive to any Fick-trained mind. We will discuss non-diffusive (or, more precisely, *fractional*) transport in Chap. 5, but it is worth to advance a few relevant features at this point.

For instance, let's examine the *average confinement time of sand* in the sandpile. This figure-of-merit is defined as the ratio of the total sand contained to the total drive strength. That is,

$$\tau_{CF} = \frac{1}{N_b p_0 L} \sum_{n=1}^{L} h_n \simeq \frac{(Z_c - N_F/2)L^2}{2 N_b p_0 L} \propto L. \tag{1.11}$$

Interestingly, the diffusive equation (Eq. 1.7) would yield a scaling $\tau_{CF} \propto L^2$ instead (see Problem 1.6). Clearly, the *SOC confinement scaling* is much less favourable.[19] In addition, the confinement time degrades as $(N_b p_0)^{-1}$, in contrast to the diffusive case, in which confinement is independent of the drive strength. That is, the faster the sand is added to the pile, the faster it will be transported out.[20] In the sandpile, this type of power degradation of confinement is closely related to the strong *profile stiffness* that the SOC steady state exhibits. By the term "stiff", it is meant that the sand profile is rather insensitive to the location or strength of the external drive.

---

[19]In fact, this apparently innocent statement can become pretty important in some cases, such as when trying to confine plasma energy in a tokamak fusion reactor. The determination of whether the confined plasma exhibits SOC or not might cost (or save) the fusion program quite a few millions of additional dollars.

[20]Power degradation is not unique to SOC systems. It may also happen in diffusive systems whenever D is a function of the external power. That is the case, for instance, in driven plasmas since D is then a function of the local temperature.

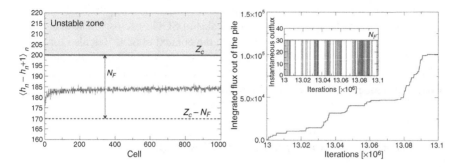

Fig. 1.10 Left: Average steady-state slope profile for a running sandpile defined by $L = 1000$; $N_F = 30$, $Z_c = 200$ $p_0 = 10^{-4}$ and rain size $N_b = 10$. The slope remains, on average, below Z_c and above $Z_c - N_F$ everywhere. In spite of being submarginal on average, the sandpile sustains significant transport. **Right:** Integrated outflux of sand coming out of the bottom of the pile for 10^6 iterations starting at iteration 1.3×10^7. The instantaneous outflux is shown in the inset, being equal to N_F when the edge is unstable (i.e., $Z_{\text{edge}} > Z_c$), and zero otherwise

In fact, the sandpile profile remains always close to Z_c by exciting avalanches, which quickly bring the profile back below threshold. A stronger drive simply means that avalanches will be triggered more frequently and be possibly larger. Instead, a diffusive system would adapt its profile shape to the external source, not being constrained by anything. Profile stiffness is in fact the reason why SOC systems can drive significant transport while apparently staying submarginal (i.e., below threshold) on average (see Fig. 1.10). This counter-intuitive behaviour simply reflects the fact that the sandpile stays quiet (and submarginal) the majority of the time, but intermittently, it becomes locally supercritical. Avalanches are then excited that drive net transport.[21]

A final characteristic worth mentioning is the significance of *extreme events* for transport in the SOC state. Extreme events can be defined here, for convenience, as those *avalanches with linear extensions involving a sizeable part of the system size*. Spatio-temporal criticality makes the contribution of extreme events to the overall SOC transport always significant, independently of how large a system is considered. This is a consequence of the divergent power-law statistics of the SOC state that endows larger avalanches with probabilities that are always meaningful. In contrast, diffusive systems exhibit Gaussian-type statistics and the probability of extreme events decreases exponentially as the system size increases. Therefore, their contribution to the overall transport becomes quickly negligible.

After this description of the features of transport in the running sandpile, it is not surprising that Fick's law turns out to be inadequate to capture most, if not all, of

---

[21]It should be noted, however, that SOC could also sustain average supermarginal profiles if the drive is strong enough to keep the profiles above marginal more often than not. SOC would however disappear if the system becomes overdriven, with profiles staying above threshold everywhere, all the time.

the features of SOC transport. Any candidate for effective transport model should be able to tackle the lack of characteristic scales of SOC systems.[22] The search for such a model is a problem of great interest in the context of SOC, but also for many other systems unrelated to SOC. This common need has stimulated the development of a new type of transport formalism, not based on Fick's law, commonly known as **fractional transport theory**. We will discuss it in detail in Chap. 5.

1.4   Overview of the First Part of This Book

In this chapter we have discussed many concepts of relevance for the understanding and investigation of complex systems. We have argued that, in our opinion, the two most defining features of complexity are self-organization and emergence. We have also shown that these features appear in systems that are widely different and may emerge in various forms. We will devote the rest of the first part of this book to describe and quantify some of these features, with a particular emphasis on the introduction of techniques that are useful to characterize them.

We will start with a discussion about statistics in Chap. 2, since most of the approaches that can reveal complex behaviour rely heavily on the use of probabilities and averaging, both from a theoretical and practical point of view. Then, we will continue with the concept of scale-invariance in Chap. 3, both in space and time. Important scale-invariant mathematical models that complex systems are often compared against, such as the fractional Brownian (fBm) or Lévy (fLm) motions, will be introduced here. Chapter 4 will discuss at length the concept of memory in the context of complex systems. Finally, Chap. 5 will review in depth the (still expanding) theory of fractional transport and its implications for the description of transport in complex systems.

The running sandpile will provide a lifeline throughout this introductory journey. At the end of each chapter, some relevant aspect of the running sandpile dynamics will be analyzed by means of some of the techniques just presented. These exercises will serve as an illustration, but also as a window through which SOC dynamics can be further explored and understood.

In most of the material included, we have tried to be mathematically correct but not excessively rigorous. There are situations in which a formula will be provided without proof, or a physical argument will be stated without deriving it in painful detail. In most of these cases, references are provided where the interested reader can find proofs and derivations. Some appendices have been included at the end of each chapter to stress some concepts, formalisms or methods of particular interest.

---

[22]Finding effective transport models for non-diffusive systems is a very important problem and a very active area of research, particularly in engineering contexts such as the development of magnetic confinement fusion reactors, or the understanding of transport in porous media that has applications in oil extraction or pollutant control, to name a few.

Footnotes are often used (some might say that excessively) to convey additional information. The reader should keep in mind that they can be ignored in a first read to avoid slowing down progress.

It might also happen that some of the material could be perceived as excessively introductory for some readers. In some cases, we will recommend them explicitly to browse quickly through the first sections of the chapter, that usually deal with basic concepts and introductory theory, and to move directly to the later sections that focus on methods and applications.

A list of proposed problems has also been included at the end of each chapter. Some of these problems will simply ask for the derivation of some formula that has been used, but not proved, in the main text. Other problems, much more interesting in our opinion, will ask the readers to build numerical codes to repeat or expand the examples that have been presented in the main text. In most of these cases, some basic guidance is often provided in the appendices included at the end of the corresponding chapter. We strongly recommend readers to try to work these exercises out. It is our firm belief that many of the subtleties of complexity can only be fully grasped after having had some first-hand experience with these systems.

Appendix 1: Fixed Points of a Dynamical System

A dynamical system is usually defined by a system of N ordinary differential equations (or simply, ODEs) that, written in matrix form, becomes:

$$\dot{\mathbf{x}} = \mathbf{F}(\mathbf{x}), \tag{1.12}$$

where $\mathbf{F}$ is an arbitrary N-dimensional non-linear function. The **fixed points** of the system are given by the solutions of $\mathbf{F}(\mathbf{x}_{\mathrm{fp}}) = 0$, that correspond to all the *time-independent* solutions of the problem. At most, N different fixed points may exist. Each fixed-point can be either **stable**, **unstable** or **neutral**, depending on whether a small perturbation away from the fixed point is attracted back to it (i.e., stable), diverges away from it (i.e., unstable) or remains close to the fixed point without either being attracted or ejected from it (i.e., neutral).

To determine their type [39], one needs to linearize the system of ODEs in the neighbourhood of each fixed point, $\mathbf{x}_{\mathrm{fp};k}$, where the index $k = 1, 2, \cdots, N_p$, being $N_p \leq N$ the number of fixed points of the system. To do this, one writes $\mathbf{x} = \mathbf{x}_{\mathrm{fp};k} + \mathbf{y}$ and assumes that the norm of the perturbation $|\mathbf{y}|$ remains small. The behaviour of the perturbation, when advanced in time, will tell the type of the fixed point. A linear advance is sufficient. Thus, one proceeds by inserting $\mathbf{x} = \mathbf{x}_{\mathrm{fp};k} + \mathbf{y}$ into Eq. 1.12, that yields the linear evolution equation of $\mathbf{y}$:

$$\dot{\mathbf{y}} = \left. \frac{\partial \mathbf{F}}{\partial \mathbf{x}} \right|_{\mathbf{x}=\mathbf{x}_{\mathrm{fp};k}} \cdot \mathbf{y}. \tag{1.13}$$

The matrix formed by taking partial derivatives of $\mathbf{F}$ and evaluating them at the fixed point is called the **jacobian matrix** at the fixed point. The solution of this equation is given by the exponential of the jacobian,

$$\mathbf{y}(t) = \exp\left(\left.\frac{\partial \mathbf{F}}{\partial \mathbf{x}}\right|_{\mathbf{x}=\mathbf{x}_{\text{fp};k}} (t - t_0) \right) \cdot \mathbf{y}(t_0). \tag{1.14}$$

Therefore, it is the eigenvalues of the jacobian at the fixed point that define the type of the fixed point. If there is *any eigenvalue with a **positive real part***, the perturbation will grow and the solution will move away from the fixed point, and the fixed point is **unstable**. If *all eigenvalues have **negative real parts***, the amplitude of the perturbation will eventually vanish, and the solution will come back to the fixed point, which is **stable**. If, on the other hand, *all eigenvalues are purely imaginary*, the norm of the perturbation remains constant over time and, although the solution never gets back to the fixed point, it does not run away from it. The fixed point is then **neutral**.

Problems

1.1 Predator-Prey Model: Definition
Prove that, if a term like $\gamma n_l - \nu n_l^2$ had been used in Eq. 1.2 of the predator-prey model to account for lynx mating and overpopulation, a finite number of lynx might exist in the absence of hares, which would make no sense in the context of the model.

1.2 Lotka-Volterra Model: Periodic Orbits

(a) Prove that the predator-prey model, if $\epsilon = 0$, admits the conserved quantity:

$$K(n_l, n_h) := \alpha \log(n_l) + \gamma \log(n_h) - \beta n_l - \delta n_h \tag{1.15}$$

(b) Write a numerical code to plot the orbits $K(n_l, n_h) = K_0$, for arbitrary K_0.

1.3 Predator-Prey Model: Fixed Points
With the help of the techniques discussed in Appendix 1 in this chapter,

(a) prove that all the fixed points of the PP model are given by Eqs. 1.3–1.5.
(b) show that the jacobians at the fixed points are given by:

$$A : \begin{pmatrix} \alpha & 0 \\ 0 & -\gamma \end{pmatrix}, \quad B : \begin{pmatrix} -\alpha & -\beta\alpha/\epsilon \\ 0 & \delta\alpha/\epsilon - \gamma \end{pmatrix}, \quad C : \begin{pmatrix} -\epsilon\gamma/\delta & -\beta\gamma/\delta \\ (\delta\alpha - \epsilon\gamma)/\beta & 0 \end{pmatrix}$$

$$\tag{1.16}$$

(c) find the eigenvalues of each jacobian.
(d) classify the type of each fixed point.

1.4 Predator-Prey Model: Phase Space

Write a computer code that numerically integrates the dynamical system given by Eqs. 1.1–1.2, starting from appropriate initial conditions. Use the code to reproduce all the phase space plots shown in Fig. 1.5.

1.5 The Running Sandpile: Building of a Cellular Automata

Write a computer program that evolves in time the one-dimensional running sandpile described in Sect. 1.3.1. The output of the program should include at least the time record of the state of each cell of the sandpile (stable or unstable), its height, the total mass confined and the outflux coming out of the bottom of the pile. This program will be the basis of several exercises proposed in later chapters.

1.6 Diffusive Equation: Global Confinement Scaling

Assume a one-dimensional system of size L driven by a source density S, uniform both in time and space. If transport in the system takes place according to the diffusive equation (Eq. 1.7) with boundary conditions: $n(L) = dn/dx(0) = 0$, show that:

(a) profiles are not stiff by finding the steady-state profile as a function of S and D;
(b) the scaling of the global confinement time, τ_{CF}, defined as the ratio of the total mass contained to the total drive strength, is $\tau_{CF} \propto L^2$;
(c) there is no power degradation if D is independent of the external source S.

References

1. Tautz, J.: The Buzz About Bees. Springer, Berlin (2007)
2. Krugman, P.R.: The Self-organizing Economy. Blackwell Publishers, London (2010)
3. Dana, K.S., Roy, P.K., Kurths, J. (eds.): Complex Dynamics in Physiological Systems: From Heart to Brain. Springer, Berlin (2009)
4. Goertzel, B.: Chaotic Logic: Language, Thought, and Reality from the Perspective of Complex Systems Science. Springer, Boston (1994)
5. Larsen-Freeman, D., Cameron, L.: Complex Systems and Applied Linguistics. Oxford University Press, Oxford (2008)
6. Murray, J.D.: Mathematical Biology. Springer, Heidelberg (1989)
7. Kaneko, K.: Life: An Introduction to Complex Systems Biology. Springer, New York (2006)
8. Dobson, I., Carreras, B.A., Lynch, V.E., Newman, D.E.: Complex Systems Analysis of Series of Blackouts: Cascading Failure, Critical Points, and Self-organization. Chaos 17, 026103 (2007)
9. Dijkstra, H.A.: Nonlinear Climate Dynamics. Cambridge University Press, Cambridge (2013)
10. Hergarten, S.: Self-organized Criticality in Earth Systems. Springer, Heidelberg (2002)
11. Drossel, B., Schwabl, F.: Self-organized Critical Forest Fire Model. Phys. Rev. Lett. 69, 1629 (1992)
12. Sheluhin, O., Smolskiy, S., Osin, A.: Self-similar Processes in Telecommunications. Wiley, New York (2007)
13. Park, K., Willinger, W.: The Internet as a Large-Scale Complex System. Santa Fe Institute Press, Santa Fe (2002)
14. Wright, Q.: A Study of War. University of Chicago Press, Chicago (1965)

15. Allison, M.A., Kelly, S.: The Complexity Advantage: How the Science of Complexity Can Help Your Business Achieve Peak Performance. McGraw Hill, New York (1999)
16. Chowdhury, D., Santen, L., Schadschneider, A.: Statistical Physics of Vehicular Traffic and Some Related Systems. Phys. Rep. 329, 199 (2000)
17. Binney, J., Tremaine, S.: Galactic Dynamics. Princeton University Press, Princeton, NJ (2008)
18. Pietronero, L.: The Fractal Structure of the Universe: Correlations of Galaxies and Clusters. Physica 144A, 257 (1987)
19. Baryshev, Y., Teerikorpi, P.: Discovery of Cosmic Fractals. World Scientific, Singapore (2000)
20. Frank, J., King, A., Raine, D.: Accretion Power in Astrophysics. Cambridge University Press, Cambridge (2002)
21. Priest, E.R.: Magnetohydrodynamics of the Sun. Cambridge University Press, Cambridge (2014)
22. Parker, E.N.: Cosmical Magnetic Fields. Oxford University Press, Oxford (1979)
23. Crosby, N.B., Aschwanden, M.J., Dennis, B.R.: Frequency Distributions and Correlations of Solar X-ray Flare Parameters. Sol. Phys. 143, 275 (1993)
24. Aschwanden, M.J.: Self-organized Criticality in Astrophysics. Springer, New York (2014)
25. Moffat, H.K.: Magnetic Field Generation in Electrically Conducting Fluids. Cambridge University Press, Cambridge (1978)
26. Akasofu, S.I., Chapman, S.: Solar-Terrestrial Physics. Oxford University Press, Oxford (1972)
27. Heikkila, W.J.: Earth's Magnetosphere. Elsevier Publishers, Amsterdam (2011)
28. Lui, A.T.Y., Chapman, S.C., Liou, K., Newell, P.T., Meng, C.I., Brittnacher, M., Parks, G.K.: Is the Dynamic Magnetosphere an Avalanching System? Geophys. Res. Lett. 27, 911 (2000)
29. Wesson, J.: Tokamaks. Oxford University Press, Oxford (2008)
30. Diamond, P.H., Hahm, T.S.: On the Dynamics of Turbulent Transport Near Marginal Stability. Phys. Plasmas 2, 3640 (1995)
31. Sanchez, R., Newman, D.E.: Topical Review: Self-Organized-Criticality and the Dynamics of Near-Marginal Turbulent Transport in Magnetically Confined Fusion Plasmas. Plasma Phys. Controlled Fusion 57, 123002 (2015)
32. Wagner, F., Stroth, U.: Transport in Toroidal Devices-the Experimentalist's View. Plasma Phys. Controlled Fusion 35, 1321 (1993)
33. Terry, P.W.: Suppression of Turbulence and Transport by Sheared Flow. Rev. Mod. Phys. 72, 109 (2000)
34. Boffetta, G., Cencini, M., Falcioni, M., Vulpiani, A.: Predictability: A Way to Characterize Complexity. Phys. Rep. 356, 367 (2002)
35. Prigogine, I., Nicolis, G.: Self-organization in Non-Equilibrium Systems. Wiley, New York (1977)
36. Chaisson, E.J.: Cosmic Evolution: The Rise of Complexity in Nature. Harvard University Press, Harvard (2002)
37. Lotka, A.J.: Elements of Physical Biology. Williams and Wilkins, London (1925)
38. Vano, J.A., Wildenberg, J.C., Anderson, M.B., Noel, J.K., Sprott, J.C.: Chaos in Low-Dimensional Lotka-Volterra Models of Competition. Nonlinearity 19, 2391 (2006)
39. Tenenbaum, M., Pollard, H.: Ordinary Differential Equations. Dover, New York (1985)
40. Chandrasekhar, S.: Hydrodynamic and Hydromagnetic Stability. Dover, New York (1981)
41. Mandelbrot, B.B.: Fractals: Form, Chance and Dimension. W H Freeman, New York (1977)
42. Feder, J.: Fractals. Plenum Press, New York (1988)
43. Bassingthwaighte, J.B., Liebovitch, L.S., West, B.J.: Fractal Physiology. Oxford University Press, New York (1994)
44. Gutenberg, B., Richter, C.F.: Magnitude and Energy of Earthquakes. Ann. Geofis. 9, 1 (1956)
45. Roberts, D.C., Turcotte, D.L.: Fractality and Self-organized Criticality of Wars. Fractals 6, 351 (1998)
46. Frisch, U.: Turbulence: The Legacy of A. N. Kolmogorov. Cambridge University Press, Cambridge (1996)
47. Ma, S.K.: Modern Theory of Critical Phenomena. W.A. Benjamin, Inc., Reading (1976)

48. Bak, P., Tang, C., Wiesenfeld, K.: Self-organized Criticality: An Explanation of the 1/f Noise. Phys. Rev. Lett. 59, 381 (1987)
49. Jensen, H.J.: Self-organized Criticality: Emergent Complex Behaviour in Physical and Biological Systems. Cambridge University Press, Cambridge (1998)
50. Chang, T.: Self-organized Criticality, Multi-fractal Spectra, Sporadic Localized Reconnections and Intermittent Turbulence in the Magnetotail. Phys. Plasmas 6, 4137 (1999)
51. Fick, A.: Uber Diffusion. Ann. Phys. 170, 59 (1885)
52. Einstein, A.: Investigations on the Theory of Brownian Movement. Dover, New York (1956)
53. Metzler, R., Klafter, J.: The Random Walk's Guide to Anomalous Diffusion: A Fractional Dynamics Approach. Phys. Rep. 339, 1 (2000)
54. Hunt, A., Ewing, R.: Percolation Theory for Flow in Porous Media. Springer, New York (2009)
55. Hwa, T., Kardar, M.: Avalanches, Hydrodynamics and Discharge Events in Models of Sand Piles. Phys. Rev. A 45, 7002 (1992)
56. Dickman, R., Muñoz, M.A., Vespignani, A., Zapperi, S.: Paths to Self-organized Criticality. Braz. J. Phys. 30, 27 (2000)
57. Boffetta, G., Carbone, V., Giuliani, P., Veltri, P., Vulpiani, A.: Power Laws in Solar Flares: Self-organized Criticality or Turbulence? Phys. Rev. Lett. 83, 4662 (1999)
58. Sanchez, R., Newman, D.E., Carreras, B.A.: Waiting-Time Statistics of Self- Organized-Criticality Systems. Phys. Rev. Lett. 88, 068302 (2002)
59. Sattin, F., Baiesi, M.: Self-Organized-Criticality Model Consistent with Statistical Properties of Edge Turbulence in a Fusion Plasma. Phys. Rev. Lett. 96, 105005 (2006)
60. Mier, J.A., Sanchez, R., Newman, D.E.: Characterization of a Transition in the Transport Dynamics of a Diffusive Sandpile by Means of Recurrence Quantification Analysis. Phys. Rev. E 94, 022128 (2016)

Chapter 2
Statistics

2.1 Introduction

Statistics is a fundamental tool to study physical phenomena in any system, be it complex or not. However, it is particularly important in the study of complex systems, for reasons that will be explained in this chapter in detail.

Any discussion on statistics must necessarily start with a definition of **probability**, of which two complementary views are often used in mathematical physics: the **temporal** and the **ensemble** approach. The former is more intuitive, since it hinges on our every day experience with measurement and repetition. The latter is more convenient from a theoretical point of view, being used often in mathematical treatments. In the **temporal approach**, the probability of an event is related to *its occurrence frequency in the system where it happens*. In the **ensemble approach**, one considers *many virtual copies of the same system*, and define the probability of an event as the *fraction of copies in which it is observed*. Clearly, preparing multiple copies of a system is usually inconvenient from an experimental point of view, making ensemble theory a tool that is useful mainly for theoreticians. A system or a process is called **ergodic** if the *temporal and ensemble approaches yield the same statistics*. In turbulence theory, such equivalence is known as *Taylor's hypothesis* [1], and it plays a central role in the theory of turbulence. Ergodic assumptions are also central to the development of statistical mechanics [2]. However, there are many systems that are not ergodic, not even in an approximate manner. In these cases, theoretical models based on the ensemble approach may not be applicable to actual data, that are often collected in the form of a multiple time series. One must then progress with care. Some of these systems do exhibit complex dynamics.

We will not discuss the basics of probability theory here, since it is a topic usually covered both at the high school and introductory college levels. Many monographs are available that provide a good introduction to the subject for those readers that might need to refresh their knowledge, or that would simply like to

© Springer Science+Business Media B.V. 2018
R. Sánchez, D. Newman, *A Primer on Complex Systems*,
Lecture Notes in Physics 943, https://doi.org/10.1007/978-94-024-1229-1_2

extend it [3, 4]. Instead, we will start with a review of the concept of **probability density function** (or **pdf**), that will be used often throughout this book. Since the level of the material reviewed is introductory, some readers may prefer to jump directly to Sect. 2.3, where the significance of specific pdfs starts to be discussed. In later sections (Sects. 2.4 and 2.5) we will also describe several techniques that are particularly useful to determine algebraic tails accurately, among other things. This is an important capability in this context, since algebraic tails are often seen in complex systems but are difficult to determine in practice due to the low probability of the events they contain.

2.2 The Probability Density Function

The probability density function is introduced to deal with *any quantity that has a continuous support*. That is the case, for instance, of the temperature in a room, of any component (or the magnitude) of the local velocity in a turbulent fluid, or of the time intervals between successive disintegrations of a radioactive material. In all these cases, the outcome of each measurement is given by a **continuous variable** that takes values within a certain range $\Omega = (a, b)$.[1]

2.2.1 Definition

Any definition of probability based on repetition becomes useless in the case of continuous variables since the total number of possible outcomes is infinite. To get around this, a function, $p(x)$, is introduced[2] such that $p(x)dx$ gives the probability P of the outcome X lying between x and $x + dx$ (assuming dx small):

$$P(X \in [x, x + dx]) = p(x)dx. \tag{2.1}$$

$p(x)$ is known as the **probability density function** (or, simply, the **pdf**). This function must satisfy a couple of conditions to ensure that $p(x)dx$ is a proper probability. First, probabilities must be always positive. It thus follows that

$$p(x) \geq 0, \quad \forall x \in \Omega. \tag{2.2}$$

---

[1]In the case of the room temperature, $a = -273.13\,°C$ and $b = +\infty$; for a component of the fluid velocity, $a = -\infty; b = +\infty$ while its magnitude varies between $a = 0$ and $b = +\infty$. Finally, waiting-times between disintegrations vary within the range defined by $a = 0$ and $b = +\infty$.

[2]We will always represent probabilities with an uppercase P. Also, we will represent the generic outcome of an observation or experiment by an uppercase X.

Table 2.1 Examples of some common probability density functions (pdfs)

Name	p(x)	Domain
Uniform	$\dfrac{1}{b-a}$	[a, b]
Gaussian	$\dfrac{1}{\sqrt{2\pi w^2}}\exp\left(-\dfrac{(x-\mu)^2}{2w^2}\right)$	$(-\infty, \infty)$
Cauchy	$\dfrac{1}{\pi}\dfrac{\gamma}{(x-\mu)^2+\gamma^2}$	$(-\infty, \infty)$
Exponential	$\dfrac{1}{\tau}\exp(-x/\tau)$	$[0, \infty)$
Laplace	$\dfrac{1}{2\tau}\exp(-|x-\mu|/\tau)$	$(-\infty, \infty)$

Secondly, the joint probability for all possible outcomes must equal one. Thus,

$$\int_{\Omega} p(x)dx = 1. \tag{2.3}$$

It is worth noting that $p(x)$ may exceed one locally since the pdf is not a probability in itself.[3] Only, $p(x)dx$ is! Some common analytical pdfs are collected in Table 2.1 and illustrated in Fig. 2.1.

2.2.2 Cumulative Distribution Function

The *cumulative distribution function* is defined as[4]:

$$\text{cdf}(x) = \int_{a}^{x} dx' p(x'), \quad x \in \Omega, \quad a = \min(\Omega). \tag{2.4}$$

It gives *the probability of the outcome being less than a certain x*. For that reason, it is also denoted by $P[X \leq x]$. Naturally, the cdf must be equal to zero at the minimum possible value for x (i.e., $x = a$), and tends to one as the maximum allowed value (i.e., b) is approached (see Fig. 2.2). The pdf can be derived from the cdf via the relation,

$$p(x) = \frac{d}{dx}\left(P[X \leq x]\right). \tag{2.5}$$

---

[3]For instance, a Gaussian pdf will exceed 1 if $w < 1/(2\pi)$. A Cauchy will exceed 1 if $\gamma < 1/\pi$.
[4]Both the **cdf** and the **sf** are very useful tools to determine the tails of pdfs in the case of power-law statistics, as will be discussed in Sect. 2.4.3.

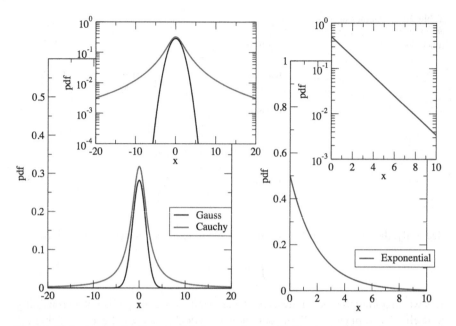

Fig. 2.1 Plots showing the Gaussian (black, $\mu = 0$, $w = 2$), Cauchy (red, $\mu = 0$, $\gamma = 1$) and Exponential (blue, $\tau = 2$) pdfs in lin-lin and log-lin scales. Note the much heavier tail of the Cauchy distribution, with respect to the Gaussian

2.2.3 Survival Function

The *survival function*[5] is defined as:

$$\mathrm{sf}(x) = \int_x^b dx' p(x').$$

(2.6)

That is, it gives the *probability of the outcome being larger than a certain* x, or $P[X \geq x]$. As a result, it is complementary to the cdf, being equal to one at the minimum possible value for x (i.e., a), and becoming zero as the maximum value (i.e., b) is approached (see Fig. 2.2). The pdf is obtained from the survival function via,

$$p(x) = -\frac{d}{dx} \left(P[X \geq x] \right).$$

(2.7)

---

[5]The survival function is also known as the **complementary cumulative distribution function** (or **ccdf**). The name *survival function* originated in biological studies that investigated the probability of a certain species surviving beyond a certain length of time.

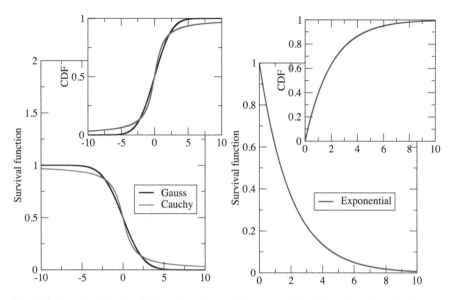

Fig. 2.2 Plots showing the cdf (above) and survival functions (below) for a Gaussian (black, $\mu = 0$, $w = 2$), Cauchy (red, $\mu = 0, \gamma = 2$) and Exponential (blue, $\tau = 2$) pdfs. It should be noted that the much heavier tail of the Cauchy distribution implies a much waker growth (for the cdf) and decrease (for the sf) than in the case of the Gaussian

A list of survival and cumulative distribution functions of some common analytic pdfs is given in Table 2.2.[6] It is also worth noting that cdf(x) + sf$(x) = 1$, since the total probability must always equal one.

2.2.4   Characteristic Function

Another important function related to the pdf is the *characteristic function*.[7] It is defined by the integral (using $\imath = \sqrt{-1}$ for the imaginary unit)

$$\phi^p(k) := \int_\Omega dx \, e^{\imath k x} p(x). \tag{2.9}$$

---

[6]The **error function** used in Table 2.2 is defined as:

$$\mathrm{erf}(x) \equiv \frac{2}{\sqrt{\pi}} \int_0^x dx \, e^{-x^2}. \tag{2.8}$$

[7]Characteristic functions are very important in the context of complex systems since, for many meaningful pdfs with power-law tails, only their characteristic function has a closed analytical expression. This makes, for instance, that comparisons against experimental data can often be done more easily using the characteristic function.

Table 2.2 Cumulative distribution and survival functions of some common pdfs

Name	cdf(x)	sf(x)
Uniform	$\dfrac{x-a}{b-a}$	$\dfrac{b-x}{b-a}$
Gaussian	$\dfrac{1}{2}\left[1+\mathrm{erf}\left(\dfrac{x-\mu}{\sqrt{2}w}\right)\right]$	$\dfrac{1}{2}\left[1-\mathrm{erf}\left(\dfrac{x-\mu}{\sqrt{2}w}\right)\right]$
Cauchy	$\dfrac{1}{2}+\dfrac{1}{\pi}\arctan\left(\dfrac{x-\mu}{\gamma}\right)$	$\dfrac{1}{2}-\dfrac{1}{\pi}\arctan\left(\dfrac{x-\mu}{\gamma}\right)$
Exponential	$1-\exp\left(-\dfrac{x}{\tau}\right)$	$\exp\left(-\dfrac{x}{\tau}\right)$
Laplace	$\begin{cases}\dfrac{1}{2}\exp\left(\dfrac{x-\mu}{\tau}\right), & x<\mu \\ 1-\dfrac{1}{2}\exp\left(\dfrac{\mu-x}{\tau}\right), & x>\mu\end{cases}$	$\begin{cases}1-\dfrac{1}{2}\exp\left(\dfrac{x-\mu}{\tau}\right), & x<\mu \\ \dfrac{1}{2}\exp\left(\dfrac{\mu-x}{\tau}\right), & x>\mu\end{cases}$

Table 2.3 Characteristic functions of some common pdfs

Name	p(x)	$\phi^p(k)$
Uniform	$\dfrac{1}{b-a}$	$\dfrac{\exp(\imath kb)-\exp(\imath ka)}{\imath k(b-a)}$
Gaussian	$\dfrac{1}{\sqrt{2\pi w^2}}\exp\left(-\dfrac{(x-\mu)^2}{2w^2}\right)$	$\exp\left(\imath\mu k-\dfrac{1}{2}w^2k^2\right)$
Cauchy	$\dfrac{1}{\pi}\dfrac{\gamma}{(x-\mu)^2+\gamma^2}$	$\exp\left(\imath\mu k-\gamma|k|\right)$
Exponential	$\dfrac{1}{\tau}\exp\left(-x/\tau\right)$	$\dfrac{1}{(1-\imath\tau k)}$
Laplace	$\dfrac{1}{2\tau}\exp\left(-|x-\mu|/\tau\right)$	$\dfrac{\exp(\imath\mu k)}{(1+\tau^2k^2)}$

It is worth noting that, if the range $\Omega = (-\infty, \infty)$, the characteristic function becomes the Fourier transform of the pdf (see Appendix 1 for a brief introduction to the Fourier transform and the conventions adopted in this book), $\phi^p(k) = \hat{p}(k)$. In that case, the pdf can be trivially recovered by inverting the transform to get,

$$p(x) = \frac{1}{2\pi}\int_{-\infty}^{\infty} dk\, e^{-\imath kx}\phi^p(k). \tag{2.10}$$

If $\Omega = [0, \infty)$ or $\Omega = [a, b]$, the range can be trivially expanded to $(-\infty, \infty)$ by extending the pdf definition to be $p(x) = 0, \ \forall x \notin \Omega$, which allows the equivalence to the Fourier transform. This convention will be implicitly assumed throughout this book. A list of characteristic functions of common pdfs is given in Table 2.3.

2.2.5 Expected Values

Expected values are defined, for any arbitrary function $g(x)$ of the random variable x whose statistics are described by $p(x)$, by the weighted integral:

$$\langle g \rangle = \int_\Omega dx\, g(x)p(x), \tag{2.11}$$

that effectively sums the product of each possible value of the function, $g(x)$, with its probability of taking place, $p(x)$. Some expected values are particularly useful. For example, the characteristic function just discussed (Eq. 2.9) can be expressed as the expected value:

$$\phi^p(k) = \langle \exp(\imath k) \rangle . \tag{2.12}$$

Some of the more common *statistical measures* that characterize a process or event are also expected values. Of particular importance are the so-called **moments** and **cumulants** of a distribution.

2.2.6 Moments

The **moments** of the pdf are defined as the expected values of the *integer powers of the outcome* x:

$$m_n := \langle x^n \rangle = \int_\Omega dx\, x^n p(x), \quad n = 1, 2, \cdots \tag{2.13}$$

It can be shown that *all the information contained in the pdf is also contained in the infinite series of moments*. Therefore, the knowledge of all the moments is as good as knowing the pdf itself. Indeed, any pdf can be reconstructed from its moments. One simply needs to insert the Taylor series of the exponential,

$$e^x = \sum_{n=0}^{\infty} \frac{x^n}{n!}, \tag{2.14}$$

in Eq. 2.12 to get:

$$\phi^p(k) = \sum_{n=0}^{\infty} \frac{(\imath k)^n}{n!} m_n, \tag{2.15}$$

where we have assumed that all moments are finite. From $\phi^p(k)$, the pdf follows by using Eq. 2.10. Inversely, the characteristic function can be used to generate all finite moments of its associated pdf via:

$$m_n = (-1)^n \frac{d^n \phi^p}{dk^n}(0). \tag{2.16}$$

2.2.7   Cumulants

Cumulants are another very useful family of expected values. They are defined as:

$$c_n = \left\langle x^n \frac{\ln p(x)}{p(x)} \right\rangle = \int_\Omega dx \, x^n \ln p(x). \tag{2.17}$$

The cumulants contain, as did the moments, all the information of the pdf. Indeed, the characteristic function of $p(x)$ can be expressed in terms of its cumulants as,

$$\ln \phi^p(k) = \sum_{n=0}^{\infty} \frac{(\imath k)^n}{n!} c_n, \tag{2.18}$$

assuming that all cumulants exist. The finite cumulants can also be generated from the characteristic function using,

$$c_n = (-1)^n \frac{d^n [\ln \phi^p]}{dk^n}(0). \tag{2.19}$$

It is straightforward to derive relations between moments and cumulants. For instance, by equating Eqs. 2.19 and 2.16, the following relations follow for the lowest four (see Problem 2.3):

$$c_1 = m_1 \tag{2.20}$$

$$c_2 = m_2 - m_1^2 \tag{2.21}$$

$$c_3 = m_3 - 3m_2 m_1 + 2m_1^3 \tag{2.22}$$

$$c_4 = m_4 - 4m_3 m_1 - 3m_2^2 + 12m_2 m_1^2 - 6m_1^4 \tag{2.23}$$

These expressions provide alternative formulas to compute the cumulants, once the moments are known.

The first few cumulants efficiently condense abundant information about the shape of the pdf. In particular, the four first cumulants are often combined to define four popular statistical figures-of-merit: the *mean*, the *variance*, the *skewness* and the *kurtosis*.

Fig. 2.3 Sketch illustrating
the meaning of positive and
negative skewness of a pdf

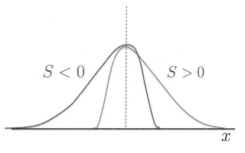

1. **Mean**. Its explicit expression is:

$$\bar{x} \equiv c_1 = m_1 = \int_\Omega dx \, xp(x).$$ (2.24)

The mean $\bar{x}$ provides the **typical value**[8] for the outcome x.

2. **Variance**. It is defined as:

$$\sigma^2 \equiv c_2 = m_2 - m_1^2 = \int_\Omega dx \, (x - \bar{x})^2 p(x).$$ (2.25)

Its square-root, σ, is known as the **standard deviation** and it is often used to characterize *the spread of all possible outcomes around the mean*.[9]

3. **Skewness**. It is defined as:

$$S \equiv \frac{c_3}{c_2^{3/2}} = \frac{1}{\sigma^3} \int_\Omega dx \, (x - \bar{x})^3 p(x).$$ (2.26)

The skewness quantifies the *asymmetry of a pdf around its mean*. It is zero for distributions that are symmetric around their mean. That is, those for which $p(\bar{x} - x) = p(\bar{x} + x)$. The geometrical meaning of positive and negative skewness is illustrated in Fig. 2.3.

---

[8]The mean should not be confused with the **most probable value**, that is that value of x where $p(x)$ reaches its maximum. It is also different from the **median**, that is the value x_m, for which $P(X > x_m) = P(X < x_m)$.

[9]The standard deviation is often used to estimate the error of a measurement, with the mean of all tries providing the estimate for the measurement. In this case, it is implicitly assume that all errors are accidental, devoid of any bias [5].

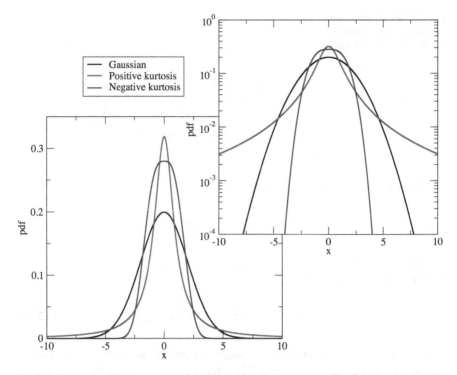

Fig. 2.4 Illustration of the meaning of positive and negative kurtosis. The Gaussian (or normal) distribution has zero kurtosis and is shown in black ($\mu = 0$, $w = 2$). In red, a Cauchy distribution is shown ($\mu = 0$, $\gamma = 1$); it has an infinite (positive) kurtosis. The blue curve, on the other hand, decays as $p(x) \sim \exp(-x^3)$ for large x and has a negative kurtosis. Left box is in lin-lin scale; right box, in lin-log scale

4. **Kurtosis**. It is defined as[10]:

$$\kappa \equiv \frac{c_4}{c_2^2} - 3 = \frac{1}{\sigma^4} \int_\Omega dx \, (x - \bar{x})^4 p(x) - 3. \tag{2.27}$$

The kurtosis κ quantifies, for symmetric distributions, the *peakedness of a pdf around its mean* and the *significance of its tail* relative to the Gaussian pdf. As the kurtosis gets large and positive, so does the significance (also referred to as *heaviness* or *fatness*) of the tail, that decays more slowly than the Gaussian pdf (see Fig. 2.4, in red). On the other hand, as it gets more negative, the central part of the pdf will be flatter than for a Gaussian pdf, but the tails will fall much faster (Fig. 2.4, in blue).

---

[10]Note that our definition of kurtosis corresponds to what in some textbooks is called the **excess kurtosis**. In those cases, the kurtosis is defined instead as $\kappa := \langle (x - \mu)^4 \rangle / c_2^2$ which, for a Gaussian pdf yields a value of three. The **excess kurtosis** is then defined as $\kappa_e := \kappa - 3$, so that a positive κ_e means a heavier tail than a Gaussian and a negative value, the opposite.

Table 2.4 Cumulants of some common pdfs

Name	p(x)	Mean	Variance	Skewness	Kurtosis
Uniform	$\dfrac{1}{b-a}$	$\dfrac{a+b}{2}$	$\dfrac{(b-a)^2}{12}$	0	$-\dfrac{6}{5}$
Gaussian	$\dfrac{1}{\sqrt{2\pi w^2}} \exp\left(-\dfrac{(x-\mu)^2}{2w^2}\right)$	μ	w^2	0	0
Cauchy	$\dfrac{1}{\pi}\dfrac{\gamma}{(x-\mu)^2+\gamma^2}$	μ	∞	0	∞
Exponential	$\dfrac{1}{\tau}\exp(-x/\tau)$	τ	τ^2	2	6
Laplace	$\dfrac{1}{2\tau}\exp(-|x-\mu|/\tau)$	μ	$2\tau^2$	0	3

The values of these statistical figures-of-merit for some common analytical pdfs[11] have been collected in Table 2.4. It is worth noting that, in some cases, the cumulants may not exist because of lack of convergence of the integral that defines them.

2.3 Significance of Specific Pdfs

One of the first things scientists often do when examining experimental or numerical data is to look at their pdf. The shape of these pdfs can sometimes offer a first hint about the nature of the underlying dynamics of the system under study. **Gaussian statistics** are usually associated with non-complex dynamics, lack of long-term memory and diffusive transport. They have been extensively used to interpret experimental data, to estimate errors in measurements and virtually in almost every situation in which statistics appear, not limited to the physical sciences by any means. In fact, they are so familiar to us that the Gaussian pdf has earned the name of **normal distribution**. **Power-law statistics**, on the other hand, are often found in situations where complex dynamics are at play.[12] There are many other possibilities,

---

[11]The skewness and kurtosis are particularly useful to detect departures of Gaussian behaviour in experimental data, due to the fact that the cumulants of the Gaussian pdf satisfy that $c_n = 0, n > 2$ (see Problem 2.4). These deviations are also referred to as *deviations of normality*, since the Gaussian is also known as the *normal distribution*!

[12]One must however remain cautious and avoid taking any of these general ideas as indisputable evidence. Such connections, although often meaningful, are not bullet-proof and sometimes even completely wrong. Instances exist of complex systems that exhibit Gaussian statistics, and non-complex systems with power-law statistics. It is thus advisable to apply as many additional diagnostics as possible in order to confirm or disregard what might emerge from the statistical analysis of the system. Knowing as much as possible about the underlying physics also helps.

though. Some of them convey their own significance, such as the **exponential** or the **log-normal** distributions. We will discuss some of these pdfs in detail in this section.

2.3.1   Gaussian and Lévy Pdfs: Additive Processes

Why are Gaussians (or near-Gaussians) found so often in natural processes? The traditional answer is provided by the **central limit theorem** (CLT), the most popular member of a series of mathematical theorems stating that, under certain conditions, the pdf of the outcomes of performing certain operations on random variables will tend towards some member of just a few families of **attractor distributions**. The existence of such theorems has important practical implications. First, it explains why just a few pdfs, amongst the many possible ones, seem to be observed so often in nature. Secondly, these theorems identify the most reasonable pdfs to choose from when trying to model a certain process.

The standard **central limit theorem** identifies the attractor distributions for the outcomes of *the addition of a sufficiently large number (N) of independent random variables*.[13] We will not dwell here on the mathematical details of the CLT. We refer the interested reader to the abundant bibliography in the matter [4, 6, 7]. It suffices to say that, loosely speaking, the CLT states that, *if the means and variances of all independent variables are finite and respectively equal to μ and σ^2, the attractor pdf will be a **Gaussian pdf**, with mean μ and variance given by σ^2/N. Many phenomena in nature can be thought of as resulting from the addition of many, approximately independent, microscopic events. These processes are usually referred to as **additive processes**. One example is provided by most transport processes, since the observed macroscopic displacements over a certain amount of time are usually the result of many microscopic displacements.[14]

Most textbooks present the CLT along the lines just described. However, the CLT can also be formulated for cases in which *the requirement is dropped of a finite variance for the distributions being added*. In that case, the CLT states that the attractor pdf is a member of the **Lévy family of pdfs** [6] that is selected depending on "how fast the variances of the individual random variables diverge", among other things. What is it meant by saying that the variance diverges faster or slower? Let's consider the case in which the individual pdfs behave asymptotically as[15],

$$p(x) \sim x^{-s}, \quad x \to \infty; \quad 1 < s < 3. \tag{2.28}$$

---

[13]A *random variable* is the way we refer to the unpredictable outcome of an experiment or measurement. Mathematically, it is possible to give a more precise definition, but in practical terms, its unpredictability is its more essential feature.

[14]For example, Gaussian distributions are also well suited to model unbiased measurement errors, since they are always assumed to be the result of many additive, independent factors [5].

[15]$s > 1$ is needed to make sure that the pdf be integrable and normalizable to unity. We also assume $x > 0$, but the same argument could be done for $x < 0$ by changing $x \to -x$.

Their variance would then scale as,

$$\sigma^2 = \int^\infty dx\, x^2 p(x) \sim \int^\infty dx\, x^{2-s} \sim x^{3-s}\big|^\infty, \tag{2.29}$$

which clearly diverges whenever $s < 3$. The closer s is to one, "the faster" the divergence is. The value of the exponent s is one of the elements that determines the member of the Lévy family that becomes the attractor.

Lévy distributions should often be expected in complex systems for the same reasons that Gaussians are so prevalent in other systems. Indeed, since the statistics in complex systems are often scale-invariant (i.e., power-law like), any additive processes taking place in them will tend to exhibit Lévy statistics if their power-law tails are sufficiently heavy.

Let's define the Gaussian and Lévy distributions in more detail.

2.3.1.1 Gaussian Distribution

The Gaussian pdf[16] is defined as (see Fig. 2.1, left panel):

$$N_{\mu,\sigma^2}(x) := \frac{1}{\sqrt{2\pi\sigma^2}} \exp\left(-\frac{(x-\mu)^2}{2\sigma^2}\right), \quad \Omega = (-\infty, \infty). \tag{2.30}$$

Its mean is $c_1 = \mu$ and its variance $c_2 = \sigma^2$. The width of the Gaussian is usually defined as $w = \sqrt{\sigma^2}$. All the moments of the Gaussian are finite and can be expressed in terms of μ and σ^2. Another important characteristic is that its cumulants all vanish for $n > 2$. That is, all above the variance are identically zero.

Finally, its characteristic function is:

$$\phi^N_{[\mu,w]}(k) = \exp\left(\imath\mu k - \frac{1}{2}\sigma^2 k^2\right). \tag{2.31}$$

2.3.1.2 Lévy Distributions

Lévy distributions [7] span a four-parameter family, $L_{[\alpha,\lambda,\mu,\sigma_L]}(x)$, with the ranges of their defining parameters being: $0 < \alpha < 2$, $|\lambda| \le 1$, $\sigma_L > 0$ and $|\mu| < \infty$. An analytic expression does not exist for them, except in a few cases. Their

---

[16]We adhere to the convention of representing the Gaussian with the uppercase letter N, a consequence of its being called the *normal* distribution.

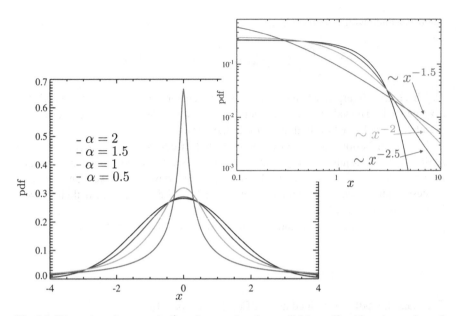

Fig. 2.5 Illustration of symmetric ($\lambda = 0$, $\sigma_L = 1$ and $\mu = 0$) Lévy pdfs with various values of the parameter $\alpha \in (0, 2)$. As it can be appreciated, the pdfs exhibit an asymptotic power law tail that becomes heavier, the smaller α is. The case with $\alpha = 2$ (in black) corresponds to the Gaussian pdf [Left box is in lin-lin scale; right box, in lin-log scale]

characteristic function can however be written in closed form [7]:

$$
\phi^L_{[\alpha,\lambda,\mu,\sigma_L]}(k) = \begin{cases} \exp\left\{\imath\mu k - \sigma_L^\alpha |k|^\alpha \left[1 - \imath\lambda \mathrm{sgn}(k) \tan\left(\frac{\pi\alpha}{2}\right)\right]\right\}, \alpha \neq 1 \\[2em] \exp\left\{\imath\mu k - \sigma_L |k| \left[1 + \frac{2\imath\lambda}{\pi}\mathrm{sgn}(k)\log|k|\right]\right\}, \quad \alpha = 1 \end{cases}
$$

$$(2.32)$$

The meanings of the four parameters, α, λ, σ and μ are:

- *Parameter α.* Its range is $0 < \alpha < 2$. This parameter sets the **asymptotic behaviour of the tail of the distribution at large** x (see Fig. 2.5). In fact, all Lévy pdfs satisfy that[17]:

$$
L_{[\alpha,\lambda,\mu,\sigma_L]}(x) \sim C_\alpha \left(\frac{1 \pm \lambda}{2}\right) \sigma_L^\alpha |x|^{-(1+\alpha)}, \quad x \to \pm\infty \qquad (2.33)
$$

---

[17]For that reason, CLT states that the average of random variables with individual pdfs exhibiting tails $p(x) \sim x^{-s}$ will be attracted towards a Lévy distribution with $\alpha = s - 1$ if $s < 3$. The value of λ of the attractor distribution will depend on the level of asymmetry of the individual pdfs.

where the constant is ($\Gamma(x)$ is Euler's gamma function),

$$C_\alpha = \begin{cases} \dfrac{\alpha(\alpha - 1)}{\Gamma(2 - \alpha)\cos(\pi\alpha/2)}, & \alpha \neq 1 \\[4mm] \dfrac{2}{\pi}, & \alpha = 1 \end{cases} \qquad (2.34)$$

Therefore, the tail of a Levy becomes heavier, the smaller α is. In addition, all moments diverge for $n \geq \alpha$. Therefore, all Lévy pdfs *lack a finite variance*. Furthermore, all distributions with $\alpha \leq 1$ also lack a finite mean!

- *Parameter* λ. Its range is $|\lambda| \leq 1$. This parameter quantifies the degree of **asymmetry** of the distribution (see Fig. 2.6), since all Lévy pdfs satisfy that

$$L_{[\alpha,\lambda,\mu,\sigma_L]}(x) = L_{[\alpha,-\lambda,\mu,\sigma_L]}(-x). \qquad (2.35)$$

Thus, the only symmetric Lévy pdfs are those with $\lambda = 0$. The degree of asymmetry increases with the magnitude of λ, while its sign tells whether the distribution leans more to the left (negative x's) or to the right (positive x's) side.[18]

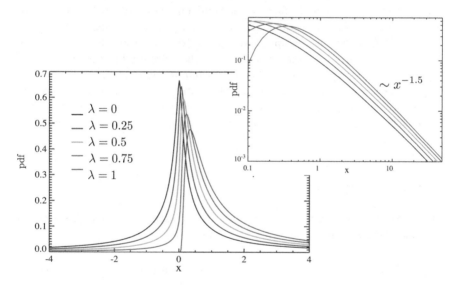

Fig. 2.6 Illustration of (right)-asymmetric ($\alpha = 0.5$, $\sigma_L = 1$ and $\mu = 0$) Lévy pdfs with various values of the parameter $\lambda \in [0, 1]$. As it can be appreciated, the pdfs exhibit all the same asymptotic power law tail, since α is the same. The degree of right-asymmetry becomes stronger, the closer to unity λ is. For $\lambda = 0$, it is symmetric. On the other hand, the case with $\lambda = 1$ (in pink) corresponds to an extremal Lévy pdf (see discussion in text), that is defined only for $x \geq 0$ [Left box is in lin-lin scale; right box, in lin-log scale]

---

[18]The case of maximum asymmetry (i.e., when $|\lambda| = 1$) is particularly interesting, and will be discussed separately later in this chapter. We will also refer to it often in Chap. 5, while discussing fractional transport in the context of continuous time random walks.

- *Parameter* σ_L. Its range is $\sigma_L > 0$. The parameter σ_L is called the **scale parameter** because of the following property:

$$L_{[\alpha,\lambda,\mu,\sigma_L]}(ax) = L_{[\alpha,\text{sgn}(a)\lambda,a\mu,|a|\sigma_L]}(x), \quad \forall a. \tag{2.36}$$

That is, changing σ_L only affects the "width" of the distribution. It is worth noting that this width is somewhat different from the usual Gaussian width, that is given by the square-root of the variance, which diverges here. It can also be shown that, for all Lévy pdfs with $\mu = 0$, their finite moments verify:

$$\langle |x|^p \rangle \propto \sigma_L^p, \quad p < \alpha. \tag{2.37}$$

so that one can define an *effective width* as,

$$w := \langle |x|^p \rangle^{1/p} \propto \sigma_L, \quad p < \alpha. \tag{2.38}$$

- *Parameter* μ. Its range is $|\mu| < \infty$. The parameter μ is called the **translation parameter**. The reason for this name is related to a property of the Fourier transform. In particular, the fact that the Fourier transform of any function whose independent variable is shifted by some arbitrary Δ is the same as that of the unshifted function, but multiplied by a phase $\imath \Delta k$ (see Appendix 1, Eq. 2.127). Therefore, the $\imath \mu k$ term that appears in the characteristic function of the Lévy pdf (Eq. 2.33) simply displaces the distribution to the right (if $\mu < 0$) or left ($\mu > 0$) for $x = 0$. The parameter μ is also somewhat related to the usual mean. In fact, μ gives the mean of all Lévy distributions with $\lambda = 0$ and $\alpha > 1$. For non-symmetric distributions (i.e., $\lambda \neq 0$) with $\alpha > 1$, the mean depends however both on μ and σ_L.

Analytical Expressions for Lévy Pdfs

There are only two Lévy distributions that have a closed analytical expression: (1) the **Cauchy distribution** ($\alpha = 1, \lambda = 0$; see Fig. 2.1),

$$\text{Cau}_{[\sigma_L]}(x) \equiv L_{[1,0,0,\sigma_L]}(x) = \frac{\sigma_L}{\pi(x^2 + \sigma_L^2)}, \quad x \in (-\infty, \infty) \tag{2.39}$$

and (2) the **Lévy distribution**,[19] ($\alpha = 1/2, \lambda = 1$),

$$\text{Lev}_{[\sigma_L]}(x) \equiv L_{[1/2,1,0,\sigma_L]}(x) = \left(\frac{\sigma_L}{2\pi}\right)^{1/2} \frac{1}{x^{3/2}} \exp(-\sigma_L/2x), \quad x \in [0, \infty) \tag{2.40}$$

---

[19] Although the name Lévy distribution is used, in this book, for any member of the Lévy family of pdfs, in many statistical contexts the name is reserved to this specific choice of parameters. This is kind of unfortunate, and can lead to confusions sometimes.

Extremal Lévy Distributions

The attentive reader may have noted that the analytical expression for $L_{[1/2,1,0,\sigma_L]}(x)$ (see Eq. 2.40) is only defined for $x \geq 0$. This distribution is an example of a particularly important subfamily of Lévy distributions known as **extremal Lévy distributions**. These subfamily includes all Lévy pdfs with $\alpha < 1$ and $\lambda = \pm 1$. What makes them interesting (see Fig. 2.6) is that the range of these pdfs only spans the interval $[0, \pm\infty)$.[20] This property makes extremal Lévy distributions with $\lambda = 1$ well suited as attractor pdfs for processes that involve variables that can only take positive values, such as waiting times, when they lack a finite mean. We will come back to these distributions again in Chap. 5.

Gaussian Pdf as a Limiting Case of the Lévy Family

A final comment about Lévy distributions worth making is that the Gaussian pdf can be obtained as a limiting case of the Lévy characteristic function (Eq. 2.32).[21] Indeed, in the limit $\alpha \to 2^-$, $\lambda \to 0$, the Lévy characteristic function reduces to:

$$\phi^L_{[2,0,\mu,\sigma_L]}(k) = \exp\left\{\imath\mu k - \sigma_L^2 k^2\right\} = \phi^N_{[\mu,\sqrt{2}\sigma_L]}(k), \tag{2.41}$$

that is the characteristic function of a Gaussian pdf with mean μ and with variance $\sigma^2 = 2\sigma_L^2$ (see Eq. 2.31). We will use this notion extensively in Chap. 5, when discussing fractional transport.

2.3.2 Log-Normal and Log-Stable Pdfs: Multiplicative Processes

We discussed previously that the CLT explains, in the case of additive processes, the attraction of their statistics towards the Gaussian or Lévy pdfs. However, sometimes one encounters processes of a multiplicative nature instead. Multiplicative processes are the consequence of the *accumulation of many small percentage changes*, instead of many small increments. In this situation, one should not expect to see any attraction towards Gaussian or Lévy statistics, but to something else.

Let's consider, for instance, the case of a highly communicable epidemic. It all starts with a few (maybe even just one) infected people, the so-called *patients zero*. We will build a very simple model for the propagation of the epidemic by assuming that the average probability that an infected person transmits the disease to a healthy

---

[20]For $\alpha > 1$, the Lévy pdfs with $\lambda = \pm 1$ do cover the whole real line, but it can be shown that they vanish exponentially fast as $x \to \mp\infty$ [7].

[21]In fact, some authors include the Gaussian as the only non-divergent member of the Lévy family.

one is p, and that the average number of people met by a normal person during a normal day is n. The increment in the number of infected people in a single day will be then given by,

$$\Delta N_{\text{infected}} \simeq pn N_{\text{infected}} \longrightarrow \frac{\Delta N_{\text{infected}}}{N_{\text{infected}}} \simeq pn. \qquad (2.42)$$

That is, the increment in the number of infected increases with the number of infected during the epidemic, but the percentage change in the number of infected remains equal to pn.

Within the context of our simple model for the epidemic, the number of infected after the first $(k + 1)$-th days can be estimated as,

$$N_{\text{infected}}^{k+1} = N_{\text{infected}}^{k} + p_k n_k N_{\text{infected}}^{k} - \epsilon_k N_{\text{infected}}^{k} = (1 + p_k n_k - \epsilon_k) N_{\text{infected}}^{k}. \qquad (2.43)$$

That is, the model simply adds, to the number of infected that already exists, the number of newly infected people during the k-th day and subtracts the number of people that has overcome the disease that day or has died from it. Evidently, p_k is the infection probability in the k-th day, n_k is the number of people met that day and $0 \leq \epsilon_k \leq 1$ is the probability of a sick person becoming healthy again or dying. The reason why we do not use constant values for these coefficients in the model is that any of these quantities may fluctuate from one day to another for various reasons.[22]

It is now straightforward to connect the number of infected at the $(k + 1)$-th day to the number of patients zero by applying Eq. 2.43 recursively to get:

$$N_{\text{infected}}^{k+1} = \left[\prod_{i=1}^{k} \eta_i \right] \cdot N_{\text{infected}}^{0}, \quad \eta_i = 1 + p_i n_i - \epsilon_i \geq 0, \qquad (2.44)$$

where N_{infected}^{0} is the number of people initially infected. It is clear that the statistics of N^k are not those of an additive process, but of a multiplicative one related to k copies of the η random variable, that contains the details about the infection rate, degree of curability of the disease, etc. So the question becomes, which are the attractor distributions for these kind of multiplicative processes? The answer is either the log-normal or the log-stable distributions. Let's see why.

2.3.2.1   Log-Normal Distribution

The **log-normal distribution** is the pdf of any random variable X whose associated random variable $Z = \ln X$ follows a Gaussian distribution [8]. More precisely, if

---

[22]These reasons may include, for instance, the varying strength of the immune system of the people encountered, their personal resistance to that particular contagion, the specific weather of that day or whether the day of interest is a working day or part of a weekend.

Z is distributed according the normal distribution $N_{[\mu,\sigma^2]}(z)$, then X is distributed according to:

$$\text{LogN}_{\mu,\sigma^2}(x) = \frac{1}{x\sigma\sqrt{2\pi}}\exp\left(-\frac{(\ln x - \mu)^2}{2\sigma^2}\right), \qquad (2.45)$$

that is the usual definition of the log-normal distribution (see Fig. 2.7). Its cumulants become also a function of the mean and variance of the underlying Gaussian. It is straightforward to prove (see Exercise 2.9) that its mean is $c_1 = \exp(\mu + \sigma^2/2)$ and its variance is $c_2 = (\exp(\sigma^2) - 1)\exp(2\mu + \sigma^2)$.

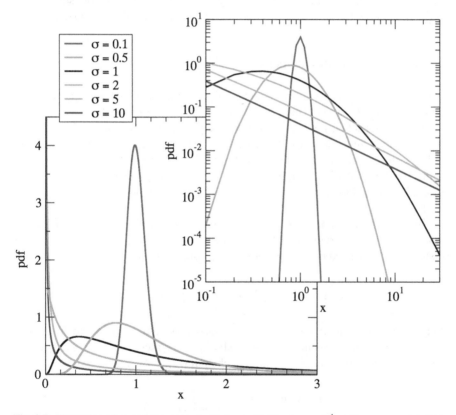

Fig. 2.7 Illustration of several log-normal pdfs ($\mu = 0$). Clearly, an x^{-1} scaling range exists that can be made arbitrary large by making $\sigma \gg 1$. At values of $\sigma < 1$, the extension of the x^{-1} scaling is almost negligible. Instead, the pdf has a maximum at a position $0 < x_0 < 1$ that moves towards 1 as σ tends to zero [Left box is in lin-lin scale; right box, in lin-log scale]

Why do log-normal distributions appear as attractor pdfs for multiplicative processes? One just needs to take the logarithm[23] of Eq. 2.44,

$$\log\left(N_{\text{infected}}^{k+1}/N_{\text{infected}}^{0}\right) = \ln\left(\prod_{i=1}^{k}\eta_i\right) = \sum_{i=1}^{k}\ln\eta_i, \tag{2.46}$$

to see that the new process $Y_{k+1} \equiv \ln\left(N_{\text{infected}}^{k+1}/N_{\text{infected}}^{1}\right)$ is an additive processes. Therefore, if each of the individual processes being added (i.e., each $\ln\eta_i$) has well defined characteristic scales (i.e., a finite variance), the central limit theorem predicts that the statistics of their sum, Y_{k+1}, will necessarily fall within the basin of attraction of a Gaussian pdf. It naturally follows from here that the statistics of $N_{\text{infected}}^{k+1}$ will then follow the log-normal distribution associated to that Gaussian.

2.3.2.2 Log-Stable Distributions

It is often the case in complex systems, however, that characteristic scales cannot be defined (see Chap. 1). Therefore, it may be the case that the logarithms of the random variables that form the multiplicative process (i.e. $\ln\eta_i$ in Eq. 2.46) lack a finite variance. In those cases, the basin of attraction to which their sum (i.e., Y_{k+1} in Eq. 2.46) would belong cannot be that of the Gaussian law, but one associated to some member of the Lévy family of pdfs. Which pdf plays a role analogous to that of the log-normal distribution in these cases? The *log-stable distribution* that will be defined next.

The **log-stable distribution** is the pdf followed by any random variable X whose associated random variable $Z = \ln X$ follows a Lévy distribution [7]. The log-stable distribution associated to the Lévy pdf, $L_{[\alpha,\lambda,\mu,\sigma_L]}(z)$ is given by,

$$\text{LogSt}_{[\alpha,\lambda,\mu,\sigma_L]}(x) = \frac{1}{x}L_{[\alpha,\lambda,\mu,\sigma_L]}(\ln x). \tag{2.47}$$

Regretfully, since most Lévy laws lack closed analytic forms, the log-stable distributions must be calculated numerically in most cases. A notable exception is the log-Cauchy pdf, associated to the Cauchy pdf (Eq. 2.39), that is given by:

$$\text{LogCau}_{[\sigma_L]}(x) = \frac{\sigma_L}{x\pi\left(\ln^2(x) + \sigma_L^2\right)}. \tag{2.48}$$

Interestingly, log-stable distributions have become recently quite popular to model the statistics of various complex systems such as the world stock market [9] or the transition to turbulence in fluids [10], thus pointing to the relevance of

---

[23]We have normalized $N_{\text{infected}}^{k+1}$ to the initial number infected for simplicity.

multiplicative processes in those systems. The physical mechanisms behind those processes are easy to point out, being related to the presence of cascades in either volatile stocks and option prices or in turbulence.

2.3.3 Weibull, Gumbel and Frechet Pdfs: Extreme Value Pdfs

It has already been mentioned that heavy-tailed power-law pdfs often appear in the context of complex systems. Due to the divergent character of those tails, the relatively large probability (as compared to normal distributions) of the events that are found near their end, also known as **extreme values** or **extreme events**, becomes particularly relevant.[24]

Can one say anything about the statistics of such extreme events? The answer is yes. Another limit theorem exists that provides attractor pdfs for the statistics of extreme values. Let's say that one has N independent random variables, X_i, with $i = 1, \cdots, N$, all distributed according to a common pdf, $p_{X_i}(X) = f(X)$, $\forall i$. An **extreme value** will be defined as the maximum value of each outcome of the N variables,

$$X_{ev} = \max(X_1, X_2 \cdots X_N). \tag{2.49}$$

It turns out that the statistics of such extreme values fall within the basin of attraction of a one-parameter family of pdfs, whose cumulative distribution function is [11]:

$$H_{[\xi,\mu,\sigma_E]}(x) = \exp\left(-\left[1 + \left(\frac{x-\mu}{\sigma_E/\xi}\right)\right]^{-1/\xi}\right), \quad \xi \in (-\infty, \infty). \tag{2.50}$$

Here, the essential parameter is ξ, since μ and σ_E act as translation and rescaling parameters, very similar to their analogous Lévy counterparts. Three different families of pdfs are derived from this cumulative distribution function: the **Gumbel pdf**, the **Frechet pdf** and the **Weibull pdf**. We discuss them separately in what follows. Criteria are also provided that determine the pdf to which the statistics of the extreme values of a particular parent pdf will end up attracted to.

---

[24]In many cases, extreme events have such a long-term influence on the system that, in spite of their apparently small probability, they dominate the system dynamics. One could think, for instance, of the large earthquakes that happen in the Earth's crust, or of the global economical crisis that affect the world economy, and the impact that may have on our daily life for many years after.

2.3.3.1   The Gumbel Pdf

By taking the limit $\xi \to 0$ of Eq. 2.50 and differentiating it with respect to x (see Eq. 2.4), one obtains the **Gumbel pdf** (see Fig. 2.8, in black):

$$\mathrm{Gu}_{[\mu,\sigma_E]}(x) = \frac{1}{\sigma_E} \exp\left\{-\left[\left(\frac{x-\mu}{\sigma_E}\right) + \exp\left\{-\left(\frac{x-\mu}{\sigma_E}\right)\right\}\right]\right\}, \quad x \in (-\infty, \infty),$$

$$(2.51)$$

that decays exponentially for large (positive and negative) values of x. There is a practical criterion provided by *Gnedenko* [11] to decide whether a particular parent distribution, $p(x)$, will have its extreme values distributed according to a Gumbel pdf. One just needs to check whether the survival function (see Eq. 2.6) associated with $p(x)$ verifies,

$$\lim_{x \to \infty} \frac{\mathrm{sf}(x)}{\mathrm{sf}(cx)} = 0, \quad \forall c > 0. \tag{2.52}$$

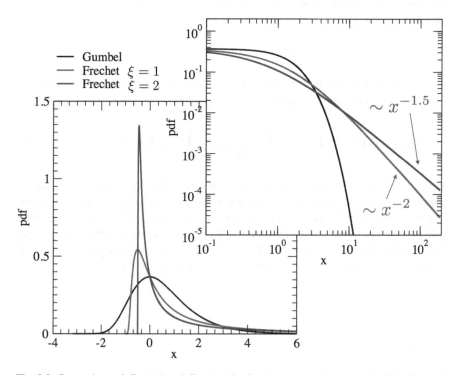

Fig. 2.8 Comparison of Gumbel and Frechet distributions ($\mu = 0$, $\sigma_E = 1$). The Gumbel distribution is defined for any x and decays exponentially for large positive and negative values of the exponent. Frechet pdfs, on the other hand, are defined only for $x > \mu - \sigma_E/\xi$ and exhibit a power-law decay with exponent $1 + 1/\xi$ [Left box is in lin-lin scale; right box, in lin-log scale]

Here, c is an arbitrary positive constant. Examples of pdfs contained within this basin of attraction are the **Gaussian pdf** (Eq. 2.30) and the **exponential pdf**, that will be discussed soon (Sect. 2.3.4).

2.3.3.2 The Frechet Pdf

For $\xi > 0$, the resulting pdf is the **Frechet pdf** that is given by:

$$\text{Fr}_{[\xi,\mu,\sigma_E]}(x) = \frac{1}{\sigma_E}\left(1 + \frac{x-\mu}{\sigma_E/\xi}\right)^{-(1+1/\xi)} \exp\left[-\left(1 + \frac{x-\mu}{\sigma_E/\xi}\right)^{-1/\xi}\right]. \tag{2.53}$$

This pdf is defined only over the domain,

$$x \in \left[\mu - \frac{\sigma_E}{\xi}, \infty\right). \tag{2.54}$$

The Frechet pdf exhibits algebraic tails for large (positive) values of x (see Fig. 2.8). Again, Gnedenko gives us a criterion to decide whether a certain parent pdf, $p(x)$, has extreme values with statistics that will tend to the Frechet pdf [11]. One just need to check whether the survival function associated with $p(x)$ verifies,

$$\lim_{x\to\infty} \frac{\text{sf}(x)}{\text{sf}(cx)} = c^k, \quad k > 0, \quad \forall c > 0. \tag{2.55}$$

Any survival function that has an algebraic tail, $\text{sf}(x) \sim x^{-k}$ verifies this condition. Thus, its related pdf will see its extreme values converging to the Frechet law with parameter $\xi = 1/k$. In particular, **Lévy pdfs** with a tail index α will have their extreme values converging to Frechet pdfs with parameter $\xi = 1/\alpha$.

2.3.3.3 The Weibull Pdf

For $\xi < 0$, the resulting extreme pdf is the **Weibull pdf**. It is given as:

$$\text{We}_{[\xi,\mu,\sigma_E]}(x) = \frac{1}{\sigma_E}\left(1 - \frac{x-\mu}{\sigma_E/|\xi|}\right)^{(1/|\xi|-1)} \exp\left[-\left(1 - \frac{x-\mu}{\sigma_E/|\xi|}\right)^{1/|\xi|}\right]. \tag{2.56}$$

The Weibull pdf is also defined over a restricted domain,

$$x \in \left(-\infty, \mu + \frac{\sigma_E}{|\xi|}\right], \tag{2.57}$$

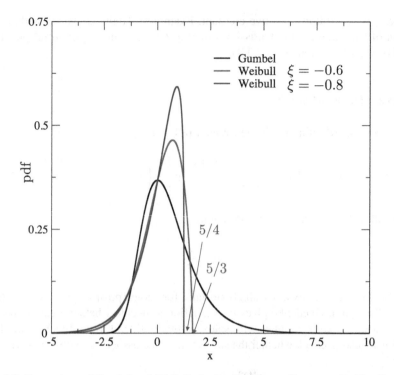

Fig. 2.9 Comparison of Gumbel and Weibull distributions ($\mu = 0$, $\sigma_E = 1$). The Gumbel distribution is defined for any x and decays exponentially for large positive and negative values of the exponent. Weibull pdfs, on the other hand, are defined only for $x < \mu + \sigma_E/|\xi|$. The cutoff point moves towards zero as $|\xi|$ gets larger [Left box is in lin-lin scale; right box, in lin-log scale]

and has a short positive tail, due to the cutoff value at $x_c = \mu + \sigma/|\xi|$ (see Fig. 2.9). The parent pdfs that fall within the basin of attraction of the Weibull pdf are those whose survival function verify [11],

$$\lim_{x \to 0^-} \frac{\text{sf}(cx + w)}{\text{sf}(x + w)} = c^k, \quad k > 0, \quad \forall c > 0. \tag{2.58}$$

The value of w is given by the equation, $\text{sf}(w) = 0$. That is, w is the maximum value that x can take. One example of a pdf whose extreme values follow the Weibel pdf is the **uniform pdf**, defined over any finite interval.

2.3.4   Exponential and Related Pdfs: Poisson Processes

Another pdf often found in nature is the **exponential pdf** (see Fig. 2.1, right panel),

$$E_\tau(x) := \tau^{-1} \exp(-x/\tau), \quad \Omega = [0, \infty), \tag{2.59}$$

in which x can only take positive values.[25] Thus, it is often related to the statistics of time intervals or distances, that are naturally positive. For instance, x could be time intervals between disintegrations of a radioactive material, or between the occurrence of mutations in a DNA strand. Or the distance between defects in a solid.

As in the case of the Gaussian, it is only natural to ask for the reasons that make exponential pdfs to appear so often. It turns out that the answer is related to a mathematical concept known as a **Poisson process** [12]. It is defined as *any process in which the time at which an event happens does not depend in any way of the past history of the process*. The fulfilment of this condition implies, among other things, that the statistics of the time intervals between successive events (i.e., a *waiting-time*) must follow an **exponential pdf**. Since many processes in nature behave like Poisson processes, exponential pdfs are very abundant. They are somewhat more rare in complex systems, where memory of past events is relevant in the dynamics. For that reason, an absence of exponential pdfs in waiting-time statistics can sometimes be used to detect possible complex dynamics.[26]

In what follows, we will discuss the statistics of the waiting-times of a Poisson process, the distribution of the number of events taking place in a given lapse of time, known as the **Poisson pdf**, as well as other related distributions, such as the **Gamma pdf** or the **stretched exponential**.

2.3.4.1 Exponential Pdf

Let's determine the distribution, $p(w)$, that is followed by the waiting times w of a Poisson process. We will assume that an event has just happened at time $t = 0$. From what we have said about Poisson processes, it should be clear that the probability of the next waiting-time being larger than w (given by the survival function $sf(w) = P[W \geq w]$) must be equal to the probability of having to wait for longer than w *after having already waited for an amount of time $\hat{w}$ following the*

---

[25] A related pdf, the **Laplace pdf**, can be defined for both positive and negative values:

$$\mathrm{Lp}_\tau(x) = \frac{1}{2\tau} \exp(-|x|/\tau), \quad \Omega = (-\infty, \infty). \tag{2.60}$$

However, the Laplace pdf does not play any similar role to the exponential in regards to time processes. It is somewhat reminiscent of the Gaussian distribution, although it decays quite more slowly as revealed by its larger kurtosis value (see Table 2.4).

[26] One must be careful, though. There are examples of complex systems with exponential statistics. For instance, a randomly-driven running sandpile exhibits exponential statistics for the waiting-times between avalanches. This is a consequence of the fact that avalanches are triggered only when a grain of sand drops, which happens randomly throughout the sandpile. It is only when sufficiently large avalanches are considered that non-exponential waiting-time statistics become apparent, revealing the true complex behaviour of the sandpile, as we will see in Sect. 4.5.

event.[27] Mathematically, this idea is expressed by[28]:

$$\frac{P[W \geq w + \hat{w}]}{P[W \geq \hat{w}]} = P[W \geq w] \tag{2.61}$$

where the left-hand-side is the probability of having to wait for at least w for the next event to happen after having *already* waited for $\hat{w}$. The right-hand-side is the probability of having to wait for at least w after $t = 0$, when the previous event happened. We can rewrite this relation as,

$$P[W \geq w + \hat{w}] = P[W \geq \hat{w}] \cdot P[W \geq w], \tag{2.62}$$

whose only solution is the exponential function[29]:

$$P[W \geq w] = \exp(aw). \tag{2.64}$$

The pdf associated with this survival function is (Eq. 2.7),

$$p(w) = -\frac{d}{dw}\left(P[W \geq w]\right) = a\exp(-aw) = E_{a^{-1}}(w), \tag{2.65}$$

which yields the exponential pdf (Eq. 2.59; see Fig. 2.1). The parameter a gives the inverse of the average waiting time of the process, as can be easily checked by calculating the expected value (see Eq. 2.24):

$$\langle w \rangle = \int_0^\infty dw\, wp(w) = \int_0^\infty dw\, aw\exp(-aw) = \frac{1}{a}. \tag{2.66}$$

2.3.4.2 Poisson and Binomial Discrete Distributions

Many other properties of a Poisson process can be easily derived. For instance, *the probability of k events occurring during a lapse of time t* is given by the discrete

---

[27]This follows from the fact that the triggering of events in a Poisson process is independent of the past history of the process. Thus, what we may have already waited for is irrelevant and cannot condition how much more we will have to wait for the next triggering!

[28]The conditional probability of B happening assuming that A has already happened, $P(A|B)$, is given by: $P(A|B) = P(A \cap B)/P(A)$, where $P(A \cap B)$ is the joint probability of events A and B happening, and $P(A)$ is the probability of A happening [4].

[29]Indeed, note that due to the properties of the exponential function, it follows that:

$$P[W \geq w + \hat{w}] = \exp(a(w + \hat{w})) = \exp(aw) \cdot \exp(a\hat{w}) = P[W \geq \hat{w}] \cdot P[W \geq \hat{w}]. \tag{2.63}$$

Poisson distribution[30]:

$$\text{Po}_\tau(X = k, t) = \frac{1}{k!}\left(\frac{t}{\tau}\right)^k \exp(-t/\tau), \tag{2.67}$$

where $\tau = a^{-1}$ is the mean waiting-time of the Poisson process and X is the number of events triggered in time t.[31]

One way to derive the discrete Poisson distribution is as a limit of another discrete distribution, the **binomial distribution**. The binomial distribution gives the probability of succeeding k times in a total of n trials, assuming that the probability of success of each trial is p:

$$B(k, n; p) = \binom{n}{k} p^k (1 - p)^{n-k}, \quad \text{with} \quad \binom{n}{k} = \frac{n!}{k!(n-k)!} \tag{2.68}$$

The Poisson distribution is obtained in the limit of $n \to \infty$ and $p \to 0$, but keeping the product np finite. The result is (see Problem 2.6),

$$B(k, n; p) \longrightarrow \frac{(np)^k}{k!} \exp(-np). \tag{2.69}$$

To conclude the derivation, we note that np gives the average number of successful trials during the whole process. In the notation of the continuous process, the number of events in a lapse of time t is given by t/τ, if τ is the average waiting time. Thus, Eq. 2.67 is obtained by making the substitution $np \to t/\tau$.

2.3.4.3 The Gamma Pdf

Another popular pdf, the **Gamma pdf** is the continuous version of the Poisson discrete distribution. It can obtained by replacing k by a continuous argument β, and by including an additional $1/\tau$ factor, needed to ensure the proper normalization of the pdf.[32] The Gamma pdf depends on two parameters, $\beta > 0$ and $\tau > 0$:

$$\text{Gamma}_{[\beta,\tau]}(t) = \frac{1}{\Gamma(\beta)}\left(\frac{t^{\beta-1}}{\tau^\beta}\right)\exp(-t/\tau), \quad t \geq 0. \tag{2.70}$$

---

[30]Discrete means that the possible outcomes are countable; in this case, the number of events k, that can only take integer values. One consequence of being discrete is that the distribution gives actual probabilities, so that it is no longer a density of probability.

[31]Note that the Poisson distribution is a discrete distribution, not a pdf, and is normalized to unity only when summing over all possible values of k. Can the reader prove this?

[32]Remember that $\Gamma(x)$ becomes the usual factorial function, $\Gamma(k + 1) = k!$, for integer k.

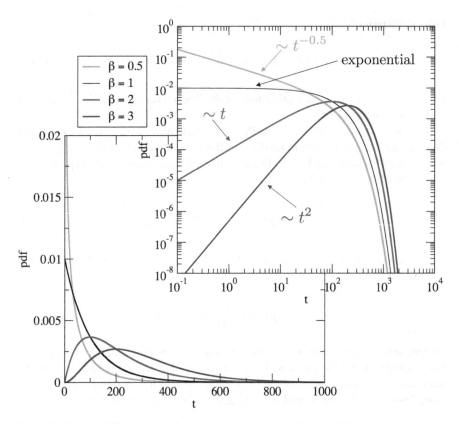

Fig. 2.10 Gamma pdf for several values of the shape parameter β ($\tau = 100$). A power-law region exists for $t < \tau$ with exponent $\beta - 1$, that can be positive or negative. At larger values of t, an exponential decay dominates the scaling [Left box is in lin-lin scale; right box, in lin-log scale]

β is often called the **shape parameter** (see Fig. 2.10) and τ, the **scale parameter**. Note that, for $\beta = 1$, the usual exponential pdf is recovered. Its mean is given by $c_1 = \beta\tau$, whilst its variance is $c_2 = \beta\tau^2$. Its skewness is given by $S = 2/\sqrt{\beta}$, whilst its kurtosis is $\kappa = 6/\beta$. Its characteristic function is,

$$\phi_{[\beta,\tau]}^{\text{Gamma}}(k) = \frac{1}{(1 - \imath\tau k)^\beta}. \tag{2.71}$$

It is possible to endow the Gamma pdf with some physical meaning if $\beta > 1$ is an integer number. In that case, the Gamma pdf arises naturally when the variable t is formed by adding together $\beta - 1$ partial variables, τ_i, each of them distributed according to the same exponential distribution, $p(\tau_i) = E_\tau(\tau_i)$. This might be the

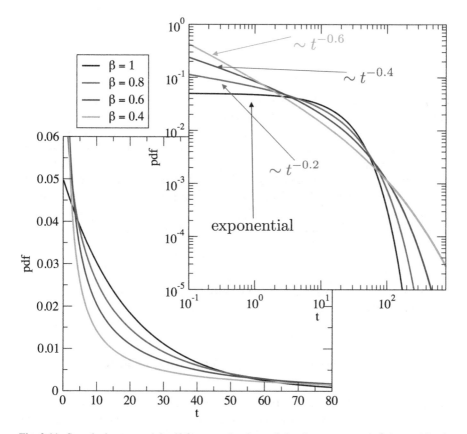

Fig. 2.11 Stretched exponential pdf for several values of the shape parameter β ($\tau = 20$). A power-law region exists for $t \lesssim \tau$ with negative exponent $\beta - 1$. The smaller β is, the longer the power-law scaling extends along t, due to the fractional exponent in the exponential [Left box is in lin-lin scale; right box, in lin-log scale]

case, for instance, of the distribution of failing times of a power system that is composed of β identical backup batteries, each of them with the same exponential distribution of failing times.[33]

2.3.4.4   The Stretched Exponential Pdf

Another interesting pdf somewhat related to the exponential is the one-parameter family of **stretched exponential pdfs** (see Fig. 2.11), that is defined over the range

---

[33]This is, in fact, a nice example of a physical system in which the appearance of a power law in a pdf has nothing to do with any complex behaviour, and the exponential decay at large t is not a finite size effect.

$\beta \in (0, 1]$ as:

$$e_{[\beta,\tau]}(t) = \frac{\beta t^{\beta-1}}{\tau^\beta} \exp(-(t/\tau)^\beta), \quad t \geq 0. \tag{2.72}$$

For $\beta = 1$, the usual exponential pdf is recovered. For $\beta < 1$, a power-law scaling region appears at to a timescale of the order of τ, beyond which the exponential term dominates. Due to the fractional exponent in the exponential, the lower β is, the more extended the power scaling region becomes. In contrast to the Gamma pdf, we do not know of any derivation of the stretched exponential pdf from physical arguments. Stretched exponentials are often used phenomenologically.[34] We will also use them in Chap. 5, during the discussion of fractional transport, due to some of their interesting properties.

The cumulative distribution function of this family is given by:

$$\mathrm{cdf}^{\varepsilon}_{[\beta,\tau]}(t) = \exp(-(t/\tau)^\beta), \tag{2.73}$$

popularly known as the **stretched exponential**, from which the whole family of pdfs borrows its name.

2.4 Techniques for the Practical Estimation of Pdfs

We believe that the ability to estimate pdfs from numerical or experimental data should always be in the bag of tricks of any scientist or engineer. Some people might however argue that, given any set of data, the estimation of its mean and variance is sufficient for most practical purposes, and that one rarely needs to make the effort to estimate the full pdf.[35] For instance, in order to quantify the resistance of an airfoil with respect to vertical forcing, one could apply multiple times the same external forcing (say, setting some prescribed weight on the tip) and measure each time the vertical displacement of the tip of the wing. The computation of the mean is trivial. One just needs to add all values and divide the result by the number of measurements and, for the variance, to subtract the mean from each measurement, square the result, add them all, and divide by the number of measurements.[36] With these two numbers, one can quantify the expected resistance and its typical variability.

---

[34]It was first introduced by R. Kohlrausch to describe the discharge of a capacitor at the end of the nineteenth century, and is often used in the context of relaxation processes.

[35]This idea is rooted in the perceived prevalence of Gaussian pdfs, whose cumulants higher than the variance all vanish.

[36]It is worth to point out the relation between these prescriptions, that we all learned in high school, and Eqs. 2.24 and 2.25, that require knowledge of the full pdf. Given a series of data $\{x_i, \ i = 1, 2, \cdots N\}$, these prescriptions for the mean and variance are given by:

There are however situations in which having a full knowledge of the pdf is important. This is certainly the case in the context of complex systems, in which the shape of the tail of the pdf may give us a hint of the kind of underlying dynamics we are dealing with. For instance, in turbulent fluids and plasmas, the departure away from diffusive effective transport can often be monitored as a transition from near-Gaussian to heavy-tailed statistics for quantities such as fluctuations, velocities and fluxes (see Chap. 5). The accurate determination of these tails is however a delicate business due to the fact that events that contribute to them happen with a much smaller probability than those contributing to central part of the distribution.[37]

In this section, we will discuss several methods to estimate pdfs from experimental data in ways that try to minimize these problems as much as possible. We will start, however, by discussing the standard method of estimating pdf that is based on the idea of **data binning**.

2.4.1 Constant Bin Size Method

The standard method to construct a pdf from a set of data proceeds as follows. Suppose that we have made N measurements of a certain quantity, X:

$$X^* = \{X_1, X_2, X_3, \cdots, X_N\}. \tag{2.75}$$

We can then estimate the pdf by first building a set of M **bins** that spans the whole domain covered by the data: $[\min(X^*), \max(X^*)]$. The number of bins, M, must be chosen much smaller than N, so that a reasonable number of values will presumably fall within each bin. We define each bin, b_i, as the interval:

$$b_i = [\min(X^*) + i\Delta b, \min(X^*) + (i+1)\Delta b], \quad i = 1, M. \tag{2.76}$$

$$\bar{x} := \frac{1}{N} \sum_{i=1}^{N} x_i, \quad \sigma^2 := \frac{1}{N} \sum_{i=1}^{N} x_i^2 - \bar{x}^2. \tag{2.74}$$

It is simple to show that the same formulas are obtained if one introduces **the numerical pdf obtained from the data series** in Eqs. 2.24 and 2.25. Let's take the mean. The sum over all data points that appears in its definition can be rewritten as a sum over all possible different values of x, if we multiply each value by the number of times it appears. Since the sum is normalized to N, the total number of data points, the sum over all different possible values would become just a sum of the products of each value and its probability. That is, precisely, what Eqs. 2.24 and 2.25 express. Can the reader prove the same equivalence for the variance?

[37]The central values are also the ones that contribute more strongly to the values of means and moments, except in cases in which these moments diverge!

where the size of each bin, Δb is defined by:

$$\Delta b = \frac{\max(X^*) - \min(X^*)}{M - 1}. \tag{2.77}$$

The size of all bins is the same, giving the method its name: the **constant bin size** (or CBS) method.

Next, one simply counts the number of elements of set X^* fall within each bin. Let's call this number N_i. Then, the probability of one measurement having fallen in bin b_i is simply given by,

$$P_i = \frac{N_i}{N}, \tag{2.78}$$

and, subsequently, the probability density function takes a value at that bin given by,

$$P_i = p_i \Delta b \longrightarrow p_i = \frac{N_i}{N \Delta b}. \tag{2.79}$$

It is typical to assign the value of the pdf just obtained to the mid-point of the bin. Thus, the numerical pdf is composed of all the pairs of the form,

$$p_{\text{CBS}}(X^*) \tag{2.80}$$
$$= \left\{ (x_i, p_i) = \left(\min(X^*) + \left(i + \frac{1}{2} \right) \Delta b, \frac{N_i}{N \Delta b} \right), \quad i = 1, 2 \cdots, M \right\}$$

The whole procedure has one free parameter M, the number of bins. The performance of CBS relies on choosing the bin size wisely. A sufficient number of values of X should fall within each bin, so that the probability can be measured accurately. As an illustration, Fig. 2.12 shows the different pdfs computed with CBS method for data distributed according to the Gaussian $N_{0,1}(x)$ using different M values (shown in the upper frame of Fig. 2.13). If M is to small, we can barely see the Gaussian shape of the pdf. On the other hand, if we make it too large (see $M = 500$ in the figure), the number of points per bin quickly becomes too small as we move away from zero in either direction. The result is a much more noisy pdf and a worse accuracy at the tail, where the density of points per bin will be the smallest. This is an intrinsic problem of CBS. Despite of this, the method works pretty well for the Gaussian pdf examined here.

The story becomes very different if the pdf has a heavy tail. For instance, we will consider the case of noise distributed according to a Cauchy pdf, $L_{[1,0,0,1]}(x) = 1/\pi(x^2 + 1)$. The algebraic tail implies that the probability of the very large events remains always non-negligible (see lower frame in Fig. 2.13). This fact has a deleterious effect on the CBS method, as can be seen in Fig. 2.14. The method fails to reproduce the pdf properly, even with 500 bins! To see how badly the tail of the pdf is captured, we have included an inset in the figure showing the pdf in

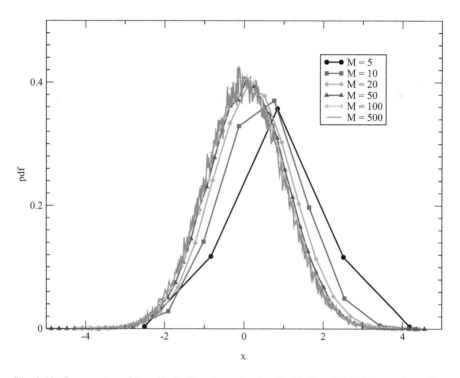

Fig. 2.12 Computation of the pdf of a Gaussian noise signal with $N = 100,000$ data points using the CBS method with $M = 5, 10, 20, 50, 100$ and 500 bins. At the smallest bin sizes, the shape of the Gaussian is not well reproduced. At the largest ones, the small number of points per bin away from the center of the pdf introduces noise

log-log format. One can see that the x^{-2} scaling of the Cauchy is barely captured with 500 bins. More bins would be needed to capture it properly. But note that in practical situations, we will rarely have 100,000 data points at our disposal. Thus, increasing the number of bins is often not an option. Clearly, if we are interested in reconstructing the tails of power-law distributions, we need to do better than CBS.

2.4.2 Constant Bin Content Method

The first thing one can do to improve the accuracy at the tail region is to abandon the notion of using bins of a constant size, and to make them larger as we advance into the tail region. In this way, as the bin size increases, more data values will fall within the bin, and the accuracy will improve. Some people use **logarithmic binning** to achieve this. That is, they use bins with sizes that increase logarithmically as one advances towards the tail. However, there is a more optimal strategy: to force the

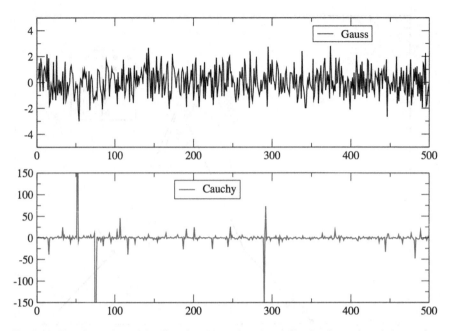

Fig. 2.13 First 500 points of the Gaussian (upper frame) and Cauchy (lower frame) time series. The heavy tail of the Cauchy pdf, decaying as $p(x) \sim x^{-2}$ is apparent in the larger presence of 'large' (i.e., extreme) events

bins to adjust their size in order to contain always the *same number of data values*. We will refer to this method as the **constant bin content** (or CBC) method.

The method can be implemented as follows. Let's consider again the sequence of values of the variable X given by:

$$X^* = \{X_1, X_2, X_3, \cdots, X_N\} \tag{2.81}$$

First, we will reorder the data in increasing order,

$$X^*_{inc} = \{\tilde{X}_1, \tilde{X}_2, \tilde{X}_3, \cdots, \tilde{X}_N, \quad \text{with } \tilde{X}_i < \tilde{X}_j \text{ if } i < j\}. \tag{2.82}$$

Next, let's consider a reasonable number of events per bin, say R. Of course, this number cannot be too large, since we only have N data values at our disposal. Thus $R \gg 1$, but keeping $R \ll N$. Since the total number of bins will be $M = N/R$, it is advisable to choose R so that M is an integer. The first bin, b_1, will then correspond to the interval:

$$b_1 = [\tilde{X}_1, \tilde{X}_R] \rightarrow \Delta b_1 = \tilde{X}_R - \tilde{X}_1, \tag{2.83}$$

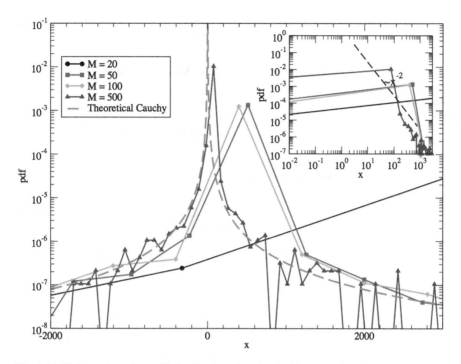

Fig. 2.14 Computation of the pdf of a Cauchy noise signal with $N = 100,000$ data points using the constant bin size method (CBS) with $M = 20$, 50, 100 and 500 bins. Here, the constant size method fails miserably. Only with the largest number of bins gets the pdf closer to the theoretical one, $L_{[1,0,0,1]}(x) = 1/\pi(x^2 + 1)$, shown with a (brown) dashed curve

being Δb_1 its width. It will be centered at the mid-value,

$$x_1 = \frac{\tilde{X}_1 + \tilde{X}_R}{2}. \tag{2.84}$$

The probability of a value falling in this first bin is then given by,

$$P_1 = \frac{R}{N}, \tag{2.85}$$

that will be equal to the probability of falling inside any other bin, since all of them contain the same number of values. However, the value of the probability density function at that first bin will be different from any other bin, since their sizes will be different. Indeed, it is given by,

$$P_1 = p_1 \Delta b_1 \rightarrow p_1 = \frac{P_1}{\Delta b_1} = \frac{R/N}{(\tilde{X}_R - \tilde{X}_1)}. \tag{2.86}$$

It is rather straightforward to extend the algorithm to the remaining bins. Indeed, for the i-th bin we will have that:

$$b_i = [\tilde{X}_{(i-1)R+1}, \tilde{X}_{iR}] \rightarrow \Delta b_i = \tilde{X}_{iR} - \tilde{X}_{(i-1)R+1}, \tag{2.87}$$

with the bin centered at:

$$x_i = \frac{\tilde{X}_{(i-1)R+1} + \tilde{X}_{iR}}{2} \tag{2.88}$$

and its corresponding value of the pdf is,

$$p_i = \frac{R/N}{(\tilde{X}_{iR} - \tilde{X}_{(i-1)R+1})}. \tag{2.89}$$

Thus, the numerical pdf computed with the CBC method is given by,

$$p_{\text{CBC}}(X^*) \tag{2.90}$$

$$= \left\{ (x_i, p_i) = \left(\frac{\tilde{X}_{(i-1)R+1} + \tilde{X}_{iR}}{2}, \frac{R/N}{(\tilde{X}_{iR} - \tilde{X}_{(i-1)R+1})} \right), \quad i = 1, 2 \cdots, M \right\}$$

It might happen, however, that some of the values could be found in the sequence multiple times. That requires to modify the CBC algorithm a bit. First, we need to reorder the non-repeated data to form an **strictly decreasing** sequence,

$$X_{inc}^* = \left\{ (\bar{X}_1, r_1), (\bar{X}_2, r_2), \cdots, (\bar{X}_P, r_P), \quad \text{with } \bar{X}_i < \bar{X}_j \text{ if } i < j \right\} \tag{2.91}$$

Here, P is the number of non repeated values and r_i is the number of times the i-th distinct value is repeated. It must necessarily happen that,

$$N = \sum_{k=1}^{P} r_k. \tag{2.92}$$

Then, we should choose the number of bins as $M = P/R$, where R is a reasonable number of (non-repeated) events to define each bin. Then, the procedure is identical to the one described earlier, except for the fact that the probability for each bin will be calculated using,

$$p_i = \frac{\sum_{i=1}^{R} r_i}{N(\bar{X}_{iR} - \bar{X}_{(i-1)R+1})}, \tag{2.93}$$

instead of Eq. 2.89. Clearly, both expressions become identical if $r_i = 1$, $\forall i$.

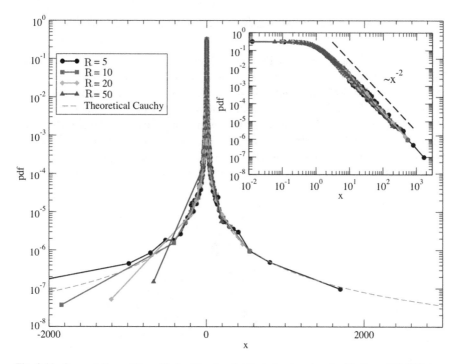

Fig. 2.15 Computation of the pdf of a Cauchy-distributed noise signal with $N = 100,000$ data points using the constant bin content method (CBC) with $R = 5, 10, 20$ and 50 points per bins. The method works pretty well, even at the lowest number of points-per-bin, with the pdf quite close to the theoretical one, $L_{[1,0,0,1]}(x) = 1/\pi(x^2 + 1)$, shown with a (brown) dashed curve. The tail behaviour is also very well captured, as shown in the inset

In order to test the CBC method, we will apply it to the same Cauchy-distributed data that made the CBS method look so bad.[38] The results are shown in Fig. 2.15, where the number of points-per-bin R is varied between 5 and 50. Clearly, the CBC method works pretty well, even for the smallest number of points per bin chosen. The tail of the Cauchy distribution, $p(x) \sim x^{-2}$ is resolved almost over three decades, as it is shown in the inset of the figure. It is worth noting that, as R is increased, the part of tail resolved is shortened. Therefore, it is important to keep R meaningful but not too large.

---

[38]For Gaussian-distributed data, it must be said that both methods work pretty well.

2.4.3 Survival/Cumulative Distribution Function Method

There are times when even the CBC method is unable to resolve the tails of a pdf. This is usually the case if the number of data points available is very small. Say that one has $N = 500$ data points. Is that sufficient to tell whether their pdf has a power-law tail or not? Assuming a minimum number of data values per bin of $R = 5$, that gives us 100 bins, which might seem enough to define the tail. However, when trying to apply it to 500 values from the same Cauchy-distributed time series we used previously, it does not work very well (see the upper frame of Fig. 2.16). Clearly, almost no point is obtained in the tail region (much less a full decade!).

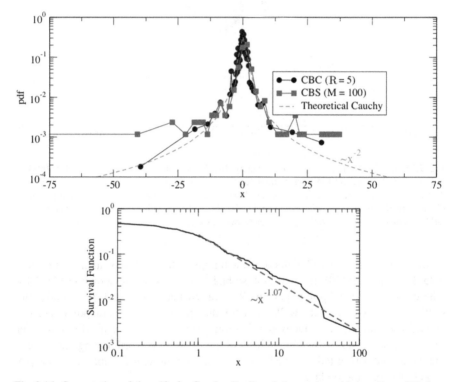

Fig. 2.16 Computation of the pdf of a Cauchy-distributed time series with just $N = 500$ data points using CBC method with $R = 5$ points per bins and the CBS method with $M = 100$ bins. In the lower frame, the survival function of the data is plotted, in log-log scale, for the positive part of the data. It exhibits a power-law tail that decays as $\sim x^{-1}$ for more almost two decades

CBS does even worse,[39] since the tail seems to be flat. Is there anything better one can do that does not rely on using bins?

The answer is of course yes. Even for that small amount of data, one can still compute the survival (Eq. 2.6) and the cumulative distribution functions (Eq. 2.4) quite accurately. It will be remembered that $sf(x)$ and $cdf(x)$ give, respectively, the probability of the outcome of a measurement being larger or smaller than x. Once known, one can get the pdf from either of them by simple differentiation (Eqs. 2.7 and 2.5). But in many cases of interest the pdf is not needed.[40] One just wants to know whether the pdf has a power-law tail, and to estimate its exponent. The $sf(x)$ and $cdf(x)$ can give us that. The argument goes like this. We have previously said that the positive tail of a Lévy function of index α scales as,

$$p(x) \sim x^{-(1+\alpha)}, \quad x \to \infty \tag{2.94}$$

Therefore, its survival function will have to verify that,

$$sf(x) = \int_x^\infty p(x')dx' \sim x^{-\alpha}, \quad x \to \infty \tag{2.95}$$

It is thus sufficient with computing the survival function from some given data and look at the tail! It it scales like Eq. 2.95, with $\alpha \in (0, 2)$, the positive side of the pdf of our data is probably close to a Lévy with that tail index. For analogous reasons, if one interested instead in determining the tail for large negative x's, one should then use the cumulative distribution function.

Let's see how to estimate $sf(x)$ from our set of data,[41]

$$X^* = \{X_1, X_2, X_3, \cdots, X_N\} \tag{2.96}$$

Say we want to estimate the survival function because we are interested in the tail for large, positive X. The first step is to reorder the data to form an **strictly decreasing** sequence,

$$X_{dec}^* = \{\bar{X}_1, \bar{X}_2, \bar{X}_3, \cdots, \bar{X}_N, \quad \text{with } \bar{X}_i > \bar{X}_j \text{ if } i < j\}. \tag{2.97}$$

Note that $\min(X^*) = \bar{X}_N$ and $\max(X^*) = \bar{X}_1$, after the reordering. Also, we will assume for now that no repeated values exist, so that the sequence is indeed strictly decreasing (we will see how to remove this constraint shortly). Then, we have that

---

[39]CBS tends to give flatter tails because, in most bins out there, there is either one point or no point at all. Since one only plots the bins with non-zero values, the tail seems artificially flatter than it actually is.

[40]These methods will in fact yield a very noisy pdf, since numerical derivatives magnify errors!

[41]To build the cumulative distribution function, $cdf(x)$, the procedure is identical to the one discussed here for the survival function, but the initial reordering of the data should be in increasing order!

the probability of X being larger or equal than say, $\bar{X}_4$, is trivially given by $4/N$, since there is only 4 numbers greater or equal than $\bar{X}_4$ in the sequence: $\bar{X}_1, \bar{X}_2, \bar{X}_3$ and $\bar{X}_4$. Thus, by this simple reordering and without any binning we can trivially construct the survival function as:

$$\text{sf}_{\text{no-rep}}(X^*) = \left\{ (x_i, S_i) = \left(\bar{X}_i, \frac{i}{N} \right), \quad i = 1, 2 \cdots, N \right\} \tag{2.98}$$

It might happen, as in the CBC case discussed in Sect. 2.4.2, that some values may be repeated in the ordered sequence. That requires to modify the algorithm slightly. Let's assume that the repetition index of each distinct value, $\bar{X}_i$ is given by the index r_i. Then, we reorder the sequence decreasingly as:

$$X^*_{inc} = \left\{ (\bar{X}_1, r_1), (\bar{X}_2, r_2), \cdots, (\bar{X}_P, r_P), \quad \text{with } \bar{X}_i \geq \bar{X}_j \text{ if } i < j \right\} \tag{2.99}$$

Here, P is the number of non repeated values. It must necessarily happen that,

$$N = \sum_{k=1}^{P} r_k. \tag{2.100}$$

It is still very easy to compute the probability of X being larger or equal than $\bar{X}_4$ in the case of the existence of repeated values. It is simply: $(r_1 + r_2 + r_3 + r_4)/N$. The survival function is then,

$$\text{sf}_{\text{rep}}(X^*) = \left\{ (x_i, S_i) = \left(\bar{X}_i, \frac{1}{N} \sum_{k=1}^{i} r_k \right), \quad i = 1, 2 \cdots, P \right\} \tag{2.101}$$

We proceed now to test the survival function method on the $N = 500$ points of Cauchy-distributed data that the CBC and CBS methods could not handle. The resulting survival function is plotted in the lower frame Fig. 2.16 in a log-log scale. Only the positive part of the independent variable is included. The survival function exhibits a power-law decay with an exponent close to -1, that is consistent with a Cauchy pdf, that decays as $p(x) \sim x^{-2}$. This is quite remarkable, given the very low number of points available. It is thus clear that the survival function method[42] should be a very useful method to use in experimental situations, where data points are often scarce.

---

[42]Or the cumulative distribution function method, if one is interested in estimating the tail for negative values of the outcome.

2.5 Techniques to Compare Experimentally Obtained Pdfs with Analytical Forms

Once a probability density function has been estimated using any of the methods described in Sect. 2.4, it is often the case than one would like to know whether the result is close to some analytical form. That is, is the pdf obtained from some data closer to a Gaussian pdf? Or to a Lévy law? To a Gamma pdf? Or to any other? The answer is usually obtained in two steps. First, one estimates the values of the parameters defining these analytical forms (say, μ and σ^2 for a Gaussian pdf; or α, λ, σ_L and μ, for a Lévy law) that best reproduce the experimental pdf. Secondly, one quantifies which of all these analytical forms (if any) provides a more adequate model for the data. The first part is usually done using some kind of parameter estimation method [13]. We will describe two (very popular) such methods: **maximum likelihood estimation** and **minimum chi-square estimation**. To carry out the second part, we will rely on one of the so-called **goodness-of-fit** tests [14]: the **Pearson's chi-square test**.

2.5.1 Maximum Likelihood Estimators

One of the more widely used methods to estimate the parameters that define any analytical pdf from experimental data is **maximum likelihood estimation** (MLE) [15]. The method was made popular by statistician and biologist Ronald A. Fisher in the 1920s [16]. The idea is simple. Let's assume that we have a set of data,

$$X^* = \{X_1, X_2, X_3, \cdots, X_N\} \tag{2.102}$$

for which we have estimated its pdf, $p(X^*)$, perhaps by using any of the methods described in Sect. 2.4. We will then hypothesize that this pdf could be well represented by the analytical pdf $\hat{p}(x| p_1, p_2, \cdots, p_{N_e})$. However, we need to estimate the collection of parameters p_i, that uniquely define the pdf, from the data. How do we proceed?

First, we will estimate the **likelihood**, L, that the data set were actually drawn from the analytical pdf. Assuming that the data are statistically independent,[43] we

---

[43]Indeed, remember that $p(A \cup B) = p(A)p(B) + p(A \cap B)$. Thus, the probability of A and B happening equals the product of their individual probabilities only if events A and B are independent.

can estimate it as the product[44]:

$$L(X^* | p_1, p_2, \cdots, p_{N_e}) = \prod_{i=1}^{N} \hat{p}(X_i | p_1, p_2, \cdots, p_{N_e}). \tag{2.103}$$

The next step would be to estimate the collection of parameters that makes L maximum. However, it is more convenient to work with the so-called **log-likelihood**,

$$\log \left[L(X^* | p_1, p_2, \cdots, p_{N_e}) \right] = \sum_{i=1}^{N} \log \left[\hat{p}(X_i | p_1, p_2, \cdots, p_{N_e}) \right]. \tag{2.104}$$

The properties of the log-likelihood towards maximization are identical to those of the likelihood, since the logarithm is a monotonously increasing function. Thus, their maxima happen for the same parameter values. Using logarithms has the additional advantage of dealing with sums, instead of products, which simplifies taking derivatives. The parameter values sought can then be obtained by solving,

$$\frac{\partial}{\partial p_i} \left\{ \log L \left[(X^* | p_1, p_2, \cdots, p_{N_e}) \right] \right\} = 0, \quad \forall i = 1, \cdots, N_e \tag{2.105}$$

The resulting collection of parameter values, $(p_1^*, p_2^*, \cdots p_{N_e}^*)$ defines the **maximum likelihood estimator** of the pdf from the data.

2.5.1.1 MLE Example: The Gaussian Pdf

Let's work out the maximum likelihood estimator in the case of the Gaussian pdf (Eq. 2.30). There is only two parameters (i.e., $N_e = 2$) to be determined, the mean μ and the variance σ^2. The likelihood for N values is then given by the sum,

$$L^{\text{Gaussian}}(X^* | \mu, \sigma^2) = \prod_{i=1}^{N} \left\{ \frac{1}{\sqrt{2\pi\sigma^2}} \exp\left(-\frac{(X_i - \mu)^2}{2\sigma^2} \right) \right\} \tag{2.106}$$

$$= \left(\frac{1}{\sqrt{2\pi\sigma^2}} \right)^N \exp\left(-\frac{\sum_{i=1}^{N}(X_i - \mu)^2}{2\sigma^2} \right),$$

---

[44]The likelihood is not the same as the probability, since it is a product of probabilities densities, not of probabilities. That is, one would need to multiply each $p(X_i | p_1, \cdots, p_{N_e})$ by dX_i to get a probability. In fact, the likelihood is related to the **joint pdf** of the data set. That is, the probability of each of the values being between X_i and $X_i + dX_i$ simultaneously.

The log-likelihood is then,

$$\log\left[L^{\text{Gaussian}}(X^*|\mu,\sigma^2)\right] = -\frac{N}{2}\log(2\pi\sigma^2) - \frac{1}{2\sigma^2}\sum_{i=1}^{N}(X_i - \mu)^2. \tag{2.107}$$

Then, we simply look for the extremal values. First,

$$\frac{\partial}{\partial\mu}\left(\log\left[L^{\text{Gaussian}}(X^*|\mu,\sigma^2)\right]\right) = \frac{1}{\sigma^2}\left(\sum_{i=1}^{N}X_i - N\mu\right) = 0$$

$$\longrightarrow \hat{\mu} = \frac{1}{N}\sum_{i=1}^{N}X_i. \tag{2.108}$$

Then, for the variance, we obtain:

$$\frac{\partial}{\partial\sigma^2}\left(\log\left[L^{\text{Gaussian}}(X^*|\mu,\sigma^2)\right]\right) = -\frac{N}{2\sigma^2} + \frac{1}{2\sigma^4}\sum_{i=1}^{N}(X_i - \mu)^2 = 0$$

$$\longrightarrow \hat{\sigma}^2 = \frac{1}{N}\sum_{i=1}^{N}(X_i - \mu)^2. \tag{2.109}$$

To conclude, we replace $\mu \to \hat{\mu}$ in the expression for $\hat{\sigma}^2$. The final expression for the estimator of the variance is[45]:

$$\hat{\sigma}^2 = \frac{1}{N}\sum_{i=1}^{N}X_i^2 - \frac{1}{N^2}\sum_{i=1}^{N}\sum_{j=1}^{N}X_iX_j. \tag{2.110}$$

2.5.1.2   MLE Example: The (Bounded from Below) Power-Law Pdf

In later chapters of this book we will discuss power-law statistics at length.[46] It is thus convenient to calculate explicitly the maximum likelihood estimators in this

---

[45]It is because of this type of substitution that maximum likelihood estimators are said to be *biased* and some authors prefer other approaches [13]. This is a consequence of the fact that, in general, the estimator for a function of some parameter is usually not exactly identical to the evaluation of the function at the estimator for that parameter. Or, in other words, that $\hat{f}(x) \neq f(\hat{x})$.

[46]Power-laws are often (but not always!) related to complex dynamics (see Chap. 3).

case. We will consider, following [17], the following analytical pdf:

$$\hat{p}(x) = \left(\frac{a-1}{x_{\min}^{1-a}}\right) x^{-a}, \quad x \geq x_{\min} \geq 0. \tag{2.111}$$

The need for a lower bound $x_{\min}$ comes from the fact that, otherwise, the analytical pdf would not be normalizable for $a \geq 1$. The value of $x_{\min}$ is usually determined by direct inspection of the data.

Following the same steps as we did for the Gaussian pdf, but using the power-law pdf instead, one easily finds that the likelihood for N values becomes,

$$L^{\text{power}}(X^*|a) = \left(\frac{a-1}{x_{\min}}\right)^N \prod_{i=1}^{N} \left(\frac{X_i}{x_{\min}}\right)^{-a}, \tag{2.112}$$

and the log-likelihood is simply:

$$\log\left[L^{\text{power}}(X^*|a)\right] = n\log(a-1) - n\log x_{\min} - a \sum_{i=1}^{N} \log\left(\frac{X_i}{x_{\min}}\right). \tag{2.113}$$

The value of the maximum likelihood estimator for a is then obtained by maximizing the log-likelihood. The result is (see Exercise 2.9) :

$$\hat{a} = 1 + N \left[\sum_{i=1}^{N} \log\left(\frac{X_i}{x_{\min}}\right)\right]^{-1}. \tag{2.114}$$

It must be said, however, that finding maximum likelihood estimators can be rather involved for many analytical pdfs. Exact expressions are only available for a few cases (see Table 2.5, for some examples). In most situations, the estimation must other be done numerically. This is not difficult and often implies to solve just a few coupled nonlinear equations[47] (see Problem 2.10). Unfortunately, MLE methods are difficult to apply to one of the most important pdfs in the context of complex systems: Lévy distributions. The reason is that only the characteristic function of Lévy pdfs has an analytical expression (Eq. 2.32), not their pdf. Therefore, constructing their likelihood using Eq. 2.103 is rather involved. Although this could certainly be accomplished numerically [19], we will soon propose a different

---

[47]Usually, one needs to use some numerical algorithm to search for zeros of nonlinear equations. For instance, a Newton method [18].

Table 2.5 Maximum likelihood estimators for some common analytical pdfs

$\hat{p}(x)$	Parameters	MLE
Uniform	$\dfrac{1}{b-a}$	$\hat{a} = \min(X^*); \; \hat{b} = \max(X^*)$
Gaussian	Eq. 2.30	$\hat{\mu} = \dfrac{1}{N} \sum_{i=1}^{N} X_i; \; \hat{\sigma}^2 = \dfrac{1}{N} \sum_{i=1}^{N} (X_i - \hat{\mu})^2$
Exponential	Eq. 2.59	$\hat{\tau} = \dfrac{1}{N} \sum_{i=1}^{N} X_i$
Laplace[a]	$\dfrac{1}{2\tau} \exp\left(-\dfrac{|x-\mu|}{\tau}\right)$	$\hat{\mu} = \mathrm{median}(X^*); \; \hat{\tau} = \dfrac{1}{N} \sum_{i=1}^{N} |X_i - \hat{\mu}|$
Gamma	Eq. 2.70	$\hat{\beta} \simeq \dfrac{1}{2} \left[\log\left(\dfrac{1}{N} \sum_{i=1}^{N} X_i\right) - \dfrac{1}{N} \sum_{i=1}^{N} \log(X_i) \right]^{-1}$ $\hat{\tau} = \dfrac{1}{\hat{\beta}} \left(\dfrac{1}{N} \sum_{i=1}^{N} X_i\right)$
(Bounded) power law	Eq. 2.111	$\hat{a} = 1 + N \left[\sum_{i=1}^{N} \log\left(\dfrac{X_i}{x_{\min}}\right) \right]^{-1}$

[a] The median is the value m for which $P(X \leq m) = P(X \geq m)$

technique (not based on maximum likelihood) to estimate Lévy parameters from an arbitrary number of data (see Sect. 2.5.3).

2.5.2 Pearson's Goodness-of-Fit Test

Once the parameters of the model pdf have been estimated from the data, one needs to quantify whether the model is a good representation or not. There are several methods available in the literature to do this.[48] Here, we will describe one of the most popular ones: **Pearson's goodness-of-fit test** [13]. The method, made popular by Karl Pearson at the beginning of the twentieth century, relies on another analytical pdf, the **chi-square distribution**.

---

[48]The interested reader may also take a look at the **Kolmogorov-Smirnov test** [14], or at one of its refinements, the **Anderson-Darling test** [20]. Both of them are also used frequently to provide goodness-of-fit figures of merit.

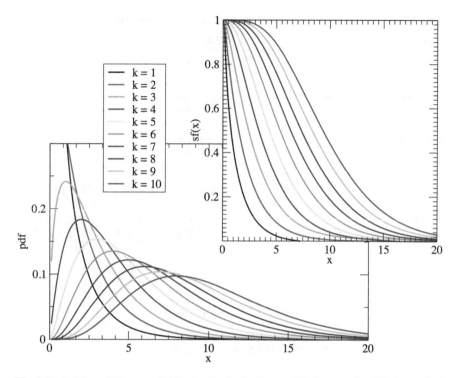

Fig. 2.17 Left box: Chi-square distribution for the first integer k indices; on the right, its survival function, $\mathrm{sf}(x) = P[X \geq x]$ (see Eq. 2.6)

2.5.2.1 Chi-Square Distribution

The **chi-square pdf of integer order** k is defined as[49]:

$$p_k^{\text{chi-sq}}(x) = \frac{1}{2^{k/2}\Gamma(\frac{k}{2})} x^{\frac{k}{2}-1} e^{-\frac{x}{2}}, \quad x \geq 0. \tag{2.115}$$

Its mean is $\mu = k$, and its variance is $\sigma^2 = 2k$. Figure 2.17 shows the chi-square pdf for the first ten integer values of k, together with its respective survival functions

---

[49]The attentive reader may have noticed that the chi-square distribution is a member of the family of Gamma pdfs (Eq. 2.70), that we described in Sect. 2.3.4.

(see Eq. 2.6), given by[50]:

$$\mathrm{sf}^{\mathrm{chi\text{-}sq}}(x) = P_k^{\mathrm{chi\text{-}sq}}[X \geq x] = 1 - \frac{\Gamma_{x/2}\left(\frac{k}{2}\right)}{\Gamma\left(\frac{k}{2}\right)}, \tag{2.117}$$

that will prove very important in this context.

Why is the chi-square distribution relevant for the comparison of data and models? The reason relies on the following property. Let's consider the random variable,

$$\chi^2 = \sum_{i=1}^{k} Y_i^2, \tag{2.118}$$

where each Y_i is an independent, random variable distributed according to a Gaussian distribution $N_{[0,1]}(y)$ (see Eq. 2.30). Then, the statistics of χ^2 follow the chi-square pdf of order k [13].

2.5.2.2 Pearson's Test of Goodness-of-Fit

How does one use the chi-square distribution to construct a goodness-of-fit test? The trick is to consider the hypothesis that the model pdf, $\hat{p}(x)$, that we have inferred (using N_e maximum likelihood estimators for N data, for instance) is actually an exact representation of the process, and that any deviation from the model values found in the experimental pdf (that is, $p(X_i)$ ($i = 1, 2, \cdots, N$) determined using any of the methods presented in this chapter) is *only due to accidental errors*. Since accidental errors are pretty well described by Gaussian statistics with a zero mean,[51] the strategy to follow is to normalize the differences between model and experimental pdfs at each point (in order to approximate the unit variance condition), square them and sum them. The resulting quantity should follow the statistics of a chi-square distribution.

---

[50] $\Gamma_x(a)$ is the **incomplete gamma function**, that is defined as:

$$\Gamma_x(a) = \int_0^x ds\, s^{a-1} e^{-s}. \tag{2.116}$$

[51] In contrast, systematic errors have a non-zero mean, introducing a bias in the quantity being measured.

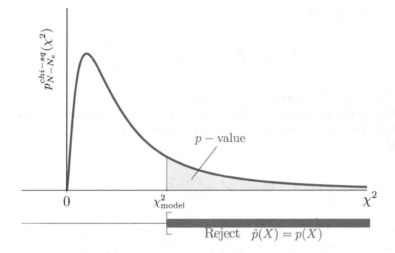

Fig. 2.18 Illustration of Pearson's test. Once a value for χ^2_{model} is obtained from the differences between the model and experimental pdfs using Eq. 2.119, the associated p-value is determined by estimating the area shown in blue under the chi-square distribution or order $N - N_e$. That is, by estimating the probability of observing a value at least as extreme as the observed χ^2_{model} in a chi-squared distribution of order $N - N_e$. In order for the model pdf to be a significant representation of the data, the resulting p-value should be at least in the range (0.05–0.10)

Pearson's test does precisely this. It chooses a normalization of the differences between model and experimental pdfs so that the variable of interest becomes:

$$\chi^2_{model} := \sum_{i=1}^{N} \frac{[p(X_i) - \hat{p}(X_i)]^2}{\hat{p}(X_i)}, \tag{2.119}$$

Since the number of independent quantities[52] in this sum is, roughly, $N - N_e$, one would expect that the statistics of χ^2_{model} be given by the chi-square distribution of order $N - N_e$.

Once the order of the chi-squared pdf has been established, one needs to quantify how good the obtained for χ^2_{model} in terms of supporting (or invalidating) the hypothesis of $\hat{p}$ being an exact representation of the statistics of the process. This is accomplished by computing the so-called p-**value** of the fit (see Fig. 2.18). This figure-of-merit is defined as *the probability of observing a value at least as extreme as χ^2_{model} in a chi-squared distribution of order $N - N_e$*. Or, in other words, using the survival function of the chi-square distribution (Eq. 2.117),

$$p\text{-value} = P^{chi\text{-}sq}_{N-N_e}\left[\chi^2 \geq \chi^2_{model}\right] = sf^{chi\text{-}sq}_{N-N_e}\left(\chi^2_{model}\right). \tag{2.120}$$

---

[52]The number of independent values, also known as the number of degrees of freedom, is given by the number of data points N minus the number of parameters that define the model pdf, N_e.

Table 2.6 Maximum allowed value for χ^2_{model} for selected p-values for the chi-square distributions of orders one to ten

$N - N_e$	χ^2_{model}									
1	0.004	0.02	0.06	0.15	0.46	1.07	1.64	2.71	**3.84**	6.64	10.83
2	0.10	0.21	0.45	0.71	1.39	2.41	3.22	4.60	**5.99**	9.21	13.82
3	0.35	0.58	1.01	1.42	2.37	3.66	4.64	6.25	**7.82**	11.34	16.27
4	0.71	1.06	1.65	2.20	3.36	4.88	5.99	7.78	**9.49**	13.28	18.47
5	1.14	1.61	2.34	3.00	4.35	6.06	7.29	9.24	**11.07**	15.09	20.52
6	1.63	2.20	3.07	3.83	5.35	7.23	8.56	10.64	**12.59**	16.81	22.46
7	2.17	2.83	3.82	4.67	6.35	8.38	9.80	12.02	**14.07**	18.48	24.32
8	2.73	3.49	4.59	5.53	7.34	9.52	11.03	13.36	**15.51**	20.09	26.12
9	3.32	4.17	5.38	6.39	8.34	10.66	12.24	14.68	**16.92**	21.67	27.88
10	3.94	4.87	6.18	7.27	9.34	11.78	13.44	15.99	**18.31**	23.21	29.59
p-Value	0.95	0.90	0.80	0.70	0.50	0.30	0.20	0.10	**0.05**	0.01	0.001

Due to the properties of the survival function (Eq. 2.6), the p-value tends to zero as $\chi^2_{\text{model}} \to \infty$. That is, when the model and the data are as far apart as possible. On the other limit, the p-value becomes one when model and data are identical. That is, if $\chi^2_{\text{model}} \to 0$. As a result, the p-value of any model should lie within 0 and 1. The closer to 1 it is, the better the model is. In fact, it is widely accepted that the p-value should be at least in the range (0.05–0.10) for the hypothesis to have any significance. In the case of several competing models, the one yielding the largest p-value would thus be the most significant.

On a technical note, finding the p-value requires the evaluation of the survival function of the chi-square distribution of the proper order (see Eq. 2.117), that must be done numerically.[53] If the order of the chi-square distribution needed is small, there are tables that make the estimation of the p-value much easier. For instance, Table 2.6 shows, for the chi-square distributions with orders one to ten, the values of χ^2_{model} for ten different p-values. In bold, one can see the maximum allowed value of χ^2_{model}, for each order, beyond which the hypothesis $\hat{p}(x) = p(X)$ should be rejected if one accepts 0.05 as the minimum acceptable p-value.[54]

2.5.3 Minimum Chi-Square Parameter Estimation

It turns out that the chi-square distribution can also be used to perform parameter estimation for pdfs, in a similar vain to maximum likelihood. The idea is quite

---

[53]There are many publicly available packages that do it. For instance, the popular statistical package R is one of them.

[54]Similar tables can be easily found in the literature. See, for instance, [21].

simple. Similarly to what we did with Eq. 2.119, one builds the quantity:

$$\chi^2(p_1, \cdots, p_{N_e}) := \sum_{i=1}^{N} \frac{[p(X_i) - \hat{p}(X_i | p_1, \cdots, p_{N_e})]^2}{\hat{p}(X_i | p_1, \cdots, p_{N_e})}. \tag{2.121}$$

The difference is now that the model pdf depends on the undetermined set of parameters: $p_1, \cdots, p_{N_e}$. Clearly, we would like to determine the values of these parameters that make χ^2 as small as possible. One can accomplish this by solving, perhaps numerically, the set of equations:

$$\frac{\partial \chi^2(p_1, \cdots, p_{N_e})}{\partial p_i} = 0, \quad \forall i. \tag{2.122}$$

One can even estimate whether the determined parameter values provide a good representation by substituting them in Eq. 2.121, which yields a value χ^2_{model}. We should then determine the associated p-value, after assuming a chi-square distribution of order $N - N_e$. If that p-value is smaller than (0.05–0.10), the model pdf is not a good representation! And if there are several candidates, the one yielding the larger p-value would be the better representation for the data.

When should one use minimum chi-square parameter estimation instead of maximum likelihood? Well, the nice property of the chi-square estimation is that it also works fine when the two quantities being compared are not pdfs. The only thing that matters is that the differences are due to accidental errors, not what the quantities being compared are. For instance, we could compare characteristic functions instead of pdfs. This becomes particularly handy in the case of Lévy pdfs (see Problem 2.13) which, as we mentioned earlier, lack an analytical expression for their pdf, but have one for their characteristic function (Eq. 2.32). We will see other examples of use of this technique later in the book (Sect. 5.4.1).

2.6 Case Study: The Running Sandpile

The running sandpile introduced by *Hwa and Kardar* [22] constitutes a standard paradigm of self-organized criticality (see Sect. 1.3). The complexity of the sandpile behaviour can be made apparent with many diagnostics. In order to illustrate the methods introduced in this chapter, we will focus here on those that have to do with the statistics of some measurable quantity. For reasons that were already discussed in Sect. 1.3, the statistics of quantities such as the linear extension or the size of the avalanches that govern transport throughout the sandpile *in steady state* should scale as, $p(s) \sim s^{-a}$, with $1 < a < 2$, if complex dynamics are dominant.

To try to determine this exponent, we will use a numerical running sandpile with $L = 1000$ cells, critical gradient $Z_c = 200$ and toppling size $N_F = 30$. The sandpile is continuously driven by dropping on each cell $N_b = 10$ grains of sand with a probability $p_0 = 10^{-6}$. Note that p_0 is chosen to ensure that the average number of iterations between two successive avalanches, $(p_0 \cdot L)^{-1} \sim 10^3$, is of the order (or larger than) the system size. In this way, the majority of avalanches triggered in the

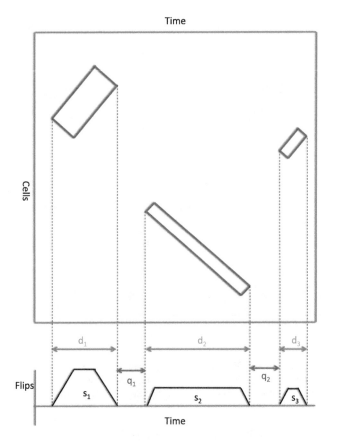

Fig. 2.19 Sketch showing the relationship between the flip time series and avalanches in the running sandpile. The evolution of the sandpile is represented in a time-space plot, with avalanches shown as red boxes that contain toppling sites. The flip time series is shown below, with time running from left to right. A trapezoidal structure in the flip series is associated with each avalanche. The duration (marked as d_i) and size (as s_i) of each avalanche are given by the length and area of each trapezoidal structure. The waiting time (q_i) between avalanches is also shown

system can have time to be finished and the system back at rest before the next rain of sand comes.[55] This is done to minimize avalanche overlapping.

The output of the numerical sandpile of interest to us will be the time series of the total number of topplings, also called **flips**, per iteration. That is, the time record of the number of cells that undergo a toppling in the sandpile. From this time series, it is easy to calculate both the duration and size of an avalanche. To understand how, Fig. 2.19 shows a sketch of the time evolution of the running sandpile. The upper part of the plot shows the domain of the sandpile in the vertical direction,

---

[55]The longest avalanches in the absence of overlapping will be of the order of the system size, since propagation from one cell to the next takes one iteration.

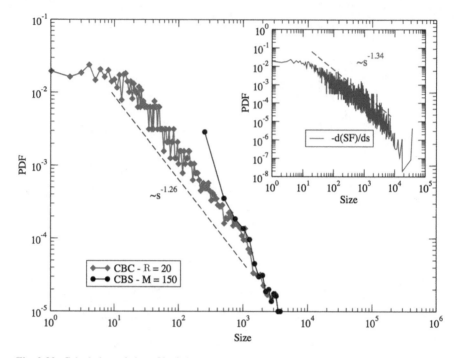

Fig. 2.20 Calculation of the pdf of the avalanche sizes for the $L = 1000$ running sandpile described in the text. Three methods have been used: constant bin size (with $M = 150$ bins), constant bin content (with $R = 20$ points-per-bin) and the survival function method (shown in the inset). The number of size values available is $N = 3186$

whilst time runs from left to right. Several possible avalanches are shown in the figure as red boxes containing the cells that are being toppled as time advances. Each avalanche starts at a given location and time, and then propagates by causing the toppling of neighbouring cells both outwards and inwards.[56] Eventually, the avalanche reaches a stopping point and dies away. The corresponding flip time series is shown in the lower part of the figure. Topplings only happen when an avalanche is active somewhere in the pile. Each of the trapezoidal figures in the flip time series is associated with one avalanche. The length and the area of these trapezoids give the duration and size of the avalanche.[57] We can also obtain the waiting-times between avalanches from the distance between two successive trapezoidal structures.

After running the sandpile to steady state, we have advanced its state for a million additional iterations. In this period, the flip time-series happens to contain 3186 avalanches. The results of applying the three methods previously discussed to the time series of their sizes are shown in Fig. 2.20. Clearly, the CBS method with $M =$

---

[56] Although note that the transport of sand in the sandpile always happens down the slope!

[57] This association certainly becomes more obscure when avalanches overlap either in space and time. Can the reader explain why? This is the reason for choosing the running sandpile parameters so carefully. In particular, N_b and p_0.

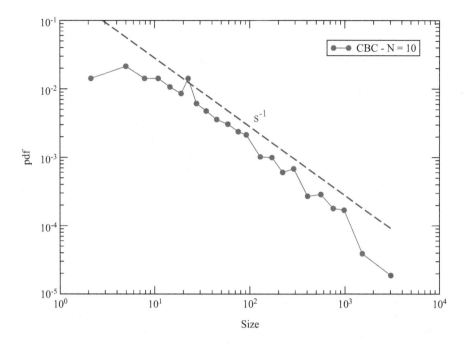

Fig. 2.21 Calculation of the pdf of the avalanche sizes for the $L = 1000$ running sandpile described in the text using the CBC method with $R = 10$ points-per-bin and $N = 300$ data points

150 bins works pretty badly. The reason is similar to the case of the Cauchy example worked out previously: the existence of a rather heavy tail, $p(s) \sim s^{-1.3}$, that extends for over three decades. This region is however captured very clearly by the CBC method using just $R = 20$ points-per-bin. The survival method, on the other hand, works also relatively well as shown in the inset of Fig. 2.20.

Finally, to prove the robustness of the CBC method, we have recalculated the pdf of the avalanche sizes but using only the first $N = 300$ values in the time-series of the sizes. That is, using ten times less values than what we used to produce the results in Fig. 2.20. The result of using CBC with $R = 10$ points-per-bin is shown in Fig. 2.21. Clearly, the exponent of the resulting tail is much less accurate, but the existence of a power-law region is still apparent over three decades with a decay exponent close to one. That is pretty remarkable, in our opinion, which should make CBC the first choice if a sufficient number of points is available. When this is not the case, and CBC is not a realistic option, the survival function (or cumulative distribution function) methods should at least be able to give us an idea of whether there is a meaningful tail or not in the data.

2.7 Final Considerations

In this chapter we have discussed some of the fundamentals of the statistical analysis
of data. The techniques presented are quite useful in practical cases, in particular
the CBC method and the survival/cumulative distribution function methods. We
encourage all readers to put them to practice using synthetic data first, in order to
better hone their skills. We have included, in Appendix 2, some methods to generate
time series with prescribed statistics, including Gaussian, Lévy and exponentially
distributed data. There are other methods available, but it seems to us that these
ones are particular simple to implement. We will use these methods again in later
chapters to explore other features of time series, relevant to the analysis of complex
dynamics.

Appendix 1: The Fourier Transform

The Fourier representation has a very large number of applications in Mathematics,
Physics and Engineering, since it permits to express an arbitrary function as a linear
combination of periodic functions. Truncation of the Fourier representation, for
example, is the basis of signal filtering. Many other manipulations (modulation,
integration, etc.) form the basis of modern communications. In Physics, use of the
Fourier representation is essential for the understanding of resonances, oscillations,
wave propagation, transmission and absorption, fluid and plasma turbulence and
many other processes. In Mathematics, Fourier transforms are often used to simplify
the solution of differential equations, among many other problems.

The **Fourier transform** of a (maybe complex) function $f(x)$ is defined as[58]:

$$F[f(x)] := \hat{f}(k) = \int_{-\infty}^{\infty} dx\, f(x)\exp(\imath kx), \qquad (2.123)$$

where $\imath = \sqrt{-1}$, k is known as the **wave number**.[59] Note that, if $f(x)$ is real, then
$\hat{f}(k)$ is Hermitian, meaning that $\hat{f}(-k) = \hat{f}^*(k)$, where the asterisk represents the
complex conjugate.

---

[58] We adopt here a convention for the Fourier transform that is common in the fields of probability
and random processes, from which many results will be referred to in Chaps. 3–5. This convention
has the benefit of converting the characteristic function into the Fourier transform of the pdf (see
Sect. 2.2.4). In other contexts, however, the positive exponential is often reserved for the inverse
transform, with the negative exponential used in the direct transform. Regarding normalization, it
is also common to find texts that use $2\pi\imath kx$ in the exponent of the exponentials (instead of just $\imath kx$),
so that the $1/2\pi$ prefactor of the inverse transform disappears. In other cases, the exponent is kept
without the 2π factor, but both direct or inverse transforms have a $1/\sqrt{2\pi}$ prefactor.

[59] We will always represent the Fourier transform with a hat symbol, $\hat{f}$. Whenever the real variable
represents a time, t, the wave number is referred to as a **frequency**, and represented as ω.

The **inverse of the Fourier transform** is then obtained as,

$$f(x) = \mathrm{F}^{-1}[\hat{f}(k)] := \frac{1}{2\pi} \int_{-\infty}^{\infty} dk \, \hat{f}(k) \exp(-\imath kx). \tag{2.124}$$

This expression is also referred to as the **Fourier representation** of $f(x)$.

Some common Fourier transform pairs are collected in Table 2.7. However, it must be noted that the existence of both the Fourier and inverse transforms is not always guaranteed. A sufficient condition (but not necessary) is that both $f(x)$ and $\hat{f}(k)$ be (Lebesgue)-integrable. That is,

$$\int_{-\infty}^{\infty} dx \, |f(x)| < \infty, \qquad \int_{-\infty}^{\infty} dk \, |\hat{f}(k)| < \infty, \tag{2.125}$$

which requires that $|f(x)| \to 0$ faster than $|x|^{-1}$ for $x \to \pm\infty$, and that $|\hat{f}(k)| \to 0$ faster than $|k|^{-1}$ for $k \to \pm\infty$. But Fourier transforms may exist for some functions that violate this condition.

Fourier transforms have many interesting properties. In particular, they are **linear operations**. That is, the Fourier transform of the sum of any two functions is equal to the sum of their Fourier transforms.

Secondly, the **Fourier transform of the n-th derivative** of a function becomes particularly simple:

$$\mathrm{F}\left[\frac{d^n f}{dx^n}\right] = (-\imath k)^n \hat{f}(k), \tag{2.126}$$

as trivially follows from differentiating Eq. 2.124.

Another interesting property has to do with the **translation of the argument**, that translates into **a multiplication by a phase** in Fourier space:

$$\mathrm{F}[f(x \pm x_0)] = \exp(\mp \imath k x_0)\hat{f}(k). \tag{2.127}$$

Particularly useful properties are the following two theorems [23]. One is **Parseval's theorem**,

$$\int_{-\infty}^{\infty} dx \, f(x) g^*(x) = \frac{1}{2\pi} \int_{-\infty}^{\infty} dk \, \hat{f}(k) \hat{g}^*(k) \tag{2.128}$$

where the asterisk denotes complex conjugation. When $f = g$, this becomes,

$$\int_{-\infty}^{\infty} dx \, |f(x)|^2 = \frac{1}{2\pi} \int_{-\infty}^{\infty} dk \, |\hat{f}(k)|^2. \tag{2.129}$$

Table 2.7 Some useful Fourier transforms

$f(t)$	$\hat{f}(k)$	Restrictions
1	$2\pi\delta(k)$	
$\delta(x)$	1	
$H(x+a) - H(x-a)$	$\dfrac{\sin(ak)}{k}$	
$\dfrac{\sin(ax)}{x}$	$2\pi\left[H(k+a) - H(k-a)\right]$	
$\lvert x\rvert^{-a}$	$\Gamma(1-a)k^{a-1}$	$0 < a < 1$
$\dfrac{1}{x^2 + a^2}$	$\dfrac{\pi\exp(-a\lvert k\rvert)}{a}$	$a > 0$
$\exp(-a\lvert x\rvert)$	$\dfrac{2a}{a^2 + k^2}$	$a > 0$
$H(x)\exp(-ax)$	$\dfrac{1}{a - \imath k}$	$a > 0$
$\exp(-ax^2)$	$\sqrt{\dfrac{\pi}{a}}\exp\left(-\dfrac{k^2}{4a}\right)$	$a > 0$
$H(x)\,x^b\exp(-ax)$	$\dfrac{\Gamma(b+1)}{(a - \imath k)^{b+1}}$	$b > -1, a > 0$
$\lvert x\rvert^b\exp(-a\lvert x\rvert)$	$\Gamma(b+1)\dfrac{\left[(a - \imath k)^{b+1} + (a + \imath k)^{b+1}\right]}{(a^2 + k^2)^{b+1}}$	$b > -1, a > 0$

$H(x)$ represents the Heaviside step function (i.e., $H(x) = 1$, for $x > 0$, and $H(x) = 0$ for $x < 0$) and $\delta(x)$ is the usual Dirac delta function. $\Gamma(x)$ is the usual Gamma function

$|\hat{f}(k)|^2$ is usually known as the **power spectrum** of $f(x)$.[60]

The second important theorem is the **convolution theorem**. It states that the convolution of two functions, defined as:

$$c_{[f,g]}(x) = \int_{-\infty}^{\infty} f(x')g(x-x')dx' \qquad (2.130)$$

satisfies that its Fourier transform is given by:

$$\hat{c}_{[f,g]}(k) = \hat{f}(k)\hat{g}(k). \qquad (2.131)$$

assuming that the individual Fourier transforms, $\hat{f}(k)$ and $\hat{g}(k)$ do exist.

In the context of *scale-invariance*, it is useful to note that:

$$F[f(ax)] = \frac{1}{|a|}\hat{f}\left(\frac{k}{a}\right). \qquad (2.132)$$

Similarly, the **Fourier transform of a power law** (see Table 2.7) is also very useful. In particular, it can be shown that, the Fourier transform of $f(x) = |x|^{-a}$, with $0 < a < 1$, is given by another power-law.[61] Namely,

$$F(|x|^{-a}) = \Gamma(1-a)k^{a-1}, \qquad (2.133)$$

being $\Gamma(x)$ Euler's gamma function. Furthermore, it can be shown that if a function $f(x) \sim |x|^{-a}$ with $0 < a < 1$ for $x \to \infty$, then its Fourier transform $\hat{f}(k) \sim k^{a-1}$ for $k \to 0$.

Appendix 2: Numerical Generation of Series with Prescribed Statistics

One of the most commonly used method generate time series with a prescribed set of statistics is the **inverse function** method. It is based on passing a time series u

---

[60]The name *power spectrum* was made popular in the context of turbulence, where f usually represents a velocity or velocity increment. The power spectrum then roughly quantifies the "*amount of energy*" contained between k and $k + dk$. For that reason, this theorem is often interpreted in Physics as a statement of the conservation of energy. In other context, the interpretation may vary. In information theory, for instance, Parseval's theorem represents the conservation of information.

[61]It should be noted that $|x|^{-a}$ with $0 < a < 1$ is an example of a function that is not Lebesgue integrable and has a Fourier transform. Its Fourier transform, $\hat{f}(k) = \Gamma(1-a)k^{a-1}$, however diverges for $k \to 0$ as a result.

whose values follow a uniform distribution[62] in $[0, 1]$ through an invertible function. That is, we generate a collection of values using:

$$x = F^{-1}(u). \tag{2.134}$$

The series of values for x is then distributed according to the pdf:

$$p(x) = \frac{dF}{dx}. \tag{2.135}$$

To prove this statement, let's consider the cumulative distribution function associated with $p(x)$ (Eq. 2.4) at $x = a$, that verifies:

$$\text{cdf}(a) = P(X \le a) = P\left(F^{-1}(U) \le a\right) = P(U \le F(a)) = F(a). \tag{2.136}$$

The fourth step follows from the fact that the cumulative function being an increasing function of its argument. The last step is due to the fact that the cumulative distribution function of the uniform distribution is simply its argument! If $F(x) = \text{cdf}(x)$, then Eq. 2.135 follows after invoking Eq. 2.5.

Generating time series with prescribed statistics is thus reduced to knowing the inverse of the cumulative distribution function of the desired distribution.[63] For instance, in the case of the **exponential pdf**, $E_\tau(x) = \tau^{-1} \exp(-x/\tau)$, the sought inverse is,

$$\text{cdf}(x) = 1 - \exp(-x/\tau) \implies \text{cdf}^{-1}(x) = -\frac{\log(1-u)}{\tau}. \tag{2.137}$$

Thus, one can generate a series distributed with an exponential pdf of mean value τ by iterating:

$$e_\tau = \frac{\log(1-u)}{-\tau}, \tag{2.138}$$

using uniformly-distributed $u \in [0, 1]$.

Regretfully, many important distributions do not have analytical expressions for their cdfs. Much less of their inverse! This is the case, for instance, of the Gamma pdf (Eq. 2.70). Another important example is the **Lévy distributions**, that only have an analytic expression for their characteristic function (Eq. 2.32). Thus, the inverse function method is useless to generate this kind of data. Luckily, a formula exists, based on combining two series of random numbers, u distributed uniformly in $[-\pi/2, \pi/2]$, and v distributed exponentially with unit mean, that is able to generate

---

[62]Most programming languages provide with intrinsic functions to generate series of data with a uniform distribution $u \in [0.1]$. Thus, we will not explain how to do it here!

[63]This inverse is known as the **quantile function** in statistics.

a series of l values distributed according to prescribed Lévy statistics. For $\alpha \neq 1$ the formula reads [24, 25]:

$$l_{[\alpha,\lambda,\mu,\sigma]} = \mu + \frac{\sigma \sin(\alpha u + K_{\alpha,\lambda})}{[\cos(K_{\alpha,\lambda})\cos(u)]^{1/\alpha}} \left[\frac{\cos((1-\alpha)u - K_{\alpha,\lambda})}{v} \right]^{1/\alpha - 1} \tag{2.139}$$

with $K_{\alpha,\lambda} = \tan^{-1}(\lambda \tan(\pi\alpha/2))$. For the case $\alpha = 1$, one should use instead [25]:

$$l_{[1,\lambda,\mu,\sigma]} = \mu + \frac{2\sigma}{\pi} \left\{ \left(\frac{\pi}{2} + \lambda u \right) \tan u - \lambda \left[\log \left(\frac{v \cos u}{\sigma \left(\frac{\pi}{2} + \lambda u \right)} \right) \right] \right\}. \tag{2.140}$$

These formulas reduce to much simpler expressions in some cases. For instance, in the **Gaussian case** ($\alpha = 2, \lambda = 0$), Eq. 2.139 reduces to:

$$g_{[\mu,\sigma]} = \mu + 2\sigma\sqrt{v}\sin(u) \tag{2.141}$$

whilst for the **Cauchy distribution** ($\alpha = 1, \lambda = 0$), Eq. 2.140 reduces to:

$$c_{[\mu,\sigma]} = \mu + \sigma \tan(u). \tag{2.142}$$

Problems

2.1 Survival and Cumulative Distribution Functions
Calculate explicitly the cumulative distribution function and the survival function of the Cauchy, Laplace and exponential pdfs.

2.2 Characteristic Function
Calculate explicitly the characteristic function of the Gaussian, exponential and the uniform pdfs.

2.3 Cumulants and Moments
Derive the relations between cumulants and moments up to order $n = 5$.

2.4 Gaussian Distribution[64]
Show that for a Gaussian pdf, $N_{[0,\sigma^2]}(x)$, all cumulants are zero except for $c_2 = \sigma^2$. Also, show that only the even moments are non-zero and equal to $m_n = \sigma^n(n-1)!!$.

2.5 Log-Normal Distribution
Calculate the first four cumulants of the log-normal distribution defined by Eq. 2.45. Show that they are given by: $c_1 = \exp(\mu + \sigma^2/2)$; $c_2 = (\exp(\sigma^2) - 1)\exp(2\mu + \sigma^2)$; $S = (\exp(\sigma^2) + 2)\sqrt{\exp(\sigma^2) - 1}$ and $K = \exp(4\sigma^2) + 2\exp(3\sigma^2) + 3\exp(2\sigma^2) - 6$.

---

[64]The double factorial is defined as $n!! = n(n-2)(n-4)\cdots$.

2.6 Poisson Distribution

Derive the limit of the binomial distribution $B(k, n; p)$ for $n \to \infty$ and $p \to 0$, keeping $np = \lambda$, and prove that it is the Poisson discrete distribution (Eq. 2.67).

2.7 Ergodicity

Consider the process defined by $y(t) = \cos(\omega t + \phi)$, where ω (the frequency) is fixed, but with a different values of ϕ (the phase) for each realization of the process. Under what conditions are the temporal and ensemble views of the statistics of y equivalent? Or, in other words, when does the process behave ergodically?

2.8 Numerical Estimation of pdfs

Write a code that, given a set of experimental data, $\{x_i, \ i = 1, 2, \cdots N\}$, calculates their pdf using each of the methods described in this chapter: the CBS, CBC and the survival/cumulative distribution function methods.

2.9 Maximum Likelihood Estimators (I)

Calculate the maximum likelihood estimators of the Exponential pdf (Eq. 2.59) and the bounded power-law pdf (Eq. 2.111) using the methodology described in Sect. 2.5.

2.10 Maximum Likelihood Estimators (II)

Prove that, in the case of the Gumbel distribution (Eq. 2.51), the determination of the maximum likelihood estimators for μ and σ requires to solve numerically the pair of coupled nonlinear equations [26]:

$$\hat{\mu} = -\hat{\sigma} \log \left[\frac{1}{N} \sum_{i=1}^{N} \exp(-X_i/\hat{\sigma}) \right] \tag{2.143}$$

$$\hat{\sigma} = \frac{1}{N} \sum_{i=1}^{N} X_i - \frac{\sum_{i=1}^{N} X_i \exp(-X_i/\hat{\sigma})}{\sum_{i=1}^{N} \exp(-X_i/\hat{\sigma})} \tag{2.144}$$

Write a numerical code that solves this equation using a nonlinear Newton method [18].

2.11 Synthetic Data Generation

Write a code that implements Eq. 2.139 to generate Lévy distributed synthetic data with arbitrary parameters. Use it to generate at least three different sets of data, and assess the goodness of Eq. 2.139 by computing their pdf.

2.12 Running Sandpile: Sandpile Statistics

Write a code that, given a flip time-series generated by the running sandpile code built in Problem 1.5, is capable of extracting the size and duration of the avalanches contained in it, as well as the waiting-times between them. Use the code proposed in Problem 2.8 to obtain the pdfs of each of these quantities.

2.13 Advanced Problem: Minimum Chi-Square Parameter Estimation

Write a code that, given $\{x_i, \ i = 1, 2, \cdots N\}$, calculates the characteristic function of its pdf, $\phi^{\text{data}}(k_j)$, with k_j being the collocation points in Fourier space. Then, estimate the parameters of the Lévy pdf that best reproduces the data by minimizing the target function, $\chi_k^2(\alpha, \lambda, \mu, \sigma)$, that quantifies the difference between the characteristic function of the data and that of the Lévy pdf (Eq. 2.32):

$$\chi_k^2(\alpha, \lambda, \mu, \sigma) = \sum_{i=1}^{N} \frac{\left| \phi^{\text{data}}(k_j) - \phi^L_{[\alpha,\lambda,\mu,\sigma]}(k_j) \right|^2}{\left| \phi^L_{[\alpha,\lambda,\mu,\sigma]}(k_j) \right|}, \tag{2.145}$$

To minimize χ_k^2, the reader should look for any open-source subroutine that can do local optimization. We recommend using, for example, the *Levenberg-Marquardt* algorithm [18].

References

1. Taylor, G.I.: The Spectrum of Turbulence. Proc. R. Soc. Lond. 164, 476 (1938)
2. Balescu, R.: Equilibrium and Non-equilibrium Statistical Mechanics. Wiley, New York (1974)
3. Tijms, H.: Understanding Probability. Cambridge University Press, Cambridge (2002)
4. Feller, W.: An Introduction to Probability Theory and Its Applications. Wiley, New York (1968)
5. Taylor, J.R.: An Introduction to Error Analysis: The Study of Uncertainties in Physical Measurements. University Science Books, New York (1996)
6. Gnedenko, B.V., Kolmogorov, A.N.: Limit Distributions for Sums of Independent Random Variables. Addison-Wesley, New York (1954)
7. Samorodnitsky, G., Taqqu, M.S.: Stable Non-Gaussian Processes. Chapman & Hall, New York (1994)
8. Aitchison, J., Brown, J.A.C.: The Log-Normal Distribution. Cambridge University Press, Cambridge (1957)
9. Carr, P., Wu, L.: The Finite Moment Log-Stable Process and Option Pricing. J. Financ. 53, 753 (2003)
10. Kida, S.: Log-Stable Distribution in Turbulence. Fluid Dyn. Res. 8, 135 (1993)
11. Kotz, S., Nadarajah, S.: Extreme Value Distributions: Theory and Applications. Imperial College Press, London (2000)
12. Cox, D.R., Isham, V.: Point Processes. Chapman & Hall, New York (1980)
13. Mood, A., Graybill, F.A., Boes, D.C.: Introduction to the Theory of Statistics. McGraw-Hill, New York (1974)
14. Chakravarti, I.M., Laha, R.G., Roy, J.: Handbook of Methods of Applied Statistics. Wiley, New York (1967)
15. Jaynes, E.T.: Probability Theory: The Logic of Science. Cambridge University Press, Cambridge (2003)
16. Aldrich, J.: R. A. Fisher and the Making of Maximum Likelihood (1912–1922). Stat. Sci. 12, 162 (1997)
17. Clauset, A., Shalizi, C.R., Newman, M.E.J.: Power-Law Distributions in Empirical Data. SIAM Rev. 51, 661 (2009)
18. Press, W.H., Teukolsky, S.A., Vetterling, W.T., Flannery, B.P.: Numerical Recipes in Fortran 90. Cambridge University Press, Cambridge (1996)

19. Brorsen, B.W., Yang, S.R.: Maximum Likelihood Estimates of Symmetric Stable Distribution Parameters. Commun. Stat. 19, 1459 (1990); Nolan, J.: Levy Processes, Chap. 3. Springer, Heidelberg (2001)
20. Stephens, M.A.: EDF Statistics for Goodness of Fit and Some Comparisons. J. Am. Stat. Assoc. 69, 730 (1974)
21. Abramowitz, M., Stegun, I.A.: Handbook of Mathematical Functions. National Bureau of Standards, Washington, DC (1970)
22. Hwa, T., Kardar, M.: Avalanches, Hydrodynamics and Discharge Events in Models of Sandpile. Phys. Rev. A 45, 7002 (1992)
23. Bracewell, R.N.: The Fourier Transform and Its Applications. McGraw-Hill, Boston (2000)
24. Chambers, J.M., Mallows, C.L., Stucka, B.W.: Method for Simulating Stable Random Variables. J. Am. Stat. Assoc. 71, 340 (1976)
25. Weron, R.: On the Chambers-Mallows-Stuck Method for Simulating Skewed Stable Random Variables. Stat. Probab. Lett. 28, 165 (1996)
26. Forbes, C., Evans, M., Hastings, M., Peacock B.: Statistical Distributions. Wiley, New York (2010)

Chapter 3
Scale Invariance

3.1 Introduction

Scale-invariance is one of the concepts that appears more often in the context of **complexity**. The basic idea behind scale-invariance can be naively expressed as: '*the system looks the same at every scale*'. Or, '*if we zoom in (or out) our view of the system, its features remain unchanged*'. Although it is true that complex dynamics are often at work when a system exhibits scale-invariance, it is important to be aware that this is not always the case. For instance, the *random walk* [1] is a process that exhibits scale-invariance but with underlying dynamics that are far from complex in any sense of the word (see Chap. 5).

Scale-invariance, and thus also complex dynamics, is often claimed whenever **power-laws** are exhibited by the statistics of some event. The reason is that mathematical functions are referred to as **scale-invariant** (or **self-similar**[1]) if they remain invariant under *any dilation* of the independent variable(s):

$$f(\lambda x, \lambda y, \cdots) = \lambda^H f(x, y, \cdots), \quad \forall \lambda > 0, \tag{3.1}$$

for some value of the **scale-invariant exponent** H. The paradigmatic example of scale-invariance is the **power law**, $f(x) = x^a$, for any real a. Indeed,

$$f(\lambda x) = (\lambda x)^a = \lambda^a f(x). \tag{3.2}$$

---

[1]Although we will use these two terms as synonyms throughout this book, mathematicians often distinguish between *scale-invariant* functions (i.e., those that satisfy Eq. 3.1) and *self-similar functions*. The latter correspond to those that are invariant under *discrete dilations*, such as any of the mathematical fractals that can be generated by iterative procedures. Throughout this book, we will also use the term *self-similar* to refer to objects or processes that can be broken down in smaller pieces that are similar (in the best case, identical; in other cases, only approximately or in a statistical sense) to the original. In other words, we will sometimes abuse the term self-similar and consider it a *weaker (discrete, approximate or statistical) version of scale-invariant*.

© Springer Science+Business Media B.V. 2018
R. Sánchez, D. Newman, *A Primer on Complex Systems*,
Lecture Notes in Physics 943, https://doi.org/10.1007/978-94-024-1229-1_3

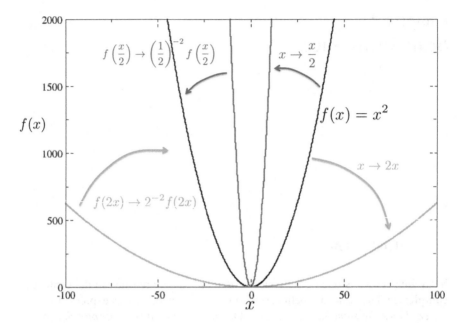

Fig. 3.1 Effect of multiplying x by λ for the power law $f(x) = x^2$. When $\lambda = 2$ (green), the graph changes as if we were zooming in into the figure; for $\lambda = 1/2$, the graph shrinks, as if we were zooming out of the figure. The original graph is recovered if one multiplies each function by λ^{-2}, given that the self-similarity exponent of the parabola is $H = 2$

Roughly speaking, the product of x and λ represents a **dilation** of that coordinate. Figure 3.1 illustrates this fact on the graph of the function $f(x) = x^2$ by showing that zooming in takes place if $\lambda > 1$, and zooming out for $\lambda < 1$. Scale-invariance simply means that we could recover the original graph by dilating the independent variable by λ^{-H}.

A famous power-law appears in the study of earthquakes. It was proposed in the 1950s that the number of earthquakes N of at least magnitude M, happening within any seismically active region, appeared to follow the scaling law:

$$N(m \le M) \propto 10^{-bM}, \quad b \sim 0.5 - 1.5, \tag{3.3}$$

known as the *Gutenberg-Richter* law [2]. This law seems to describe quite well the experimental data available for earthquake magnitudes, as measured by a seismograph. It must be noted that the magnitude of an earthquake is a logarithmic quantity. That means that an earthquake of magnitude $m + 1$ releases 10 times more energy than an earthquake of magnitude m. If we define its effective size as $s = 10^m$, the Gutenberg-Richter law becomes (see Problem 3.1),

$$p(s) \propto s^{-(b+1)}, \quad b \sim 0.5 - 1.5, \tag{3.4}$$

showing that the pdf of the effective sizes of earthquakes exhibits a power law. Equation 3.4 is scale-invariant in a sense of Eq. 3.1. In addition, the fact that the exponent b might be less than one suggests that a certain type of complex dynamics known as *self-organized criticality* (see Sect. 1.3.2) might be at play [3–5].

Similar power-law scalings appear when studying, for instance, the statistics of the energy released by solar flares [6] (see Chap. 8) or the statistics of the X-ray emissivity from accretion disks [7]. Examples are not restricted to the realm of physics and the earth sciences. For example, the mathematician and psychologist Lewis Richardson found a power-law behaviour when investigating the statistics of war casualties as early as the 1940s [8]. Similarly, the biologist Max Kleiber proposed in the 1930s that the metabolic rate, r, and the animal mass, M, were related by the famous *Kleiber law*, $r \propto M^{3/4}$ [9], whose validity is still contested to this day in spite of its many successes. In some of these cases, models resting on the idea of underlying complex dynamics have been proposed to account for the observed scale invariance.[2]

Scale-invariance if often encountered beyond the realm of pure mathematical functions. One could think, for instance, of the energy spectrum of fluctuations in fully-developed, homogeneous fluid turbulence, that decays with the famous *Kolmogorov law*, $E(k) \sim k^{-5/3}$ (k is the spatial wavenumber), over the so-called *inertial range* that separates driving and dissipative scales [10]. Objects can also be scale-invariant on their own. This is the case of many *spatial fractal patterns* [11], such as coastlines or the famous Cantor set (see Fig. 3.2), whose aspect is invariant under spatial rescaling. Some temporal processes, such as *fractal Brownian motion* [12], are *statistically invariant* objects. That is, a single realization may not be invariant under rescaling, but the average of its properties over many realizations (i.e., an ensemble average) is, as will be discussed in Sect. 3.3.

In this chapter we will define scale-invariance and introduce some popular methods to help us find out whether this object or that time process is scale-invariant. Many of these techniques try to determine whether specific properties of exact self-similar processes and objects are shared to some degree by natural patterns or by time series obtained in natural systems. In the case of spatial patterns, the mathematical models to compare with are **fractals**. Thus, we will start by discussing their properties in Sect. 3.2. In the case of time series, one often compares them against the family of stochastic processes known as **fractional Brownian/Lévy motions**, that will be introduced in Sect. 3.3. We will also go beyond pure fractals and discuss the idea of **multifractality**, both in space (Sect. 3.2.2) and time (Sect. 3.3.7). We will conclude, as always, by applying some of the techniques presented to the characterization of the scale-invariance of our test problem of choice: the running sandpile (Sect. 3.5). This analysis will let us describe in detail the idea of the **mesorange**,[3] that is a consequence of the fact that, in real systems,

---

[2]In the case of Kleiber law, it has sometimes been suggested that the observed scaling might be related to the increase in complexity of the fractal-like circulation system as animals get larger.

[3]We briefly discussed the mesorange in Chap. 1. In particular, in Sects. 1.2.3 and 1.3.2.

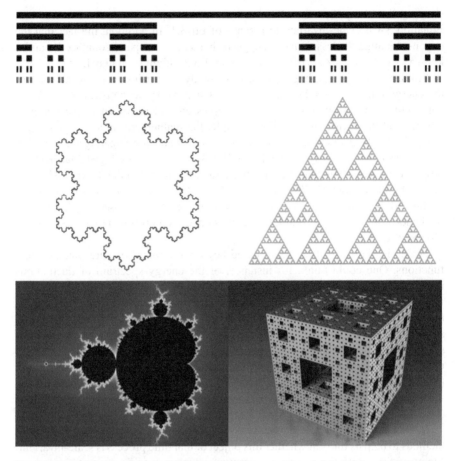

Fig. 3.2 Examples of mathematical fractals (going down, and from left to right): seventh iteration of the Cantor set (fractal dimension is 0.6309), seventh iteration of Koch's snowflake (fractal dimension is 1.2619), ninth iteration of Sierpinski's triangle (fractal dimension is 1.585), Mandelbrot's set (contour's fractal dimension is 2) and Menger's sponge (fractal dimension is 2.7268). Credits: all images from *Wikimedia Commons*; Cantor set (public domain); Koch's snowflake (A.M. Campos; public domain); Sierpinski's triangle (B. Stanislaus; CC BY-SA 3.0 license); Mandelbrot set (W. Beyer; CC BY-SA 3.0 license); Menger set (Niabot, CC BY-SA 3.0 license)

self-similarity is always restricted to a limited range of scales, in contrast to the unrestricted scale-invariance of monofractal mathematical models.

3.2   Scale-Invariance in Space

The first objects that usually come to mind when thinking of self-similarity or scale-invariance are **spatial fractals**. Loosely speaking, fractals are self-similar objects in the sense that *they are formed by parts that look very much like the whole* *object; often these parts are themselves again self-similar.* Mathematically, it is traditional to define a fractal as *a set of points, distributed within some geometrical support, whose properties upon rescaling are described by a single non-integer number—its **fractal dimension**.* In contrast, regular geometrical objects have integer dimensions—such as a straight line (dimension is 1), the circle (dimension is 2) or a pyramid (dimension is 3). The possibility of being characterized by a dimension smaller than one (i.e., fractional) is in fact the origin of the term *fractals*, first proposed by the mathematician Benoit B. Mandelbrot in the late 1960s and early 1970s [13].

In most cases, fractals are mathematical objects (see Fig. 3.2). Many are generated recursively. For instance, the **Cantor set** is obtained by repeatedly removing the central piece of all segments in the set at each iteration, after dividing them in three parts (see Problem 3.2). The iconic **Mandelbrot set**, on the other hand, is formed by all initial choices z_0 in the complex plane for which the orbit predicted by the recurrence relation, $z_{k+1} = z_k^2 + z_0$, does not escape to infinity (see Problem 3.3).

Much more interesting to the physicist and natural scientist is the fact that some natural systems seem to organize themselves by forming patterns that are reminiscent of fractals. There are many examples of naturally occurring objects that look like (approximate) fractals, at least over a relatively wide range of spatial scales. One could mention coastlines [14], blood and pulmonary vessels, bubbles in foams [15], veins in leaves and branches in trees, snow flake patterns, fault line patterns, even the rings of Saturn [16], among many others (see Fig. 3.3). It has often been suggested that complex dynamics may provide the mechanism through which these patterns are generated in nature. Be it as it may, the fact is that looking for properties characteristic of mathematical fractals is probably the best way of detecting and quantifying self-similarity in natural objects and patterns. Thus the importance of understanding fractals.

3.2.1 Fractals

The **fractal dimension** is the most relevant quantity that characterizes the structure of a fractal [13]. It is often the first quantity that one estimates to assess whether any object is a fractal or not. Various definitions of fractal dimensions exist, all related but not equivalent to each other. In what follows, we will discuss the **self-similar dimension** and the **box-counting (BC) dimension**. In fact, it is the BC dimension that we will be referring to more often. It may not be the most mathematically rigorous, but its discussion will prove particularly useful for actual applications to complex systems. In particular, to the determination of fractal properties of natural objects and patterns. In passing, we will also mention other important fractal dimensions, such as the **Haussdorf dimension**, the **entropy dimension** or the **correlation dimension**.

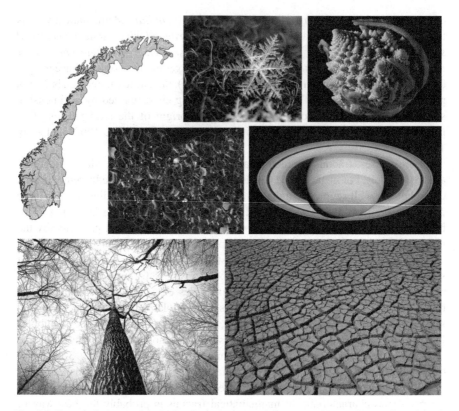

Fig. 3.3 Fractals in nature (clockwise from the upper, left corner): Norway's coastline (fractal dimension, 1.52), snowflakes, Romanesco's broccoli (1.26), Saturn's rings (∼1.7), dried-out patterns on river beds, branch systems in trees and bubbles in soap (center). [Credits: images from *Wikimedia Commons*: Norway (I. Rarelibra, public domain); snowflake (CC-A 2.0 license); Romanesco broccoli (J. Sullivan, public domain); foam (A. Karwath, CC BY-SA 2.5 license); dried riverbed (V. Panicker, CC BY-SA 2.5 license). From *Pixabay.com*: branches (public image).] Saturn (©ESA/NASA - Hubble)

3.2.1.1   The Self-Similar Dimension

The **self-similar dimension** characterizes the properties of an object that is invariant upon (often discrete) rescaling operations. It is best illustrated through an example. Let's take, for instance, the [0, 1] segment of the real line. To start, the segment is divided in three equal parts and the central part is removed. The process is then iterated, yielding the collection of points on the real line known as **the Cantor set** (see Fig. 3.2). The Cantor set is **self-similar** because *each of the parts generated at iteration $k > 0$ is identical to the original segment, but rescaled by a factor* $r = 1/3^k$. Indeed, the segments after the first iteration have a length equal to $1/3$ of the original segment, those of the second iteration have a length equal to $1/9$ of the original segment, and so forth. One can easily show (see Problem 3.2) that the number of identical parts in the set after having rescaled the Cantor set by a factor r is:

$$N(r) = r^{-D_{ss}}, \tag{3.5}$$

with $D_{ss} = \log 2 / \log 3 = 0.63$. That is, it follows a power law with exponent D_{ss}. D_{ss} is known as the **self-similar dimension**. For a general object, it is defined as:

$$D_{ss} = - \lim_{r \to 0} \frac{\log N(r)}{\log r}. \tag{3.6}$$

Why is D_{ss} referred to as a dimension? To understand why, let's calculate the length of the Cantor set after the initial segment has been rescaled by r. It is:

$$L_r = N(r) \cdot r = r^{1-D_{ss}}, \tag{3.7}$$

Therefore, the length of the set goes to zero as $r \to 0$. However, the Cantor set is not empty in that limit, even if it does not fill *the space in which is embedded* as densely as a regular object. Indeed, if we had not removed the central part of each segment at each iteration, we would have three parts in the first iteration, nine parts in the second iteration, and so forth. The number of parts would then scale as $N(r) = r^{-1}$, which would always yield a finite length at any iteration. Therefore, D_{ss} provides a measure of how densely the embedding space is filled by the parts we have kept. It is a fractional number for the Cantor set because it lies somewhere in between a regular line ($D_{ss} = 1$) and the empty space ($D_{ss} = 0$).

The same argument can be easily extended to higher-dimensional objects, and non-integer dimensions between 1 and 2 (for instance, Koch's curve shown in Fig. 3.2), or between 2 and 3 (as Menger's sponge in Fig. 3.2) are thus introduced. All of these objects also fill their embedding spaces differently from what a regular figure would do. For instance, Koch's curve fills a plane more densely than any regular curve; although less densely than a two-dimensional solid, such as a circle or a square. Menger's sponge, on the other hand, fills the three-dimensional space more densely than a hollow sphere, but less densely than a solid sphere, for instance. It is because of their non-integer dimensions that these objects are referred to under the generic name *fractals*.

3.2.1.2 The Box-Counting (BC) Dimension

The self-similar dimension can only be used with *exactly* self-similar objects. However, many objects that are not exactly self-similar are irregular enough as to fill the space in which they are embedded in a different way from what regular figures do. Thus, a less restrictive definition is needed that could be applied to these more general objects and that, if possible, should reduce to the self-similar dimension for perfectly self-similar objects. One such definition is the **box-counting (BC) fractal dimension**.

Fig. 3.4 To calculate the box-counting dimension of the coastline of Antarctica we cover it using meshes of non-overlapping squares of some prescribed size, l. Here, two such coverings are shown, one for half the square linear size than the other. It turns out that the number of squares in the covering scales as $N(l) \sim l^{-1.1}$ [17], in contrast to the $N(l) \sim l^{-1}$ that one expects for a regular curve. Credits: Antarctica satellite view (© NASA - Visible Earth)

To illustrate the BC fractal dimension, we will use an example inspired by Benoit Mandelbrot's seminal analysis of the coastline of Great Britain [14], and examine the coastline of Antarctica. At first view, coastlines are just one-dimensional curves embedded in a two-dimensional space (i.e., the map). But after closer inspection, many coastlines look kind of rough, with many irregularities. Maybe too irregular to be just a regular curve. How does one determine whether this is the case or not? One way is to cover the coastline using a mesh of non-overlapping squares of a given size, l, (see Fig. 3.4). One then counts the number of squares needed to cover the coastline. Let's call this number $N(l)$. As was done with the Cantor set, one can then estimate the length of the coastline as,

$$L(l) \simeq \sum_{i=1}^{N(l)} l_i = N(l) \cdot l \tag{3.8}$$

that assumes that each square contains inside approximately a segment of the coastline of length $l_i \simeq l$. Clearly, if the square size is too large, this is a very bad approximation. But, in the limit of small l, it should be quite close to the actual length of the coastline. That is,

$$L = \lim_{l \to 0} N(l) \cdot l. \tag{3.9}$$

This limit is however finite only if $N(l) \propto l^{-1}$. That is, the *number of boxes in the covering must be inversely proportional to their length*. This is indeed the case for a regular curve. But does it happen for the coastline of Antarctica? It turns out that it does not. A similar exercise to the one just described has been done for the Antarctic coast [17], finding that $N(l) \sim l^{-1.1}$. This scaling, when inserted in Eq. 3.9, yields an infinite length for the coastline of Antarctica! Therefore, the coastline of Antarctica behaves rather differently from a regular curve. Similar results have been obtained for many other coastlines. The first analysis of a coastline was done by

Mandelbrot for Great Britain, who found that $N(l) \sim l^{-1.25}$ [14]. Although the most remarkable result is probably that obtained for the coast of Norway, with an impressive $N(l) \sim l^{-1.52}$ [18]. It is apparent that all these coastlines have a *denser* structure than a regular curve, although not as dense as required to completely fill the two-dimensional space (the map) that contains them. Coastlines are akin to self-similar fractals, in that sense.

The **box-counting (BC) fractal dimension**, D_{bc}, quantifies how dense this structure is. It is formally defined as[4]

$$D_{bc} = -\lim_{l \to 0} \frac{\log N(l)}{\log l}, \qquad (3.11)$$

where l gives the size of the non-overlapping grid used to cover the object (see Problem 3.3).

Using the BC dimension, a **measure of the fractal content** of an object can be built that is more proper than the usual length (or area, or volume). Namely,

$$M_{bc} = \lim_{l \to 0} N(l) \cdot l^{D_{bc}}, \qquad (3.12)$$

that is now finite since $N(l) \sim l^{-D_{bc}}$. If dealing with a regular n-dimensional object, $N(l) \propto l^{-n}$ and thus, $D_{bc} = n$. Therefore, M_{bc} naturally reduces to the expression for the n-dimensional size (length, for $n = 1$, area, for $n = 2$ or volume, for $n = 3$).

3.2.2 Multifractals

The scaling properties of a fractal are described by a single exponent, its fractal dimension. However, even fractals are too simplistic models to describe the spatial distribution patterns found in some natural systems. This seems to be the case, for instance, of the distribution of certain minerals within the Earth crust, the world distribution of human disease, the spatial distribution of energy dissipation regions

---

[4]The box-counting fractal dimension is closely related to the so-called **Hausdorff dimension**, but it is less precise from a mathematical point of view. Although both dimensions coincide for many real fractals, there are documented cases in which the two yield different results. The Hausdorff dimension predates the fractal concept. It was introduced as far back as 1918, by the German mathematician Felix Hausdorff. The idea is to consider the object immersed in a metric space of dimension n. Then, one counts the number of n-dimensional balls of radius at most r, $N(r)$, required to cover the object. One then builds a **measure of the size** of the object in the form

$$M_d(r) = \sum_{i=1}^{N(r)} A(r_i) r_i^d, \qquad (3.10)$$

where $A(r)$ is a geometrical factor that depends on the metric chosen and the dimensionality. The **Hausdorff dimension** is then defined as the value of $d = d_H$ that makes that, for all $d < d_H$, the measure tends to zero. And, that for all $d > d_H$, the measure diverges. In this sense, d_H is a critical boundary between values of d for which the covering is insufficient to cover the space, and values for which the covering is overabundant.

in a turbulent flow [19], the distribution of the probability of finding an electron in three-dimensional disordered quantum systems, or the distribution of galaxies within clusters throughout the universe [20].

The objects considered in these examples, and many more, still appear to fill their embedding spaces more densely than regular objects, but they look more complicated than simple fractals. In fact, they look more like the result of having *several intertwined fractal subsets embedded within the same geometrical support, but each characterized by a different scaling exponent* [18]. This type of mathematical object does also exist and is known as a **multifractal**. However, in order to discuss multifractals, we will need to introduce new concepts such as **singularities**, the **Holder exponent**, the **singularity spectrum** and the **generalized fractal dimension**.

3.2.2.1 Multifractal Measure

We will start to illustrate the idea of multifractality by revisiting the Antarctica coastline. It will be remembered that, in order to estimate its box-counting fractal dimension, we introduced a quantitative measure of its fractal content that, for a given covering with boxes of size l, was (Eq. 3.12):

$$M(l) = N(l) \cdot l^D. \tag{3.13}$$

Here, $N(l)$ was the number of squares of size l in the covering, and D was the fractal dimension that ensured that the measure was finite when $l \to 0$. Clearly, each square in the covering contributes to the measure in the same amount since, in a fractal, the scaling properties are assumed to be the same everywhere. In a multifractal, however, it would be more reasonable to expect that different parts of the object should contribute differently. Indeed, since the fractal subsets with larger fractal dimensions are spatially denser (i.e., they fill the space around them more fully), it follows that those squares that contain larger parts of them will also be more densely filled. For the same reason, those squares that contain mostly large parts of the fractal subsets with smaller fractal dimensions will be less densely filled. How can a **multifractal measure** be built, in the same spirit that Eq. 3.13, that takes these subtleties into account?

One way to do it is to assign a **weight** to each square in the covering *that is proportional to how densely that particular square is filled*. To find this weight, we proceed to cover the i-th square in the original covering with a second covering (see Fig. 3.5) formed by a mesh of P^2 squares of linear size $l' = l/P \ll l$. Next, we count how many squares of the second covering contain some of the part of the coastline that lies within the i-th square and call this number $p_i(l)$. The weight associated to the i-th square is then given by:

$$w_i(l) = p_i(l) / \sum_{j=1}^{N(l)} p_j(l), \quad i = 1, 2, \cdots, N(l). \tag{3.14}$$

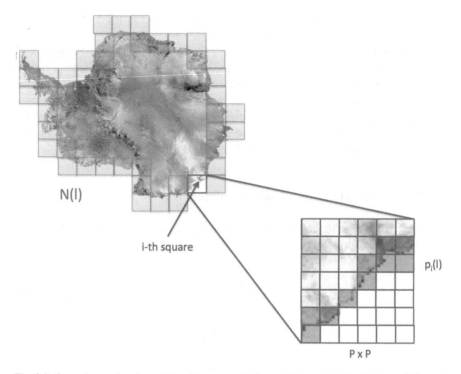

Fig. 3.5 Second covering (in red) used to compute the weights used to build the multifractal measure of the coastline of Antarctica. $p_i(l)$ is the number of squares, in the covering of the i-th square from the first covering, that contain some of the coastline. Credits: Antarctica satellite view (© NASA - Visible Earth)

The normalization ensures that the sum of all weights over all the squares in the original covering equals one. Using these weights, one defines the one-parameter **family of multifractal measures**

$$M(l, q) = \left[\sum_{i=1}^{N(l)} w_i^q \right] \cdot l^{\tau(q)}, \tag{3.15}$$

where q can take any real value. A new **family of multifractal exponents** $\tau(q)$ has also been introduced, that is determined by *requiring that $M(l, q)$ stays finite when $l \to 0$*.

It turns out that, by varying the value of q, the family of exponents $\tau(q)$ effectively quantifies the multifractal properties (if any) of the coastline. Indeed, if one considers the values:

- **q = 0**. The multifractal measure reduces to:

$$M(l, 0) = \left[\sum_{i=1}^{N(l)} 1 \right] \cdot l^{\tau(0)} = N(l) \cdot l^{\tau(0)}, \tag{3.16}$$

that is the original measure of fractal content defined in Eq. 3.12. Thus, $\tau(0)$ coincides with the BC fractal dimension of the object, D_{bc}.

- **Large, positive values of** q. The measure $M(l, q)$ will then favour contributions from squares in the covering for which the weight is large (larger, the larger q is!). That is, those coming from parts of the object where the contribution of denser subsets (i.e., larger fractal dimensions) is dominant.[5]
- **Large, negative values of** q. The same game can also be played to favour instead contributions to the measure coming from the fractal subsets with the smaller fractal dimensions. We just need to choose q large, but negative!

3.2.2.2   The Family of Multifractal Exponents $\tau(q)$

It is customary to estimate $\tau(q)$ by introducing the so-called **partition function**, $N(l, q)$, via the expression:

$$N(l, q) \equiv \sum_{i=1}^{N(l)} w_i^q \sim l^{-\tau(q)}. \tag{3.17}$$

from which the multifractal exponents can be obtained as:

$$\tau(q) = -\lim_{l \to 0} \frac{\log N(l, q)}{\log l}, \tag{3.18}$$

The shape of the function $\tau(q)$ characterizes the multifractal properties of an object. In particular, it should be noted that,

- $\tau(0)$ coincides with the box-counting fractal dimension of the object, D_{bc}.
- $\tau(1) = 0$. This follows from the normalization chosen for the weights, $\sum w_i = 1$.
- if the object is a **monofractal** (i.e., a traditional fractal that has just a single scaling exponent), then $\tau(q)$ reduces to the (decreasing) linear function of q:

$$\tau(q) = D_{bc}(1 - q). \tag{3.19}$$

This follows from the fact that, for a monofractal, all boxes in the covering are statistically identical in terms of their rescaling properties. Thus, $p_i = p$, and the weight associated to each box becomes (see Eq. 3.14):

$$w = p / \sum_{i=1}^{N(l)} p = \frac{1}{N(l)}. \tag{3.20}$$

---

[5]This simply follows from the fact that, if a number between zero and one is raised to a large positive power, the result becomes smaller the smaller the number.

The partition function then becomes, using Eq. 3.17,

$$N(l, q) = \sum_{i=1}^{N(l)} w_i^q = \sum_{1}^{N(l)} \left[\frac{1}{N(l)} \right]^q = [N(l)]^{1-q}. \qquad (3.21)$$

Since, $N(l) \sim l^{-D_{bc}}$ for a monofractal, Eq. 3.19 follows.

• for a **multifractal**, $\tau(q)$ will exhibit a more complicated (i.e., nonlinear) dependence on q.

It is traditional to define a **generalized fractal dimension** associated to the multifractal exponents $\tau(q)$. It is defined as [21]:

$$D(q) \equiv \frac{\tau(q)}{1-q}, \qquad (3.22)$$

that, for any monofractal object, naturally reduces to $D(q) = D_{bc}$, $\forall q$ [22]. For a multifractal, it is a decreasing function of q (see Fig. 3.6). Interestingly, several popular fractal dimensions are contained in $D(q)$ [23]. For instance, it becomes the

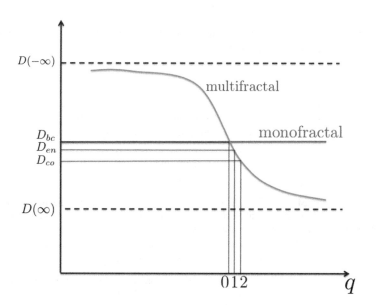

Fig. 3.6 Typical shape of the generalized dimension $D(q)$ for a monofractal (red) and a multifractal (blue). $D(0)$ is the BC fractal dimension; $D(1)$, the entropy dimension; $D(2)$, the correlation dimension. For a monofractal, all dimensions coincide with the box-counting one. For a multifractal, they are different with $D(0) \geq D(1) \geq D(2)$

entropy dimension[6] when $q = 1$. If $q = 2$, the generalized fractal dimension gives the **correlation dimension**.[7]

3.2.2.3   Local Holder Exponent: α

The estimation of the family of multifractal exponents $\tau(q)$ is not the only way in which the multifractal character of a set can be exposed and quantified. Another path uses instead the concept of the **local Holder exponent**.

To illustrate its meaning, we will return to the Antarctica coastline example. It will be remembered that, when looking for a multifractal measure, we associated a weight, w_i, to the i-th square of size l in the covering (see Eq. 3.14). For a monofractal, we argued that this weight should be independent of the square considered, since the object should scale in the same way everywhere. Thus, we set $w_i = w = N(l)^{-1} \sim l^{D_{bc}}$. However, for a multifractal object, one should expect that the weights of different squares could scale differently, since different parts of the object may contain different fractions of each of the fractal subsets that form them. Thus, we will assume instead that,

$$w_i(l) \sim l^{\alpha_i}. \tag{3.26}$$

The exponent $\alpha_i > 0$ is known as the **local Holder exponent** or, in some contexts, as the **local singularity exponent**.[8]

---

[6]One can compute the amount of information associated to the covering of size l by calculating its Shannon entropy, $S(l) = \sum_{i=1}^{N(l)} w_i \log w_i$. The **entropy dimension** is then defined after assuming:

$$S(l) \sim l^{-D_{en}} \to D_{en} = -\lim_{l \to 0} \frac{\log S(l)}{\log l}. \tag{3.23}$$

[7]The **correlation dimension** of a set of N points is computed [21] by counting the total number of pairs of points, n_p, that have a distance between them smaller than some $\epsilon > 0$. For small ϵ, the limit of the function known as the correlation integral,

$$C(\epsilon) = \lim_{N \to \infty} \frac{n_p}{N^2} \sim \epsilon^{D_{co}}, \tag{3.24}$$

where D_{co} is the correlation dimension. It can also be formulated by introducing the function $C(l) = \sum_{i=1}^{N(l)} w_i^2$, and then defined after assuming:

$$C(l) \sim l^{-D_{co}} \to D_{co} = -\lim_{l \to 0} \frac{\log C(l)}{\log l}. \tag{3.25}$$

[8]In the theory of mathematical functions, the Holder exponent appears as a way of quantifying the degree of singularity of non-differentiable functions at a given point. Indeed, a function f that is differentiable at x satisfies that, for small δ,

$$|f(x + \delta) - f(x)| \propto |\delta|^1, \tag{3.27}$$

3.2.2.4 Singularity or Multifractal Spectrum: $f(\alpha)$

The local Holder exponent of a monofractal is the same everywhere and coincides with its BC fractal dimension. In a multifractal, however, we will have a list with various values of the Holder exponent, as well as their spatial distribution, once all the squares in the covering have been characterized using Eq. 3.26. Therefore, one could easily compute the fractal dimension, $f(\alpha)$, of *each of the spatial subsets formed by all squares that share the same Holder exponent, α. $f(\alpha)$* is known as the **multifractal spectrum** or the **singularity spectrum** [24].

For a monofractal, $f(\alpha) = D_{bc}\delta(\alpha - D_{bc})$, where $\delta(x)$ is the Dirac delta function. Thus, one could quantify the degree of multifractality of an object from the shape of $f(\alpha)$. But instead of estimating it directly, that is rather involved, our next job is to find out a way to derive $f(\alpha)$ from the family of exponents $\tau(q)$, in practice more accessible thanks to the box-counting procedures we described previously.

We will assume that the family of multifractal exponents $\tau(q)$ is known (we follow here [18] closely). The idea is to replace the sum over squares in Eq. 3.17 (effectively, an integral over the space occupied by the multifractal) by a *sum over all the fractal subsets formed by all squares with identical local Holder exponent α*:

$$N(l, q) = \sum_{i=1}^{N(l)} w_i^q \sim \int \rho(\alpha)d\alpha \; l^{-f(\alpha)} \; l^{\alpha q}. \tag{3.31}$$

Let's discuss the terms appearing in this integral. First, $\rho(\alpha)d\alpha$ gives the fraction of the total space occupied by the fractal subset with Holder exponent between α and

---

which permits to define its derivative at x as the usual,

$$\frac{df}{dx}(x) = \lim_{\delta \to 0} \frac{f(x + \delta) - f(x)}{\delta} \tag{3.28}$$

On the other hand, a function with a bounded discontinuity at x (for instance, the Heaviside step function, that has a jump of one at $x = 0$), satisfies,

$$|f(x + \delta) - f(x)| \propto |\delta|^0, \tag{3.29}$$

so that its derivative at x does not exist.

Continuous, but non-differentiable functions behave between one case and the other,

$$|f(x + \delta) - f(x)| \propto |\delta|^\alpha, \quad 0 < \alpha < 1. \tag{3.30}$$

The exponent α, in this case, *quantifies the degree of the singularity of the function at x* or, in plain words, how far it is from being differentiable. It is called the **local Holder exponent**.

How does this mathematical digression justify referring to α_i in Eq. 3.26 as a local "Holder exponent"? Well, fractals are very irregular objects, usually non-differentiable. Thus, it is reasonable to expect that their local singularities could be describable by some kind of non-integer exponent. The term "Holder exponent" is then borrowed by mere association. Although it can take values larger than one. Indeed, for a monofractal we saw that $\alpha_i = D_{BC}$, that can take all non-integer values within the interval $(0, 3)$.

$\alpha + d\alpha$. Since the fractal dimension of this fractal subset is $f(\alpha)$, the contribution of this region to the partition function $N(l, q)$ must be proportional to $l^{-f(\alpha)}$. Finally, the weight associated to this fractal subset is, logically, $l^{\alpha q}$, since α is the local Holder exponent of all squares in it.[9]

Since we are interested in how this integral scales with l, we just need to keep the contribution of the value of α that maximizes the integrand. This value will be a function of q, referred to as $\alpha^*(q)$ in what follows. It is obtained by solving:

$$\frac{d}{d\alpha}[q\alpha - f(\alpha)] = 0 \quad \Rightarrow \quad \frac{df}{d\alpha}(\alpha^*(q)) = q. \tag{3.32}$$

But we do not need to worry about solving it explicitly. Instead, we can assume $\alpha^*(q)$ known and rewrite Eq. 3.17 as,

$$N(l, q) \sim l^{q\alpha^*(q) - f(\alpha^*(q))} \sim l^{-\tau(q)}. \tag{3.33}$$

It thus follows that the relation between $\tau(q)$ and $f(\alpha)$ is,

$$q\alpha^*(q) - f(\alpha^*(q)) = -\tau(q). \tag{3.34}$$

Luckily, this relation can be easily inverted by differentiating both sides with respect to q, and then using Eq. 3.32. The procedure yields the pair of equations:

$$\alpha^*(q) = -\frac{d\tau}{dq}(q) \tag{3.35}$$

$$f(\alpha^*(q)) = \tau(q) - q\frac{d\tau}{dq}(q) \tag{3.36}$$

This solves the problem completely. This pair of equations estimate, for each value of q, the dominant value of the local Holder exponent that contributes to the multifractal measure $M(l, q)$ (Eq. 3.35), as well as the fractal dimension (Eq. 3.36) of the fractal subset formed by all regions in the multifractal with that local Holder exponent. Thus, by varying q, obtaining $\tau(q)$ and using Eqs. 3.35 and 3.36, one can estimate all relevant values of α, and their related fractal dimension, $f(\alpha)$.

3.2.2.5   Properties of the Multifractal (or Singularity) Spectrum $f(\alpha)$

Let's say, to conclude, a few words about the multifractal spectrum $f(\alpha)$ that might be useful to interpret its meaning in practical cases. A plot showing a typical

---

[9]It is quite reassuring to note that, in the limiting case in which the multifractal becomes a monofractal, only one value of α contributes to the sum (i.e., $\rho(\alpha) = \delta(\alpha - D_{bc})$). On the other hand, $f(\alpha_0) = D_{bc}$, the BC fractal dimension of the monofractal. And if we take $q = 0$, we see that the integral nicely reduces to $N(l) \sim l^{-D_{bc}}$, as it should be!

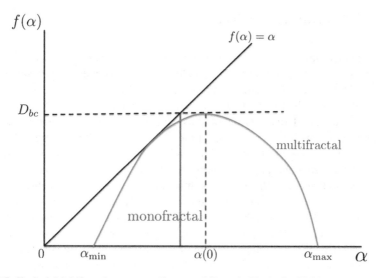

Fig. 3.7 Typical Multifractal spectrum. For a multifractal (blue), the Holder exponent varies between $\alpha_{\min}$ and $\alpha_{\max}$, with its maximum at $\alpha(0)$. For a monofractal (red), the curve becomes $f(\alpha) \simeq D_{bc}\delta(\alpha - D_{bc})$

multifractal spectrum is shown in Fig. 3.7 in order to illustrate some of its typical features:

1. First, observe that $f(\alpha) \leq D_{bc}$, $\forall \alpha$, where D_{bc} is the BC fractal dimension of the whole object, as computed from the fractal measure $M(l, 0)$.
2. Secondly, it is worth noting that $f(\alpha)$ reaches its maximum value, D_{bc}, when the Holder exponent α takes the value that solves Eq. 3.35 for $q = 0$.
3. Next, $f(\alpha) \leq \alpha$, $\forall \alpha$. This can be trivially proven by exploiting the fact that $\tau(1) = 0$.
4. Finally, $\alpha \in [\alpha_{\min}, \alpha_{\max}]$, where the limiting Holder exponents satisfy,

$$\alpha_{\min [\max]} = -\lim_{q \to +\infty [-\infty]} \frac{\tau(q)}{q}. \qquad (3.37)$$

We showed that, in the limiting case in which the suspected multifractal turns out to be a **monofractal**, $\tau(q) = D_{bc}(1 - q)$ (Eq. 3.19). Consequently, Eqs. 3.35 and 3.36 reduce to:

$$\alpha^*(q) = f(\alpha^*(q)) = D_{bc}, \quad \forall q. \qquad (3.38)$$

Thus, the multifractal spectrum reduces to $f(\alpha) = D_{bc}\delta(\alpha - D_{bc})$. That is, there is only one local Holder exponent, and it coincides with the fractal dimension of the monofractal, as we already mentioned previously.

3.3   Scale-Invariance in Time

Many scientists face soon in their careers the fact that experimental data comes, more often than not, in the form of **discrete temporal records** or **time series** (see Fig. 3.8). This used to be certainly the case in the field of turbulence. Temporal records of local velocity fluctuations were indeed much easier to obtain than a map of the spatial distribution of those fluctuations at any given time [10]. Although spatial measurements in real time are nowadays much more manageable in fluids,[10] this state of things still holds in many other fields. For instance, fluctuation measurements are still extremely difficult to make inside the kind of hot fusion plasmas that are confined in a tokamak. For that reason, most of the turbulent data available in fusion plasmas is available in the form of temporal records, the majority of them measured at the plasma edge using local Langmuir probes [25, 26]. Temporal records are also common for much more distant plasma systems, such as the Sun (see Chap. 7) and other stars, the Earth's magnetosphere (see Chap. 8) or accretion disks, to name a few. The characterization of the self-similarity properties (if any) of time series such as these is the subject of this section. This analysis is done by comparing the real time series with certain mathematical **random processes** that exhibit **exact statistical self-similarity**, in an analogous way to how one compares real spatial patterns with mathematical fractals.

Random processes are those in which, at any given time, the value of the process is a **random variable**. That is, the value varies randomly from realization

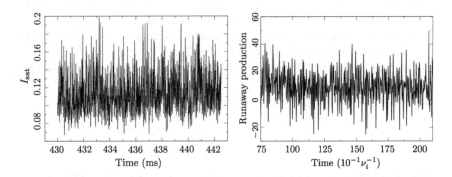

Fig. 3.8 Examples of time series from magnetically confined plasmas. **Left:** temporal evolution of the ion saturation current (these fluctuations are considered a surrogate for plasma density turbulent fluctuations) measured with a Langmuir probe at the edge of the W7-AS stellarator [27]; **right:** saturated runaway electron production obtained by a numerical Langevin code that considers the runaway dynamics in phase space in the presence of a constant electric field [28]

---

[10]Thanks in part to the advent of both laser technology and inexpensive high-speed high-resolution CCD cameras.

to realization, according to some prescribed statistics.[11] A very important role is played, in this respect, by the families of **fractional Brownian** and **fractional Lévy motions**, that we will discuss in Sects. 3.3.3 and 3.3.4. In order to understand their properties, we will first introduce several important concepts such as **statistical self-similarity**, **invariance under temporal translations** and **stationarity**. We will also rely on a new mathematical object, the **process propagator**. Although some readers may find the concept of the propagator rather complicated to grasp, we can say that gaining a good understanding of its meaning and properties will definitely pay off in the longer run, since it is by exploiting its properties that many of the methods used to quantify self-similarity in real time series are possible.[12]

3.3.1 Self-Similar Time Random Processes

Time series are discrete records obtained, for instance, by recording a continuous time process at a certain rate. A **time process**, for the purposes of this book, can be any arbitrary function of time, $y(t)$, that may or may not be continuous. If $y(t)$ is a **random variable** for every time $t > 0$, the process is called **random** or **stochastic**. If the statistics of the random variable at every time are Gaussian, the process is a **Gaussian random process** (see Fig. 3.9). Many other statistics are also possible (for instance, of the Lévy type).

A random process is **statistically self-similar**[13] with **self-similarity exponent** H, or simply H-**ss**, if it satisfies that,

$$y(t) \ =^d \ a^{-H} y\left(\frac{t}{a}\right), \quad \forall a, t \geq 0. \tag{3.39}$$

Although this definition looks similar to Eq. 3.1, that established when a mathematical function is self-similar, they are however not equivalent since the symbol "$=^d$" used in Eq. 3.39 means **equivalence in distribution**. This type of equivalence is rather different from what is meant by the more usual "$=$" sign that appears in Eq. 3.1. In the literature, the meaning of "$=^d$" is often explained by saying that the two sides it relates must have **the same statistical properties**. But what is

---

[11]If one compare fluctuation data measured at the edge of the same fusion device, but for different plasma discharges with similar conditions, they would all look like different realizations of the same random process. The same happens for the runaway production signal shown in the right frame of Fig. 3.8. The different realizations would then correspond to runs done with the same parameters but initialized with a random generator.

[12]In addition, propagators will be used heavily in Chap. 5, since they play a dominant role in the theory of fractional transport.

[13]In the rest of this book, we will often drop the adjective "*statistically*" when referring to random processes, in order to make the discussion more agile. The implication should however not be forgotten.

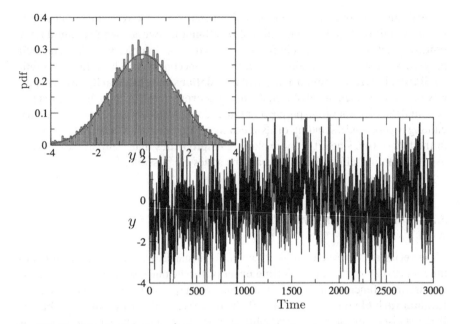

Fig. 3.9 Example of a Gaussian random process. In this case, the value of the process at any given time follows a Gaussian pdf. The process shown here is also stationary, which implies that the pdf of values at any given time is identical to the pdf of all the values taken in one realization, as will be explained later (see Sect. 3.3.5). The latter pdf is shown in the inset, together with a Gaussian fit (in red)

really meant by this statement? To answer this question more precisely, we need to introduce a new mathematical object associated to the process: its **propagator**.

3.3.2 Propagator of a Random Process

For the purposes of this book, all statistical properties of interest to us are conveyed by a single function of the random process: its **two-point probability density function** (or **two-point pdf**). We already discussed pdfs in Chap. 2. Its meaning was that, if $p(x)$ is the pdf for a certain variable x, $p(x)dx$ then gives the probability of that variable taking a value between x and $x + dx$. Similarly, the two-point pdf of a time process $y(t)$ quantifies the probability of y taking a value between y_1 and $y_1 + dy_1$ at time t_1, and a value between y_2 and $y_2 + dy_2$ at time t_2[14]:

$$p_2(y_1, t_1; y_2, t_2)dy_1 dy_2. \tag{3.40}$$

---

[14]Clearly, $p_2(y_1, t_1; y_2, t_2)$ does not exhaust all the statistical information that can be retrieved from the process. One could define a whole hierarchy of higher-order pdfs that connect up to n-points, with $n > 2$ as large as desired, all of which are also joint probability distribution functions for the process. We will not consider any these functions here, though.

Furthermore, we will always assume that the first point corresponds to the initial value of the process, y_0, at the initial time, t_0. The second point, on the other hand, may correspond to the value of the process at any $t > t_0$. Thus, we can rewrite p_2 as:

$$p_2(y_0, t_0; y, t) = G(y, t | y_0, t_0) p(y_0). \tag{3.41}$$

Here, $p(y_0)$ is the *one-point* pdf at the initial time. Or, in other words, $p(y_0)dy_0$ gives the probability of the initial value of the process being between y_0 and $y_0 + dy_0$ at t_0. This pdf, $p(y_0)$, is clearly arbitrary and independent of the dynamics of the process. It simply describes how the process has been initialized. $G(y, t | y_0, t_0)dy$, on the other hand, gives the conditional probability[15] of the process taking a value between y and dy at time t, *after assuming that the initial value at t_0 was y_0*. In contrast to $p_0(y_0)$, the function $G(y, t | y_0, t_0)$ is *intrinsic to the process*. It is often referred to as the **propagator** of the process.[16]

3.3.2.1 Information Carried by the Propagator

Very interesting information about the random process becomes accessible once the propagator is known. For instance, one could obtain all the relevant **one-point statistical information** of the process after constructing its **one-point pdf** from the propagator. Since the one-point pdf is the pdf of the values that the process takes after an arbitrary time lapse τ, it is given by:

$$p_1(y, \tau | t_0) = \int dy_0\, p(y_0) G(y, \tau + t_0 | y_0, t_0). \tag{3.43}$$

Secondly, one can also extract any **two-point statistical information** of interest for the process. One just needs to calculate the appropriate expected values (see

---

[15]It should be remembered that, in the theory of probabilities, the *conditional probability* of an event A happening, assuming that another event B has already taken place, is defined as $p(A|B) = p(A \cap B)/p(B)$. That is, it is given by the ratio of the joint probability of A and B happening, divided by the probability of B happening.

[16]The term *"propagator"* probably originates from the fact that $G(y, t | y_0, t_0)$ satisfies the following property:

$$G(y_2, t_2 | y_0, t_0) = \int dy_1 \int dt_1 G(y_2, t_2 | y_1, t_1) G(y_1, t_1 | y_0, t_0), \quad t_0 \le t_1 \le t_2. \tag{3.42}$$

This relation simply expresses that the probability of reaching the value y_2 at time t_2, after having started from the initial value y_0, is the sum of the probabilities of "propagating" the solution through all the possible intermediate values y_1 at all intermediate times t_1. It is a direct consequence of the fact that the total probability must be conserved during the evolution of the process. The propagator has played an important role in many disciplines within Mathematics and Physics. In particular, it is one of the building bricks of Richard Feynman's path-integral reformulation of Quantum Mechanics [29].

Sect. 2.2.5). For instance, consider the family of **moments** (for arbitrary real q) at time lapse τ:

$$m_q(\tau|t_0) := \langle |y(\tau+t_0)-y_0|^q \rangle \tag{3.44}$$

$$= \int dy_0\, p(y_0) \int dy |y-y_0|^q G(y, \tau+t_0|y_0, t_0).$$

These moments quantify how the process diverges from its initial value at time t_0.

Another important expected value is the so-called **autocorrelation function**, that is defined as (assuming $\tau_1 < \tau_2$),

$$C_y(\tau_1, \tau_2|t_0) := \langle y(\tau_1+t_0) \cdot y(\tau_2+t_0) \rangle \tag{3.45}$$

$$= \int dy_0\, p(y_0) \int dy_2\, y_2 \int dy_1\, y_1\, G(y_2, \tau_2+t_0|y_1, \tau_1+t_0)$$

$$G(y_1, \tau_1+t_0|y_0, t_0).$$

These quantities will play a very important role in this book. In particular, the one-point pdf and the moments are often used to test the self-similarity of a time process. In regards to the autocorrelation function, it will be shown in Chap. 4 that it gives information about the presence of long-term memory in the process.

3.3.2.2   Translationally Time-Invariant Random Processes

The properties of a random process can be expressed as conditions that its propagator must fulfil. For instance, an important family of random processes are those which are **invariant under temporal translations**. This requires that:

$$p_2(y_1, t_1; y_2, t_2) = p_2(y_1, t_1+T; y_2, t_2+T), \quad \forall T, \tag{3.46}$$

for any pair of times t_1 and t_2, and any pair of values y_1 and y_2. In simple words Eq. 3.46 means that the probability of the process taking a value y_2 at time t_2 and a value y_1 at t_1 *depends only on how much time has passed between t_1 and t_2, but not on the actual values of those times*. Therefore, p_2 must depend only on their difference, $t_1 - t_2$:

$$p_2(y_1, t_1; y_2, t_2) \rightarrow p_2(y_1; y_2|t_1-t_2). \tag{3.47}$$

If we now particularize Eq. 3.47 by assuming that t_2, y_2 correspond to the initial time (t_0) and value (y_0) of the process and $y_1 = y$ to the value at any later time $t_1 = t > t_0$, invariance under temporal translations translates, for the propagator, into:

$$G(y, t|y_0, t_0) \rightarrow G(y, t-t_0|y_0). \tag{3.48}$$

This requirement has consequences for all the quantities we defined in the previous section. In particular, the one-point pdf (Eq. 3.43) of the process becomes independent of its initial condition,

$$p_1(y, \tau|t_0) \to p_1(y, \tau) = \int dy_0 p(y_0) G(y, \tau | y_0). \tag{3.49}$$

This property actually quite useful since, if the process is always initialized from the same value, i.e. $p(y_0) = \delta(y_0 - y_0^*)$, then $p_1(y, \tau) = G(y, \tau | y_0^*)$. This property offers an easy way to estimate propagators numerically, that will be exploited in Chap. 5.

Similarly, all the q-moments defined in Eq. 3.44 become functions of the elapsed time $\tau = t - t_0$:

$$m_q(\tau|t_0) \to m_q(\tau) = \int dy_0 \, p(y_0) \int dy |y - y_0|^q G(y, \tau | y_0). \tag{3.50}$$

Finally, the autocorrelation function defined in Eq. 3.45 only depends on the values of the two *time lags*, $\tau_1 = t_1 - t - 0$ and $\tau_2 = t_2 - t_0$:

$$C_y(\tau_1, \tau_2|t_0) \to C_y(\tau_1, \tau_2) \tag{3.51}$$

$$= \int dy_0 \, p(y_0) \int dy_1 \, y_1 G(y_1, \tau_1 | y_0) \int dy_2 \, y_2 \, G(y_2, \tau_2 - \tau_1 | y_1).$$

3.3.2.3 Statistically Self-Similar Random Processes

With the help of the propagator just introduced, we can now make more precise what statistical self-similarity (i.e., Eq. 3.39) entails. A random process is H-ss if its propagator satisfies that:

$$G(y, t | y_0, t_0) = a^{-H} G \left(\frac{y}{a}, \frac{t}{a^H} \middle| y_0^*, t_0^* \right), \tag{3.52}$$

with the new initial condition given by: $y_0^* = y_0/a$ and $t_0^* = t_0/a^H$. Mathematically, this means that the propagator must be **self-affine** instead of self-similar. By self-affine it is meant that *each independent variable must be rescaled with a different factor for the original propagator to be recovered.* If Eq. 3.52 is satisfied, it can then be guaranteed that all the statistical information of interest to us in the original and the rescaled processes will be identical.

The fact that the propagator is self-affine has important consequences. For instance, the one-point pdf of the process (Eq. 3.43) also becomes self-affine,

$$p_1(y, t) = a^{-H} p \left(\frac{y}{a}, \frac{t}{a^H} \right). \tag{3.53}$$

In addition, the q-moments of the process (Eq. 3.44) will satisfy,

$$m_q(t|t_0) = \langle |y(t) - y_0|^q \rangle \propto |t - t_0|^{qH}, \quad \forall q. \tag{3.54}$$

Equations 3.53 and 3.54 will be the basis of some of the analysis techniques that will be discussed in Sect. 3.4.1 to test for self-similarity in real time series suspect of being scale-invariant.

3.3.2.4 Examples of Propagators

In order to illustrate the previous discussion, it is worthwhile to compute the propagator explicitly for a couple of simple processes.

Cosine Process

We will first consider the following **cosine process**:

$$y^{(1)}(t) = \cos(\omega(t - t_0) + \phi) \tag{3.55}$$

where the phase ϕ is fixed by the initial condition, $y(t_0) = y_0$, that we will assume distributed according to a uniform distribution $p(y_0) = 1/2$, for y_0 in $[-1, 1]$. The cosine process is random because of the randomness of the initial condition. As a result, if one looks at the value the process takes at any given time for different realizations, they would indeed be random (see Fig. 3.10). Next, we would like to find answers to the following questions. Is the cosine process invariant under time translations? Is it self-similar? To find out, we need to calculate its propagator.

The propagator of the cosine process can be easily computed by noting that its phase is set by the initial condition via,

$$\phi = \cos^{-1} y_0, \tag{3.56}$$

by rewriting next the relation between y and y_0 as,

$$y = \cos\left(\omega(t - t_0) + \cos^{-1} y_0\right) \tag{3.57}$$

and by finally realizing that the propagator must simply state the fact that these two values are the only ones that the process can connect between those times:

$$G^{(\cos)}(y, t| y_0, t_0) = \delta\left(y - \cos\left(\omega(t - t_0) + \cos^{-1} y_0\right)\right), \tag{3.58}$$

where $\delta(x)$ is Dirac's delta function. This is a consequence of the fact that, once the initial condition is fixed, the evolution of $y(t)$ is fully deterministic.

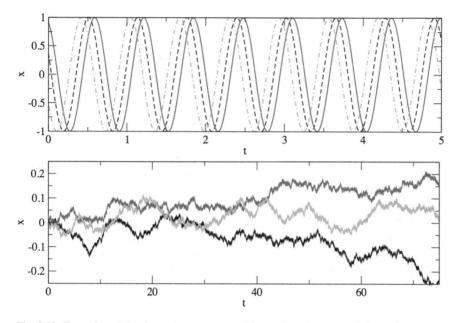

Fig. 3.10 Examples of simple random processes. **Above:** three instances of the cosine process defined by Eq. 3.55 using $\omega = 10$ and for $y_0 = -0.56$, 0.33 and 0.91. Once y_0 is chosen, the evolution is deterministic. But if one collects the values of many realizations at any given time, they behave like a random variable. **Below:** three realizations of the random time process defined by Eq. 3.64 for $y_0 = 0$. Since the process is intrinsically stochastic, the realizations diverge as time advances

We can now try to answer the questions we posed earlier. Clearly, the propagator depends only on the difference $(t - t_0)$. Therefore, the cosine process is invariant under time translations. However, it is not true that, if we make the changes $y \rightarrow y/a$ and $t \rightarrow t/a^H$, the original propagator can be recovered by doing $G \rightarrow a^{-H}G$ for any exponent H. This is particularly evident if one calculates the associated one-point pdf using Eq. 3.43. To do it, we need to compute the integral:

$$p_1^{\cos}(y, t) = \int dy_0 p(y_0)\delta\left(y - \cos\left(\omega t + \cos^{-1} y_0\right)\right). \tag{3.59}$$

To do it, we need to invoke a well-known property of Dirac's delta function. If $g(x)$ is a function that only vanishes at $x = x^*$, then,

$$\delta(g(x)) = \frac{\delta(x - x^*)}{|g'(x^*)|}. \tag{3.60}$$

In our case, we have that,

$$g(y_0^*) = y - \cos\left(\omega t + \cos^{-1} y_0^*\right) = 0 \longrightarrow y_0^* = \cos(\cos^{-1} y - \omega t). \tag{3.61}$$

Then, we can calculate,

$$|g'(y_0^*)| = \left.\frac{|\sin(\omega t + \cos^{-1} y_0)|}{\sqrt{1 + y_0^2}}\right|_{y_0 = y_0^*} = \frac{|\sin(\cos^{-1} y)|}{\sqrt{1 + \cos^2(\cos^{-1} y - \omega t)}}. \qquad (3.62)$$

We can now calculate the integral in Eq. 3.59, that yields, since $p(y_0) = 1/2$,

$$p_1^{\cos}(y, t) = \frac{1}{2|g'(y_0^*)|} = \frac{\sqrt{1 + \cos^2(\cos^{-1} y - \omega t)}}{2|\sin(\cos^{-1} y)|}. \qquad (3.63)$$

Equation 3.63 reduces to the pdf of the initial conditions, $p(y_0) = 1/2$, when $t = 0$. But its shape changes with time in a way that is not a simple rescaling, as shown in Fig. 3.11. Therefore, p_1 cannot be self-affine, and the cosine process cannot be self-similar. This conclusion was to be expected since it was clear from the start that the cosine process had a well-defined time scale: its period $T = 2\pi/\omega$.

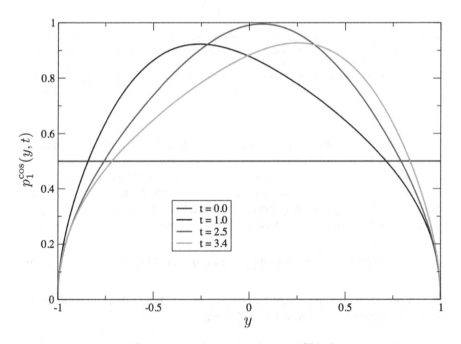

Fig. 3.11 Snapshots at different times of the one-point pdf, $p_1^{\cos}(y, t)$, for the cosine process defined in Eq. 3.55 using $\omega = 10$. It coincides with the pdf of the initial condition, $p(y_0) = 1/2$ at $t = 0$. But then, its shape changes in a way that is different from a simple rescaling. Thus, the cosine process cannot be self-similar

Random Gaussian Process

Secondly, we will consider an example of a **random Gaussian process**. In particular, we will look at Langevin's formulation of one-dimensional Brownian motion [30]:

$$y^{\text{Bm}}(t) = y_0 + \int_{t_0}^{t} ds \xi_2(s). \tag{3.64}$$

Here, $\xi_2(s)$ is a random, Gaussian noise with zero mean and variance σ_ξ^2. Due to this random noise, the resulting process becomes truly random in the sense that the values of the process at any time are unpredictable *even for realizations that start from the same initial value* (see lower frame in Fig. 3.10). Let's try to answer the same questions that were made for the cosine process. Is the process invariant under time translations? Is it self-similar?

In this case, we will not compute the propagator explicitly since we will do it in Sect. 5.2.2. We will simply advance the result (see Eq. 5.32):

$$G^{\text{Bm}}(y, t | y_0, t_0) = \frac{1}{\sqrt{2\pi\sigma_\xi^2(t - t_0)}} \exp\left(-\frac{(y - y_0)^2}{2\sigma_\xi^2(t - t_0)}\right) = N_{[y_0, \sigma_\xi^2(t-t_0)]}(y). \tag{3.65}$$

That is, the propagator is a Gaussian whose variance grows linearly with time as $\sigma^2(t) = \sigma_\xi^2(t - t_0)$. Clearly, the propagator only depends on the elapsed time $(t - t_0)$. Therefore, it is invariant under temporal translations. At the same time, it is trivial to see that the propagator of Brownian motion recovers its original form, after carrying out the transformations $y \to y/a$ and $t \to t/a^H$, simply by multiplying it by a^{-H} with $H = 1/2$. Therefore, the propagator is truly self-affine. Brownian motion, on the other hand, is self-similar with self-similarity exponent $H = 1/2$.

The same conclusion could have also been reached from the analysis of the one-point pdf of the process. For simplicity, we will assume that the process always start with the same initial condition, $y(t_0) = y_0^*$. Therefore, $p(y_0) = \delta(y_0 - y_0^*)$ and Eq. 3.43 reduces to:

$$p_1^{\text{Bm}}(y, \tau) = \frac{1}{\sqrt{2\pi\sigma_\xi^2\tau}} \exp\left(-\frac{(y - y_0^*)^2}{2\sigma_\xi^2\tau}\right) = N_{[y_0^*, \sigma_\xi^2\tau]}(y), \tag{3.66}$$

with the elapsed time being $\tau = t - t_0$. That is, the one-point pdf coincides with the propagator, as expected from a self-affine propagator.[17]

---

[17]The same result is obtained for any other choice for $p(y_0)$, as long as it is positive everywhere and normalizable to one.

3.3.2.5   Self-Similar, Translationally Time-Invariant, Random Processes

The Brownian motion we just examined belongs to a larger family of H-ss, translationally time-invariant processes, whose propagators take the general form,

$$G(y, t| y_0, t_0) = \frac{1}{(t - t_0)^H} \Phi\left(\frac{y - y_0}{(t - t_0)^H}\right),\tag{3.67}$$

for some function $\Phi(x)$, that is arbitrary except for the fact that they must satisfy that,

$$\Phi(x) \geq 0, \quad \forall x,\tag{3.68}$$

and

$$\int_{-\infty}^{\infty} dx \, \Phi(x) = 1.\tag{3.69}$$

Equations 3.68 and 3.69 ensure that the propagator, that in essence is a probability density, remains always positive (probabilities cannot be negative!) and normalizable to one.

The case of the Brownian motion corresponds to $H = 1/2$ and $\Phi(x) = N_{[0,\sigma_\xi^2]}(x)$. However, many other possibilities exist. In particular, given the clear favouritism of the Central Limit Theorem for Gaussian and Lévy pdfs (see Chap. 2), we will focus the following discussion on the two most popular members of this family: the **fractional Brownian motion** (Φ = Gaussian, $H \in (0, 1]$) and the **fractional Levy motion** (Φ = symmetric Lévy, $H \in (0, \min\{1, 1/\alpha\}]$).

It is worth saying now that, thanks to their self-similarity, all these processes share a common property. Their q-moment satisfies,

$$\langle |y(t) - y_0|^q\rangle = C_q |t - t_0|^{qH}, \quad C_q = \int dz \, |z|^q \Phi(z), \quad \forall q > -1 \mid C_q < \infty.\tag{3.70}$$

This property will be of great utility to compare real time series against translationally time-invariant, H-ss processes (See Sect. 3.4).

3.3.3   Fractional Brownian Motion

As already mentioned, Brownian motion belongs to a family of H-ss random processes known as **fractional Brownian motions (fBm)**. These processes were introduced by Benoit B. Mandelbrot in the late 60s [12], being defined as:

$$y^{\text{fBm,H}}(t) = y(t_0) + \frac{1}{\Gamma\left(H + \frac{1}{2}\right)} \int_{t_0}^{t} ds \, (t - s)^{H-1/2} \xi_2(s), \quad H \in (0, 1].\tag{3.71}$$

where t_0 is an arbitrary initial time[18] and $\Gamma(x)$ is the usual Euler's Gamma function. The self-similarity exponent H is often referred to, in this context, as the **Hurst exponent** (see also Chap. 4). For $H = 1/2$, fBm reduces to the usual Brownian motion (Eq. 3.64). For $H \neq 1/2$, on the other hand, the non-trivial kernel inside the temporal integral forces the process to exhibit interesting features that will make fBm a very important player throughout the rest of this book.

The propagator of fBm can be easily calculated [12, 31]. It is given by the Gaussian,

$$G_{\text{fBm}}^{H}(y, t | y_0, t_0) = \frac{1}{\sqrt{2\pi \hat{\sigma}_\xi^2 (t - t_0)^{2H}}} ds \, \exp\left(-\frac{(y - y_0)^2}{2\hat{\sigma}_\xi^2 (t - t_0)^{2H}}\right), \tag{3.73}$$

with $\hat{\sigma}_\xi^2 = a(H)\sigma_\xi^2$, and $a(H) = \left[(2H)^{1/2}\Gamma(H + 1/2)\right]^{-1}$. From the inspection of the propagator,[19] it follows that the process is invariant under time translations and self-similar with exponent H. The shape of the propagator is identical to that of the pdf of the noise driving it, but with a rescaled variance. As predicted by Eq. 3.70, the moments of all members of the fBm family verify that,[20]

$$m_q^{\text{fBm},H}(t|t_0) = \left\langle |y^{\text{fBm},H}(t) - y_0|^q \right\rangle \propto |t - t_0|^{qH}, \quad \forall q > -1, \tag{3.74}$$

Physically, this scaling means that the process moves away from its initial value at a faster rate as the value of H increases. This is apparent from the three realizations of fBm shown in Fig. 3.12. The values of H used are, respectively, 0.25, 0.50 and 0.75. Clearly, the process with $H < 0.5$ departs from its initial location slower (on average) than Brownian motion, whilst the opposite situation happens for $H > 0.5$. Since Brownian motion is intrinsically related to **diffusive transport** (we will discuss this connection in Chap. 5), these two behaviours are often referred to as **subdiffusion** ($H < 1/2$) and **superdiffusion** ($H > 1/2$).

---

[18]In fact, Mandelbrot preferred instead the definition [12],

$$y(t) = y(t') + \frac{1}{\Gamma\left(H + \frac{1}{2}\right)}\left[\int_{-\infty}^{t} ds \, (t - s)^{H-1/2}\xi_2(s)\right. \tag{3.72}$$

$$\left. - \int_{-\infty}^{t'} ds \, (t' - s)^{H-1/2}\xi_2(s)\right], \quad H \in (0, 1]$$

since he felt that Eq. 3.71 assigned too much significance to the initial time t_0, as will be made much clearer in Chap. 4, where we discuss memory in time processes. We have however preferred to stick with Eq. 3.71 since some t_0 must be chosen to simulate fBm processes numerically or to compare against experimental data (clearly, no code or measurement can be extended to $t \to -\infty$!). In any case, both definitions lead to very similar properties.

[19]Clearly, it satisfies Eq. 3.67 with the choice $\Phi = N_{[0,\hat{\sigma}_\xi^2]}(x)$ for any $H \in (0, 1]$.

[20]The lower limit comes from the fact that the Gaussian does not have finite moments for $q < -1$, since the integrand, $1/|x|^q$ then has a non-integrable divergence at $x = 0$.

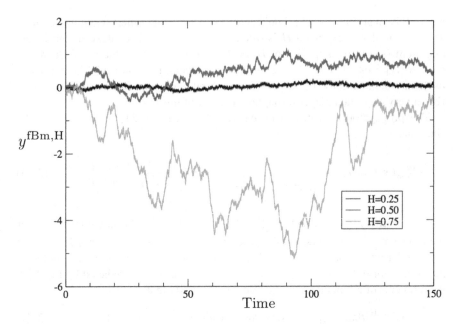

Fig. 3.12 Realizations of fBm starting at $y(0) = 0$ using self-similarity exponents $H = 0.25$ (black), $H = 0.5$ (red) and $H = 0.75$ (green). The $H = 0.5$ process is Brownian motion. For the other values of H, the process diverges from its initial value either slower ($H < 0.5$) or faster ($H > 0.5$) than Brownian motion. The former behaviour is usually referred to as subdiffusion; the second, superdiffusion

Mathematically, Eq. 3.74 also implies that one can associate a unique *local Holder exponent*, $h_s = H$, to each member of the fBm family.[21] As previously discussed (see Sect. 3.2.2; also, the discussion that will be had in Sect. 3.3.7), the local Holder exponent measures the degree of singularity of a function by looking at the scaling of the increment $|y(t + \Delta t) - y(t)| \propto |\Delta t|^{h_s}$. A differentiable function has $h_s = 1$, whilst a function with a bounded discontinuity would have $h_s = 0$. Continuous, but non-differentiable functions, such as fBm, have a value of h_s that lies in between 0 and 1. Particularizing Eq. 3.74 for $q = 1$, we find that $|y(t + \Delta t) - y(t)| \propto |\Delta t|^H$, so $h_s = H$ for fBm.[22]

---

[21] In the case of time processes, we will use the symbol h_s to refer to the Holder exponent, and reserve α for the tail-index of Levy pdfs. This is in contrast to what is usually done when carrying out multifractal analysis on spatial objects, where the symbol α is reserved for the singularity exponent. That is why we adhered to the popular criteria when discussing spatial multifractals in Sect. 3.2.2.

[22] It is somewhat curious that, although H is a measure of a global property (i.e., rescaling) and h_s measures a local property (degree of the local singularities), they do coincide for fBm. The situation is similar to what was found for spatial monofractals, where the fractal dimension (global property) and the Holder exponent (local property) were also identical. It is for this reason that it is sometimes said that H-ss processes have a monofractal character. It is also interesting to remark

3.3.4 Fractional Lévy Motion

It was discussed in Chap. 2 that the Central Limit Theorem also identified other pdfs, in addition to the Gaussian law, as preferred attractors: the heavy-tailed Lévy family (see Sect. 2.3). Since Lévy pdfs are often found in complex systems, it is logical to expect that a Lévy version of fBm should play a similar role in them. Such a family of H-ss processes does indeed exist, being known as **fractional Lévy motion (fLm)** [32]. It is defined analogously to fBm[23]:

$$y_{\alpha,H}^{fLm}(t) = y(t_0) + \frac{1}{\Gamma\left(H - \frac{1}{\alpha} + 1\right)} \int_{t_0}^{t} (t-s)^{H-1/\alpha} ds\, \xi_\alpha(s), \qquad (3.76)$$

$$\text{with} \quad 0 < H \le \max\left(1, \frac{1}{\alpha}\right).$$

ξ_α is now a zero-mean noise distributed according to a symmetric Lévy law of index α and scale factor σ_ξ. That is, according to $L_{[\alpha,0,0,\sigma_\xi]}(\xi_\alpha)$.[24] Figures 3.13 and 3.14 show realizations of fLm for various values of H and α.

The first thing to note is that, for $H = 1/\alpha$, the kernel vanishes in Eq. 3.76 and the process becomes the Lévy equivalent to Brownian motion, often referred to as a **Lévy flight**. Similarly to fBm, the motion diverges faster than a Lévy flight if $H > 1/\alpha$, and slower if $H < 1/\alpha$.[25] Secondly, all fLm traces exhibit long jumps that are associated with large noise values that contribute to the tail of each respective Lévy pdf. Clearly, the importance of these extreme jumps for the overall motion increases as the value of α decreases.

---

that the Holder exponent defined for fBm is identical to the one used in the theory of mathematical functions, whilst the one used for spatial fractals had a different definition (Eq. 3.26). As a result, the fractal dimension of fBm time traces is not equal to h_s, but given by $D = 2 - h_s = 2 - H$ (See Problem 3.5). This is correlated with the fact that, the smaller h_s, the more irregular the trace becomes, so it fills its embedding space more densely.

[23] As with fBm, another definition for fLm exists that avoids giving too much importance to the initial time. It is:

$$y_{\alpha,H}^{fLm}(t) = y(t_0) + \frac{1}{\Gamma\left(H - \frac{1}{\alpha} + 1\right)} \left[\int_{-\infty}^{t} ds\, (t-s)^{H-1/\alpha} \xi_\alpha(s) \right. \qquad (3.75)$$

$$\left. - \int_{-\infty}^{t'} ds\, (t'-s)^{H-1/\alpha} \xi_\alpha(s) \right], \quad \text{with} \quad 0 < H \le \max\left(1, \frac{1}{\alpha}\right).$$

where $t' < t$ is again an arbitrary past reference time.

[24] It is worth mentioning that the theory of fLm processes using non-symmetric Lévy distributions has not been developed very much so far, in spite of the fact that there are some physical problems where it might be useful. We will discuss one such example in Sect. 5.5, when we investigate transport across the running sandpile.

[25] However, the terms subdiffusion and superdiffusion are not used to refer to any of these behaviours, as will be discussed in Chap. 5.

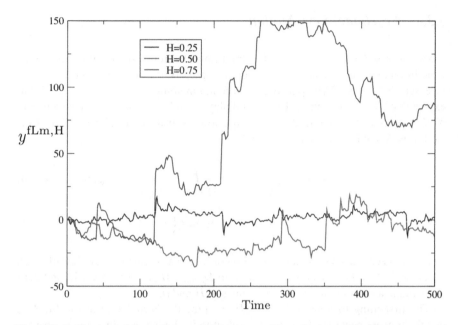

Fig. 3.13 Realizations of fLm with $\alpha = 1.5$ starting at $y(0) = 0$ using self-similarity exponents $H = 0.25$ (black), $H = 0.5$ (red) and $H = 0.75$ (blue). Here, the uncorrelated motion corresponds to $H = 1/\alpha = 0.66$. Thus, only the blue realization contains positive correlations, that cause it to diverge much faster. In contrast, the anticorrelated ones barely move away from the initial location

As with fBm, we will discuss fLms in greater detail in Chap. 5, since they are of key importance for the theory of fractional transport. Here, we just provide their propagator, that is given by [33, 34],

$$G_{\mathrm{fLm}}^{\alpha,H}(y,t|\,y_0,t_0) = \frac{1}{(t-t_0)^H} L_{[\alpha,0,0,\hat{\sigma}_\xi]}\left(\frac{(y-y_0)}{(t-t_0)^H}\right), \tag{3.77}$$

with $\hat{\sigma}_\xi = a(H,\alpha)\sigma_\xi$, and the coefficient $a(H,\alpha)$ defined as[26]:

$$a(H,\alpha) := \left[(\alpha H)^{1/\alpha}\,\Gamma\left(H - \frac{1}{\alpha} + 1\right)\right]^{-1}. \tag{3.78}$$

All fLm processes are both invariant under time translations and self-similar, as follows form the fact that the fLm propagator has the canonical form expressed in Eq. 3.67, with $\Phi = L_{[\alpha,0,0,\hat{\sigma}_\xi]}(x)$ for any $\alpha \in (0.2)$. The shape of the fLm propagator

---

[26]Note that $a(H,2) = a(H)$, the function we introduced for fBm in Sect. 3.3.3. For that reason, the limit of fLm when $\alpha \to 2$ is fBm with the same value of H.

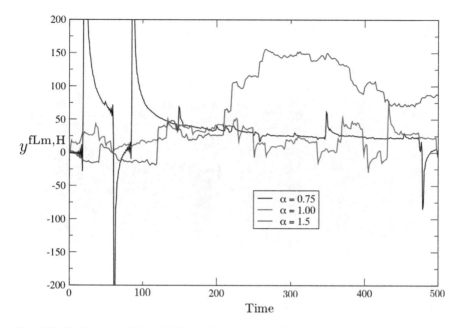

Fig. 3.14 Realizations of fLm with $H = 0.75$ starting at $y(0) = 0$ using tail exponents $\alpha = 0.75$ (black), $\alpha = 1.0$ (red) and $\alpha = 1.5$ (blue). Here, the uncorrelated motion corresponds to $H = 1/\alpha$. Thus, only the blue realization contains positive correlations. Also, note that sudden jumps, associated to the Lévy extreme events, become more dominant as α gets smaller, due to the fatter tail

also coincides with the pdf of the driving noise, as was the case of fBm, again with a different rescaling factor. However, in contrast to fBm, its moments verify,[27]

$$m_q^{\text{fLm};,\text{H}} = \left\langle |y_{\alpha,H}^{\text{fLm}}(t) - y_0|^q \right\rangle \propto |t - t_0|^{qH}, \quad -1 < q < \alpha \tag{3.79}$$

since the heavy tail of $L_{[\alpha,0,0,\hat{\sigma}_\xi]}(x)$ makes $C_q \to \infty$ in Eq. 3.70 for $q \geq \alpha$. It is also worth remarking that, because of this divergence, one can only define a local Holder exponent for fLm with $1 \leq \alpha < 2$, in which case it is again given by $h_s = H$.

3.3.5 Stationarity and Self-Similarity

Stationarity is another concept of great importance in the context of time random processes. Loosely speaking, stationarity means that *the process does not change its statistical features over time*. We will formulate this idea more precisely shortly, but

---

[27] As with the Gaussian, the lower limit comes from the fact that the symmetry Lévy does not have finite moments for $q < -1$, since $1/|y - y_0|$ then has a non-integrable divergence at $y = y_0$.

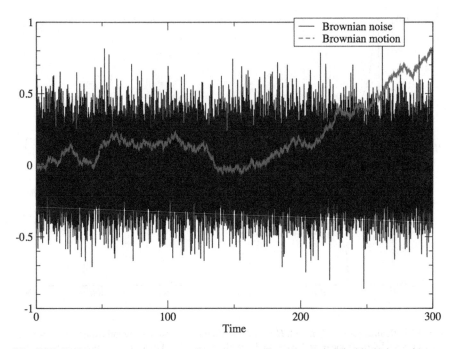

Fig. 3.15 Stationary vs. non-stationary time processes: Brownian noise (in black) is stationary. Brownian motion (in red), the integral of Brownian noise over time, is not a stationary time process

one can easily illustrate the concept by looking at the two time series in Fig. 3.15. One of them is a realization of Brownian motion (in red); the other one is the series of its *successive increments*[28] (in black). Clearly, Brownian motion is not stationary. Its mean value would be very different if estimated using the first or the second half of the data shown in the figure. The series of its increments, on the other hand, appears stationary in the sense that both mean and variance are independent of the section of the signal chosen to estimate them.

Why are stationary time processes important? The main reason is related to the practical advantages of dealing with stationary data records. Mainly, that stationary processes are usually ergodic in the sense that *temporal and ensemble statistical analysis yield similar results*. Therefore, if dealing with stationary data, one does not need a large number of different realizations to gather meaningful statistics. A very long temporal record broken up in pieces is enough, and usually much easier to obtain.

However, an interesting conundrum often appears in complex systems. Complex dynamics often abhor characteristic scales, as we have repeatedly stated. As a result, time processes taking place in them tend to exhibit self-similar features.

---

[28]Or, in other words, the noise series used to integrate Eq. 3.64.

Regretfully, it turns out that *self-similarity and stationarity are two mutually exclusive properties.*[29] The Brownian motion shown in red in Fig. 3.15, itself a $1/2$-ss process, provides a very nice illustration of this fact.

It might thus seem that, in order to be able to detect the presence of complex dynamics in natural time processes by looking for self-similarity features, we would necessarily have to deal with multiple realizations. Luckily, Fig. 3.15 already suggests one easier solution: instead of looking for self-similarity by comparing real processes to fBm/fLm, one could just *look at the increments of the natural time series and compare its properties with those of the increments of fBm/fLm processes,* that will be shown to be stationary in what follows.[30]

3.3.5.1 Definition of an Stationary Random Process

We consider that a random time process is **stationary** if it is[31]: (1) *translationally time invariant* (see Eq. 3.46); (2) its *one-point pdf is independent of the time lag τ* (see Eq. 3.49):

$$p_1(y, t|t_0) \rightarrow p_1(y); \tag{3.80}$$

and (3) its *autocorrelation function depends only on the lag difference, $\tau = \tau_2 - \tau_1$* (see Eq. 3.51):

$$C_y(\tau_1, \tau_2|t_0) \rightarrow C_y(\tau_2 - \tau_1) = C_y(\tau). \tag{3.81}$$

Clearly, it follows from the fact that the one-point pdf is independent of time (Eq. 3.80) that all the moments of the process must be also independent of time (see Eq. 3.50):

$$m_q(\tau) \rightarrow m_q, \quad \forall q. \tag{3.82}$$

Thanks to these properties, the same results should be obtained when calculating quantities from an ensemble average over a large number of realizations, or from a temporal average over a sufficiently long part of a single realization. We will take advantage of this fact in later sections (see Sect. 3.4.2).

---

[29]Except for trivially self-similar processes such as the constant process.

[30]We will also show later that, due to the translational time-invariance of both fBm/fLm, things are not so bad as stated here, and that methods exist to improve the statistics when dealing with the integrated process directly even if few (or just one) realizations are available (see Sect. 3.4.1).

[31]As we also did when discussing scale-invariance, we will only consider two-point statistical information in this book.

3.3.6   Self-Similar Processes with Stationary Increments

It is apparent that neither fBm nor fLm satisfy Eq. 3.80 (or Eq. 3.82). Thus, they are non-stationary random processes. In fact, as hinted out previously, *stationarity and self-similarity are two mutually exclusive properties*, except for trivial cases. Nevertheless, there is a famous theorem by Lamperti [35] that establishes the relation between H-ss and stationary processes. The theorem simply states that if $y(t)$ is an H-ss process, then $x(t) = e^{-tH}y(e^t)$ is a stationary one. Reversely, if $x(t)$ is a stationary process, then $y(t) = t^H x(\ln t)$ is an H-ss process.[32]

It is however possible to consider **self-similar processes with stationary increments**, often referred to as **H-sssi processes** (i.e., H-ss with stationary increments). It turns out that both fBm and fLm are two examples of H-sssi processes, as we discuss next.

3.3.6.1   Fractional Gaussian Noise (fGn)

Benoit Mandelbrot showed that the increments of fBm (previously defined in Eq. 3.71) are stationary [12]. Several increment series can be considered, depending on the chosen spacing $h > 0$:

$$\Delta_h y^{\text{fGn,H}}(t) := \frac{y^{\text{fBm,H}}(t + h) - y^{\text{fBm,H}}(t)}{h} \tag{3.83}$$

$$= \frac{h^{-1}}{\Gamma\left(H + \frac{1}{2}\right)} \left[\int_{t_0}^{t+h} (t + h - s)^{H-1/2} \xi_2(s) \right.$$
$$\left. - \int_{t_0}^{t} ds\,(t - s)^{H-1/2} \xi_2(s) \right], \qquad H \in (0, 1]. \tag{3.84}$$

The process $\Delta_h y^{\text{fGn,H}}$ will be referred to as **fractional Gaussian noise** (fGn).[33] Most interestingly, Mandelbrot also proved that *fBm is the only H-sssi process generated by (zero-mean) Gaussian noise* [12].

It turns out that fGn has some interesting properties. Some of them become particularly handy to compare it with natural stationary series. All of them are direct

---

[32]In theory, one might use Lamperti's theorem to test whether any time series is self-similar (or stationary). One would just need to apply the Lamperti's transform to it and check whether the result is stationary (or self-similar). However, due to the exponentials appearing in Lamperti's formulation, this scheme is often difficult to use in practice.

[33]It is also common in the literature to define fGn simply as $y_H^{\text{fGn}}(t+h) - y_H^{\text{fGn}}(t)$ instead, without the h^{-1} prefactor [36]. We have decided to adopt the definition that includes h^{-1} so that we can better assimilate fGn to a derivative of fBm, in spite of the latter being non-differentiable. The reason will become clearer when discussing methods to generate numerically fBm (see Appendix 1) using fractional derivatives. The only differences between one or another choice are that $\sigma_h^2 = h^{2H}\sigma_\xi^2$ in Eq. 3.86, and that the factor $r^{(H-1)}$ becomes r^H in Eqs. 3.87 and 3.88.

consequences of the properties of fBm. For instance, one can easily prove that the one-point pdf of fGn is given by:

$$p_1\left(\Delta_h y_H^{\mathrm{fGn}}\right) = N_{[0,\sigma_h^2]}(\Delta_h y), \tag{3.85}$$

i.e., a Gaussian whose variance is given by,

$$\sigma_h^2 = h^{2H-2}\sigma_\xi^2, \tag{3.86}$$

with σ_ξ^2 the variance of the Gaussian noise ξ_2. The one-point pdf of fGn is thus identical to both the pdf of the noise that drives fBm, and the Φ function that characterizes the fBm propagator. All of them are Gaussians, although with different variances.

Equation 3.86 is, in fact, an special case of a very interesting scaling property of the one-point pdf of fGn *with respect to the value of the spacing h*, that derives solely from the scale-invariant nature of fBm.[34] Under the transformation $h \to rh$, it follows that:

$$p_1(\Delta_{rh} y_H^{\mathrm{fGn}}) \equiv^d r^{(H-1)} p_1(\Delta_h y_H^{\mathrm{fGn}}), \tag{3.87}$$

This property translates, for all the convergent integer and real moments of the one-point pdf, into

$$\left\langle |\Delta_{rh} y_H^{\mathrm{fGn}}|^q \right\rangle = r^{q(H-1)} \left\langle |\Delta_h y_H^{\mathrm{fGn}}|^q \right\rangle, \quad \forall q > -1, \tag{3.88}$$

that reduces to Eq. 3.86 if $q = 2$. Equation 3.88 expresses how the moments change *when fGn is looked at in the different scales that are set by changing the value of r*. The reader will be able to see the practical usefulness of these scaling properties in Sect. 3.4 (for instance, see Eq. 3.114).

Finally, it is worth mentioning that the local Holder exponent for fGn is given by $h_s = 0$, since fGn is *discontinuous at every point*, but the discontinuity is bounded.

3.3.6.2 Fractional Lévy Noise (fLn)

Fractional Lévy motion (fLm) can also be shown to have stationary increments.[35] These increments form what is known as **fractional Lévy noise** (fLn). Similarly to

---

[34]Equation 3.87 applies for any arbitrary function Φ in Eq. 3.67, since it is only due to the scale-invariance. For example, it is also satisfied by fractional Levy noise, to be introduced next.

[35]In fact, this is one of the conditions that sets the allowed values of the self-similarity exponent H to the interval $(0, \max(1, \alpha^{-1})]$. The other one is the requirement of the propagator of the process being positive everywhere, at every time [37].

fGn, fLn will be defined, for arbitrary spacing $h > 0$, as[36]:

$$\Delta_h y_{\alpha,H}^{fLn}(t) = \frac{h^{-1}}{\Gamma\left(H - \frac{1}{\alpha} + 1\right)} \left[\int_{t_0}^{t+h} ds\,(t + h - s)^{H-1/\alpha}\xi_\alpha(s) \right. \tag{3.89}$$

$$\left. - \int_{t_0}^{t} ds\,(t - s)^{H-1/\alpha}\xi_\alpha(s)\right], \quad 0 < H \leq \max\left\{1, \frac{1}{\alpha}\right\}.$$

In the case of fLm, however, it is no longer true that Eq. 3.89 is the only H-sssi processes that can be generated by (zero-mean) Lévy noise [36]. fLm shares some of the properties of fGn, although adapted to the fact that its moments diverge for orders equal or larger α. In particular, its one-point pdf is given by:

$$p_1\left(\Delta_h y_{\alpha,H}^{fLn}\right) = L_{[\alpha,0,0,\sigma_h]}(\Delta_h y), \tag{3.90}$$

whose scale factor is given by,

$$\sigma_h = h^{H-1}\sigma_\xi. \tag{3.91}$$

Again, the one-point pdf has the same form that those of the noise driving the fLm and the Φ function that characterizes the fLm propagator. They all are symmetric Lévy pdfs of tail-index α, although each with a different scale factor.

In addition, the one-point pdf of fLn, due to the scale-invariance of fLm, satisfies an analogous property to fBm under the transformation of the spacing $h \to rh$:

$$p(\Delta_{rh} y^{fLn}) \equiv^d r^{(H-1)} p(\Delta_h y^{fLn}). \tag{3.92}$$

This property translates, for all convergent integer and real moments, into the following scaling rule:

$$\left\langle|\Delta_{rh} y_{\alpha,H}^{fLn}|^q\right\rangle = r^{q(H-1)}\left\langle|\Delta_h y_{\alpha,H}^{fLn}|^q\right\rangle, \quad -1 < q < \alpha, \tag{3.93}$$

that will be shown to be extremely useful in practical cases in Sect. 3.4.

We conclude our discussion of fLn by saying that, as was the case for fGn, the local holder exponent of fLn (for $1 < \alpha < 2$) is $h_s = 0$, since it is discontinuous at every point, with a bounded discontinuity.

3.3.7 Multifractal Time Random Processes

There are many natural time processes that look irregular, but not in the homogeneous way in which fBm or fLm look irregular. Instead, signals such as time records of the variability of the human heart rate (see Fig. 3.16) or of the neural activity of the human brain exhibit very different looking singularities over time [38, 39].

---

[36] Again, if one adopts the definition of fLn without the h^{-1} prefactor, $\sigma_h = h^H \sigma_\epsilon$ instead in Eq. 3.91. Also, the factor $r^{(H-1)}$ becomes r^H in Eqs. 3.92 and 3.93.

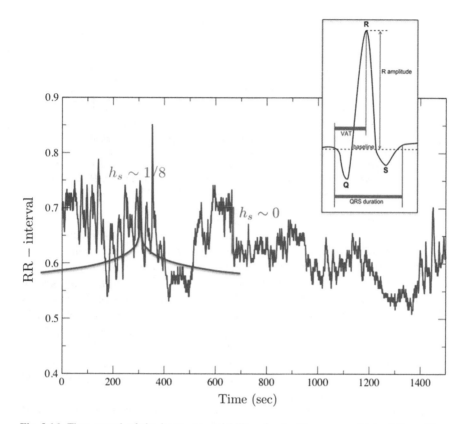

Fig. 3.16 Time record of the heart-rate variability of a healthy person obtained from *Phys-ionet.org*. The variability is measured using the interval of time between successive *R* points (see inset on the top right) of the heartbeat signal as measured in a typical electrocardiogram (ECG). Different singularities are apparent in the signal as time advances, suggesting that the nature of the process could be multifractal. Credits: inset from *Wikimedia Commons* (M. Häggström, public domain)

Some of them look more like sudden jumps or steps, while others often look more like spikes or cusps. Mathematically, a different value of the Holder exponent can be associated to each of these singularities (see Fig. 3.17). Therefore, it is impossible to capture the nature of these time process with **monofractal** models such as fBm or fLm, whose singularities are all of the same type. That is, those with Holder exponent $h_s = H$, their Hurst exponent.

There has been some efforts to build random processes, based on generalizations of fBm, to overcome this limitation. One is **multifractional Brownian motion (mBm)** [40], and its prescription is rather simple. It is defined similarly to fBm (Eq. 3.71) but with *H* becoming a function of time, $H(t) \in (0, 1]$, $\forall t > t_0$:

$$y^{\text{mBm}}(t) = y(t_0) + \frac{1}{\Gamma\left(H(t) + \frac{1}{2}\right)} \int_{t_0}^{t} (t - s)^{H(t)-1/2} ds\, \xi_2(s). \qquad (3.94)$$

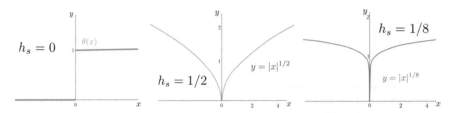

Fig. 3.17 Examples of functions that exhibit singularities at $x = 0$ with different Holder exponents. Heaviside's step function (left) has a Holder exponent of $h_s = 0$, since it has a bounded discontinuity of size one. The function $y = |x|^{1/2}$ (center) has a holder exponent $h_s = 1/2$. On the right, $y = |x|^{1/8}$, has a Holder exponent of $h_s = 1/8$ at $x = 0$. The smaller the Holder exponent, the spikier the singularity becomes until eventually, the function becomes discontinuous

It can be shown that, although mBm is no longer globally self-similar in time, it is still **locally self-similar** for fixed t, with a self-similarity exponent given by $H(t)$. In addition, its local Holder exponent becomes $h_s(t) = H(t)$, which is no longer constant. A similar generalization can also be carried out for fLm, known as **multifractional Lévy motion (mLm)** [41], that shares many of the properties of mBm: local self-similarity for fixed t and time-varying Holder exponent, $h(t) = H(t)$, if restricted to $1 < \alpha < 2$. Regretfully, both mBm and mLm impose a strong restriction on the variation of $H(t)$, that must be a regular function, which precludes the kind of erratic variation of the local Holder exponent observed in natural series. Other generalizations of fBm/fLm are in the process of being investigated, that could allow for the erratic variation of h_s with time [42].

We will not get into the details of mBm, mLm, or any other multifractal time process in this book. Instead, we will focus on building a tool, similar to the multifractal analysis developed for spatial fractals, that could be used to characterize multifractal features in time records.[37] The idea that was used for spatial multifractals was relatively simple (see Sect. 3.2.2). We considered coverings with N blocks of size l, and then we associated to each block a weight w_i, that measured how densely filled the block was, and that was related to a local Holder exponent via $w_i \propto l^{\alpha_i}$ (Eq. 3.26). Weights were normalized so that $\sum_{i=1}^{N} w_i = 1$. Then, we constructed the partition function, $N(l, q) = \sum_{i=1}^{N} |w_i|^q$, where q could be any real number, and we looked for scalings of the kind $N(l, q) \propto l^{-\tau(q)}$ (Eq. 3.17). From the analysis of $\tau(q)$ the multifractal character of the spatial object could be made apparent, since for a monofractal, a decreasing linear dependence was expected, $\tau(q) = D_{bc}(1 - q)$ (Eq. 3.19), with D_{bc} being its BC fractal dimension. Let's see how we can do something similar with time processes.

---

[37]The interested reader may also refer to Appendix 3, that discusses an alternative technique based on the use of wavelets to characterize multifractality in time series.

3.3.7.1 Family of Multifractal Exponents $\xi(q)$

In the case of a time random process, $y(t)$ we will consider the division of the time series in N_b non-overlapping blocks of size T to play a similar role to that of the spatial covering in fractals. Within each block i, we will define a *non-normalized weight* given by[38]:

$$w_i := \frac{|y(t_i + T) - y(t_i)|}{T} = |\Delta_T y(t_i)| \propto T^{h_s(t_i)-1}, \tag{3.95}$$

where $h_s(t_i)$ is the local Holder exponent of $y(t)$ at t_i for small T. Note that we have included a T^{-1} in the weight on purpose, so that we can use the notation we introduced for the increments of a process in Eq. 3.83. Next, we build the **partition function**, also referred to as the **structure function** in this context,[39,40] as:

$$N(T, q) := \frac{1}{N_b} \sum_{i=1}^{N_b} w_i^q \propto T^{\xi(q)}, \tag{3.96}$$

from which the family of multifractal exponents $\xi(q)$, can be obtained as:

$$\xi(q) = \lim_{T \to 0} \frac{\log N(T, q)}{\log T}. \tag{3.97}$$

The family of multifractal exponents $\xi(q)$ calculated in this way has the following properties:

- $\xi(0) = 0$. Indeed, this follows from: $N(T, 0) = 1 \propto T^0$.
- for fBm with Hurst exponent H, $\xi(q) = q(H - 1)$, $\forall q > -1$. That is, it is a decreasing linear function of q. This follows from the fact that $h_s(t_i) = H \leq 1$ for fBm. Thus, $N(T, q) \propto N_b^{-1} \left(N_b T^{q(H-1)} \right) \propto T^{q(H-1)}$.

---

[38]Non-normalized weights must be used in the case of time processes since, otherwise, the normalization, $\sum_{i=1}^{N_b} |y(t_i+T) - y(t_i)|$ would eliminate the scaling of the weight for a monofractal (fBm or fLm). The reason is that the support of the time process is a regular line (the temporal axis) instead of a fractal with dimension D. As a result, $N_b \propto T^{-1}$, instead of T^{-D}.

[39]The name structure function originates from the theory of turbulence, back in the 1940s, when Kolmogorov formulated his famous law that stated that the structure function of the turbulent velocity fluctuations scaled as $S_p(x) := \langle |V(\mathbf{r} + \mathbf{x}) - V(\mathbf{r})|^p \rangle \propto r^{\xi(p)}$, with $\xi(p) = p/3$ [43]. In fact, much of multifractal analysis for time processes was originally developed and extensively applied later to the study of fluid turbulence [19, 44, 45].

[40]The partition (or structure) function is sometimes introduced a little bit differently from how we do it here. The main difference is that the factor T^{-1} that appears in Eq. 3.95 is omitted. As a result, the generalized Hurst exponent is defined as $H(q) = \xi(q)/q$. The equations that give the singularity spectrum (Eqs. 3.102 and 3.103) then become $h_s^*(q) = d\xi/dq$ and $f(h_s^*(q)) = qh_s^*(q) - \xi(q) + 1$.

- for fLm with Hurst exponent H, it also happens that $\xi(q) = q(H-1)$ but only for $-1 < q < \alpha$, being α the tail-index of the Levy noise driving fLm. The sum of powers of the weight would diverge if q is equal or larger than α, as expected.
- a time process is multifractal[41] if $\xi(q)$ has a more complicated (i.e., nonlinear) dependence on q. It is traditional to define the **generalized Hurst exponent** as,

$$H(q) := \frac{\xi(q)}{q} + 1. \tag{3.98}$$

As expected, $H(q) = H$, $\forall q > -1$ for fBm with Hurst exponent H, and for fLm with the same Hurst exponent, as long as $-1 < q < \alpha$.

3.3.7.2   Singularity Spectrum $f(h_s)$

We will continue with the analogy with spatial fractals and derive next an expression for the **singularity spectrum**, $f(h_s)$, that gives the fractal dimension of the subset formed by all points on the temporal axis where the local Holder exponent of the time process is h_s or, in other words, the moments in which singularities with that Holder exponent happen. To do so, we rewrite the partition function as:

$$N_b \times N(T, q) = \sum_{i=1}^{N_b} w_i^q \sim \int dh_s p(h_s) T^{-f(h_s)} T^{q(h_s-1)} \sim T^{(\xi(q)-1)}. \tag{3.99}$$

The explanation for the formation of the integral is the same that we gave in the case of spatial fractals (see Sect. 3.2.2). To rewrite the sum over blocks that forms the partition function as a sum over fractal sets of points with Holder exponent h_s that have a fractal dimension $f(h_s)$, we simply have to multiply the fraction of the temporal axis occupied by each set, $\rho(h_s)dh_s$, by the contribution of the region to the integral, $T^{-f(h_s)}$, times the weight associated to it, $T^{q(h_s-1)}$. Since we are interested in the dominant part of the integral, this will come from the value of h_s, that we will call $h_s^*(q)$, that minimizes the exponent $qh_s - f(h_s)$. It is given by the solution of,

$$\frac{d}{dh_s}[q(h_s-1) - f(h_s)] = 0 \implies \frac{df}{dh_s}(h_s^*(q)) = q. \tag{3.100}$$

We do not need to find the solution to this equation now. Instead, we will assume that $h_s^*(q)$ is known, which allows us to relate $\tau(q)$ with the singularity spectrum:

$$q(h_s^*(q) - 1) - f(h_s^*(q)) = \xi(q) - 1, \tag{3.101}$$

---

[41]Some authors define multifractal as any instance in which $H(q) \neq H$, $\forall q$. Thus, fLm is then considered as multifractal, since $H(q) \neq H$ for $q \geq \alpha$. We do not adhere to this practice.

that can be easily inverted, by differentiating both sides with respect to q, and then use Eq. 3.100 to get:

$$h_s^*(q) = 1 + \frac{d\xi}{dq}(q) \tag{3.102}$$

$$f(h_s^*(q)) = q(h_s^*(q) - 1) - \xi(q) + 1. \tag{3.103}$$

This solves the problem completely. Indeed, this pair of equations estimate, for each value of q, the dominant value of the Holder exponent that contributes to the partition function, as well as the fractal dimension of the fractal subset formed by all regions in the time axis with that local Holder exponent. Thus, by varying q, obtaining $\tau(q)$ and using Eqs. 3.102 and 3.103, we can quantify the spectrum of fractal dimensions of the subsets with Holder exponent h_s that are intertwined to form the multifractal time process.

Note that $f(h_s) \leq 1$, since it cannot exceed the dimension of the embedding space, the real line, which is one. It is also reassuring to note that, in the case of fBm with H as its Hurst exponent, $h_s^*(q) = H$, $\forall q > -1$, as verified by substituting $\xi(q) = q(1 - H)$ in Eq. 3.102. Also, that the fractal dimension of the subset containing all parts of the temporal axis where $h_s = H$ (the whole axis, in the case of fBm!) is given by $f(h_s = H) = 1$, as it should be. The same conclusion applies also to fLm with H as its Hurst exponent, although only for $-1 < q < \alpha$. Therefore, as it was the case with spatial multifractals, a broad singularity spectrum is again a sign of multifractality, with its width quantifying the degree of multifractality of the time process.[42]

3.4 Techniques for the Practical Determination of Scale-Invariance

After having discussed the theoretical fundamentals of scale-invariance in both spatial objects and time processes, we move the focus next towards the practical characterization of the scale-invariance in time series.[43] Experimental data are often the result of sampling some quantity at a certain rate. The outcome is a *discrete temporal record*,

$$\{Y_k, \quad k = 1, 2, \ldots\}, \tag{3.104}$$

---

[42]One must however be careful, since the spectrum of discrete realizations of fBm and fLm also have a finite width, caused by their discreteness. We will come back to this issue in Sect. 3.4.3.

[43]For spatial objects, the box-counting procedures discussed in Sect. 3.2 are very good to calculate fractal dimensions and the multifractal spectrum. We will not illustrate them here, though.

where the index k plays the role of a discrete time. In order to decide whether the data set exhibits scale-invariance, one often tries to determine whether it resembles (or not) a self-similar process such as fBm or fLm. As we already mentioned, two main possibilities exist. Namely, one can choose to proceed by analyzing the time series Y directly or by examining instead the properties of its ordered increments.

In what follows we discuss several simple techniques that can be used to analyze both the signal and its increments, that will be assumed to be invariant under time translations, based on the examination of the **scaling properties of their moments**. Readers should be aware that other popular techniques also exist to determine self-similarity. Some of them will be discussed in Chap. 4, since they are intrinsically related to the concept of memory.[44] In addition, we have collected some basic information about other popular techniques such as **detrended fluctuation analysis** (or **DFA**) in Appendix 2 and **wavelet analysis** in Appendix 3. DFA, in particular, has the nice feature of being applicable *to both the signal and its increments*.

3.4.1 Analysis of Non-stationary Processes

In principle, the detection of scale-invariance in a process reduces to the determination of whether its propagator resembles Eq. 3.67 for some suitable function Φ and some exponent H. We could do even less[45] and just try to ensure that the moments of the process scale according to Eq. 3.70, for some value of the self-similarity exponent H.

In any case, sufficiently good statistics will be necessary. Since a self-similar process is never stationary,[46] good statistics however require access to a sufficiently large number of realizations.[47] Often, this can be a problem. This can be partially ameliorated in the case that *the process is translationally invariant in time* and the data set is long. Then, one could try to partition the time record in several pieces and consider each of them as a different realization. We will also discuss how to implement this procedure in what follows.

---

[44]In particular, we will discuss techniques that determine self-similarity through the analysis of the **autocorrelation function** of the time process (Sect. 4.4.1), its **power spectrum** (Sect. 4.4.2) or the so-called **rescaled range (R/S) analysis** (Sect. 4.4.4).

[45]The determination of Φ is often useful in itself, since it allow us to classify the process under investigation even further, perhaps relating it to fBm (if Φ = Gaussian) or fLm (Φ = symmetric Lévy).

[46]One should always start by checking whether the provided dataset is stationary or not. If it is, the process cannot be self-similar, although its increments might be. We will provide some techniques to test stationarity later in this section.

[47]If the process is not stationary, temporal and ensemble averaging are no longer equivalent!

3.4.1.1 Moment Scaling

The easiest test for scale-invariance is to compute the *moments* of the time series for each value of time and see whether they scale with the elapsed time as a H-ss process would (Eq. 3.70). Referring to each available realization by the superscript r, that is assumed to run from 1 to M_r, one just needs to compute the ensemble average:

$$m_k^q := \langle |Y_k - Y_1|^q \rangle = \frac{1}{M_r} \sum_{r=1}^{M_r} |Y_k^{(r)} - Y_1^{(r)}|^q, \quad k = 1, 2, \cdots, N \qquad (3.105)$$

for $q > -1$. Here, N is the number of points available. If these moments scale as,

$$m_k^q \propto k^{e(q)}, \quad e(q) \simeq qH, \qquad (3.106)$$

for some exponent H, the data resembles that of an H-ss process. In theory, the scaling should hold for arbitrarily large q. This is not usually the case, though. Due to the finiteness of the record, the lack of statistics will make that the scaling eventually fails. Also, if the statistics of the data are Lévy-like, with tail-index α, the scaling should fail for $q \geq \alpha$.

Before showing some actual tests, it is worth discussing how to break a single time record into independent pieces to improve statistics. The idea is to take advantage of the fact that *only the first k points of the series* are used to evaluate Eq. 3.105 up to time k. The rest of the record is left unused. However, if we assume that the process is *invariant under temporal translations*, we could have also estimated that moment using the values of the process from the $(k+1)$-th to the $(2k)$-th iteration, or from the $(2k+1)$-th to the $(3k)$-th, and so forth. Therefore, one could produce more estimates of the same moment by breaking the data set into non-overlapping blocks of size k, for each value of k. Thus, we can better approximate the moment at time k by averaging over all of the blocks,

$$m_k^q := \frac{1}{M_r} \sum_{r=1}^{M_r} \left[\frac{k}{N} \sum_{b=1}^{N/k} |Y_{bk}^{(r)} - Y_{(b-1)k+1}^{(r)}|^q \right], \quad k = 1, 2, \cdots N, \qquad (3.107)$$

that remains meaningful even for $M_r = 1$, for $k \ll N$ since, clearly, block-averaging only helps to provide significant additional statistics as long as $N/k \gg 1$.

To illustrate the method, we apply it first to $M_r = 15$ independent realizations of a record with 10,000 values generated by a numerical algorithm that approximates fBm with any prescribed Hurst exponent (see Appendix 1). In this case, we use $H = 0.8$ (see Fig. 3.18). The first ten integer moments, computed using Eq. 3.107, are shown as a function of k in Fig. 3.19. It is apparent that, up to approximately $q \simeq 8$, the scaling predicted by Eq. 3.70 holds rather well, with $H \simeq 0.82 \pm 0.02$. As expected, the scaling deteriorates for higher moments due to the finiteness of

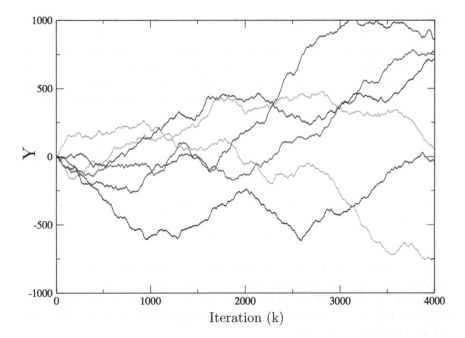

Fig. 3.18 Time traces of five of the fifteen realizations of fBm generated with a prescribed exponent $H = 0.8$ and used to estimate the Hurst exponent using the moment method

our temporal record, which misrepresents the statistics of events with very low probabilities (of the order of $1/N$ or lower). It is also apparent from Fig. 3.19 that all moments eventually deviate from the scaling for sufficiently large k due again to lack of statistics. The sooner, the higher q is.

Next, we carry the same exercise on a 10,000-long data series generated using a numerical algorithm that generates fLm with prescribed tail index and Hurst exponent (see Appendix 1). In this case, $\alpha = 1.5$ and $H = 0.8$ have been chosen (see Fig. 3.20). The moments obtained using Eq. 3.107 are shown in Fig. 3.21 for several real values of q between 1 and 3. The result is rather interesting, since only the for $0 < q < 1.5$ do the moments appear to scale with the expected Hurst exponent, $H = 0.78 \pm 0.03$. For larger values of q, the scaling exponent is a much lower $H \simeq 0.20 \pm 0.02$. As advertised earlier, this is a consequence of the lack of finite moments for $q \geq \alpha$ of the fLm with tail-index α. In theory, the fat tail makes the integral that defines C_q in Eq. 3.70 to become infinite. In practice, however, the divergence does not take place because the finiteness of the data record effectively truncates the integral to the largest value present in the part of record used to estimate the moment. Instead, C_q becomes a function of k since, as we increase the part of the record used by increasing k, even larger values will eventually appear that are never negligible in the case of Lévy statistics.

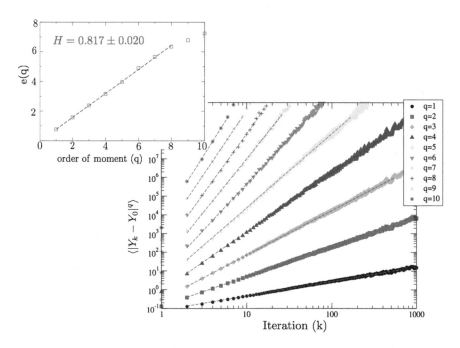

Fig. 3.19 First ten integer moments estimated using Eq. 3.107 using 15 independent realizations of fBm generated with a nominal Hurst exponent $H = 0.8$. **Above, left:** The exponents obtained from fitting the first eight moments satisfy $e(q) \sim (0.82 \pm 0.02) \cdot q$

3.4.1.2 Determination of Φ

In order to determine the shape of the function Φ that appears in Eq. 3.67 from the data, we proceed by computing the pdf of the displacement from the initial location reached by iteration k using all the realizations available. Since k is arbitrary, the statistics can be improved by including displacements for different values of k *after having rescaled them in time by multiplying each displacement by k^{-H}*, where H is the Hurst exponent obtained from the moment analysis. Since we are also assuming that the process is invariant under temporal translations, one could also break the record into non-overlapping blocks of size k, and treat them as independent realizations that can provide additional estimates for the displacement at time lapse k. The results of applying this procedure on both the generated fBm and fLm data are respectively shown in Figs. 3.22 and 3.23. We have used the temporal range $k \in [2, 500]$, as well as non-overlapping block averaging. The resulting pdfs, calculated using the CBC method (see Sect. 2.4), are very close to the Gaussian/Lévy pdfs prescribed at the generation of the synthetic data. The much larger significance of extreme events for the Lévy pdfs is apparent, however, in the fact that the tails are rather roughly resolved (see Fig. 3.23).

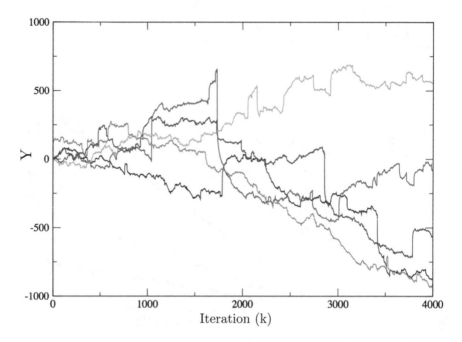

Fig. 3.20 Time traces of five of the fifteen realizations of fLm ($\alpha = 1.5$) generated with a prescribed exponent $H = 0.8$ and used to estimate the Hurst exponent using the moment method

3.4.2 Analysis of Stationary Processes

As it was mentioned previously, it is also possible determine whether a temporal process is scale-invariant by analyzing the properties of its ordered increments and testing whether they resemble those of fBm or fLm. The advantage of this approach is that, since the increments should be stationary, temporal and ensemble averaging are equivalent, which facilitates the estimation. To generate the **increments with unitary spacing** of the presumed self-similar time series, $\{Y_k,\ k = 1, 2, \cdots, N\}$, one just need to compute:

$$\Delta Y_{k+1}^{(1)} := Y_{k+1} - Y_k, \quad k = 1, 2, \cdots, N - 1. \tag{3.108}$$

The original signal could be recovered via the inverse transformation,

$$Y_k = \sum_{j=1}^{k} \Delta Y_j^{(1)}. \tag{3.109}$$

where we have defined $\Delta Y_1^{(1)} = Y_1$.

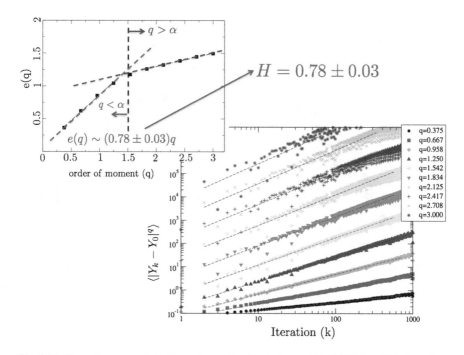

Fig. 3.21 Several q-moments between 1 and 3 estimated using Eq. 3.107 for 15 independent realizations of fLm ($\alpha = 1.5$) generated prescribing a nominal Hurst exponent $H = 0.8$. **Above, left:** The exponents obtained satisfy $e(q) = (0.78 \pm 0.03) \cdot q$, but only if restricted to $q < \alpha$

We will also need to generate **increments with a time spacing larger than one**. In analogy to how we defined them for fGn/fLn, we will compute them using:

$$\Delta Y_j^{(M)} := \frac{1}{M} \left(Y_{Mj+1} - Y_{M(j-1)+1} \right), \quad j = 1, 2, \cdots, N_b. \tag{3.110}$$

Here, $N_b = N/M$ is the number of non-overlapping blocks of size M in the series. With the help of Eq. 3.109, they can also be expressed as,

$$\Delta Y_j^{(M)} = \frac{1}{M} \sum_{k=M(j-1)+1}^{Mj} \Delta Y_{k+1}^{(1)}, \quad j = 1, 2, \cdots, N_b. \tag{3.111}$$

Thus, the increment of spacing M is equal to the *non-overlapping block-average of the original increment series using blocks of size M*.

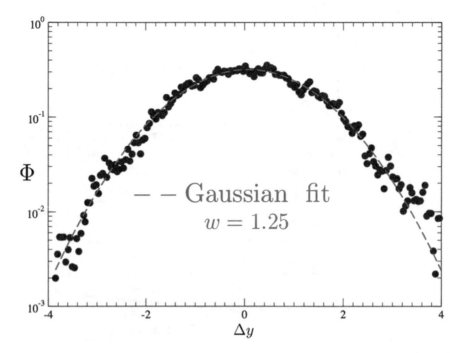

Fig. 3.22 Estimate of Φ (Eq. 3.67) for the generated fBm process. A Gaussian with width $w \simeq$ 1.25 is obtained, that should be compared with the prescribed Φ in generation, that was a Gaussian with $w \simeq 1.248$. Since the probability of large displacements decrease very quickly for fBm, the 15 realizations used are already sufficient to resolve the function quite clearly

3.4.2.1   Test of Stationarity

Once the increments have been computed for arbitrary spacing, one needs to confirm that they are indeed stationary.[48] The reason is clear, if ΔY is not stationary, Y cannot be H-sssi. However, testing for stationarity can be rather involved. Although there

---

[48]Sometimes, experimental data can be "contaminated" by slow-varying trends that mask the stationarity of a fast-varying component. For instance, one could have an stationary process superimposed with one or several periodic processes. Think, for instance, of fluctuations in a turbulent fluid which is itself rotating! In order to be able to check for the stationarity of the faster process (or the self-similarity of its integrated process), the periodic trend must be removed first. This could be done by applying some (high-pass) digital filter in order to remove the lower frequencies associated to the rotation. In other cases, the trend is due to a slowly varying external drive. This often happens when measuring turbulent fluctuations in a tokamak reactor while the plasma profiles are ramping up or down. In this case, removing the trend requires the subtraction of the *running-average* of the data, that should be calculated using a box with a size of the order of the characteristic time of the drive variation. In general, it may be difficult to differentiate between actual trends and features of the process of interest. This is, for example, one of the limitations of the DFA technique (see Appendix 2). As suggested by the previous examples, some knowledge of the dynamics is usually needed to guide our hand when dealing with trends.

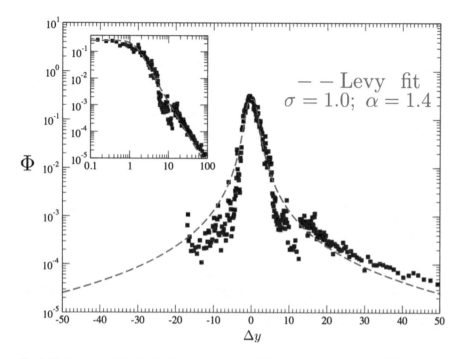

Fig. 3.23 Estimate of Φ (Eq. 3.67) for the generated fLm process. A symmetric Lévy with tail index $\alpha \simeq 1.4$ provides the best fit, whilst the prescribed Φ in generation was a Lévy law with tail-index $\alpha = 1.5$. The small number of realizations ($M_r = 15$) is the main cause for not having any that explored the left tail more deeply

are some formal tests,[49] we will assume that, for our purposes, the data is stationary if one can show a reasonable independence with time of both mean and variance. In order to check this, we proceed by splitting the data in $1 \ll N_b \ll N$ blocks of size $M = N/N_b$, and constructing the mean within each block:

$$\mu_b = \frac{1}{M} \sum_{k=1}^{M} Y_{(b-1)M+k}, \quad b = 1, 2, \cdots N_b \tag{3.112}$$

and its variance:

$$\sigma_b^2 = \frac{1}{M} \sum_{k=1}^{M} \left(Y_{(b-1)M+k} - \mu_b \right)^2, \quad b = 1, 2, \cdots N_b. \tag{3.113}$$

---

[49]For instance, one could mention the Dickey-Fuller test, the Kwiatkowski-Phillips-Schmidt-Shin test or the Phillips-Perron test, that are widely used in fields such as econometrics [46]. Most of them are however applicable only to Gaussian random processes.

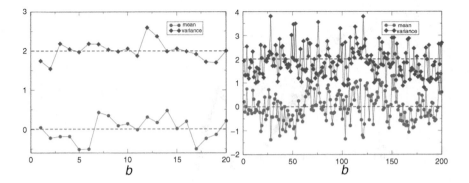

Fig. 3.24 Block-averaged means and variances calculated for one 10,000-point realization of fBm with $H = 0.8$. **Left:** using $N_b = 20$ ($M = 500$); **right:** using $N_b = 200$ ($M = 40$). Theoretical values for the mean ($= 0$) and variance ($= 2$) are shown in dashed lines

Often, a simple plot can then be used to test for (a reasonable) stationarity of the data. We have illustrated the procedure by computing mean and averages, for two different block sizes, for the non-overlapping block partition of one of the fBm realizations[50] used previously (Fig. 3.24).

3.4.2.2   Moment Scaling

Once the stationarity of the ΔY series has been established, we can test whether it behaves as the series of increments of an H-sssi process would. It is sufficient with proving that the moments of the one-point pdf of the ΔY series, upon rescaling, satisfy Eq. 3.87. Or, in other words, that for $q > 0$,

$$\frac{\langle |\Delta Y^{(M)}|^q \rangle}{\langle |\Delta Y|^q \rangle} \simeq \frac{N_b^{-1} \sum_{i=1}^{N_b} |\Delta Y_i^{(M)}|^q}{N^{-1} \sum_{i=1}^{N} |\Delta Y_i|^q} = M^{e'(q)}, \tag{3.114}$$

and that $e'(q) = q(H - 1)$, for some value of the exponent H. The advantage of dealing with increments, instead of with the integrated process, becomes now clear. Although the brackets in Eq. 3.114 would require to perform an ensemble average at time k, the stationarity of the increments have allowed us to replace the ensemble average by a temporal average over all available blocks. Thus, we can carry out the moment calculations using just one (sufficiently long) realization.

To illustrate the method, we apply it first to one of the 10,000 point long realizations of fBm that were used previously, and that were generated with $H = 0.8$ (see Fig. 3.18). The increment series ΔY^M are constructed according to Eq. 3.111.

---

[50]In the case of fLm, one should always apply the stationarity test on a finite moment. That is, using $\langle |x|^q \rangle$ with $0 < q < \alpha$!

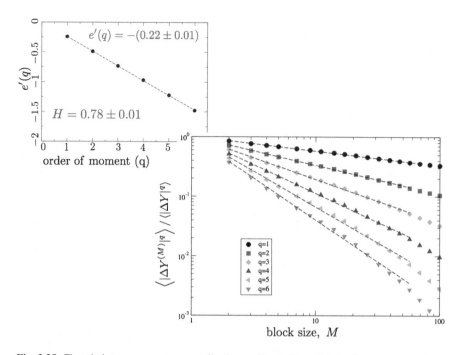

Fig. 3.25 First six integer moments, normalized according to Eq. 3.114, for the increments of one single realization of fBm with 10,000 values generated with a nominal Hurst exponent $H = 0.8$. **Above, left:** The exponents obtained from fitting the first six moments satisfy $e'(q) \sim (0.22 \pm 0.01) \cdot q$, which yields a Hurst exponent $H = 0.78 \pm 0.01$

The ratios that appear in Eq. 3.114 have been evaluated for the first six integer moments. They are shown in Fig. 3.25 as a function of the block size M. The scaling predicted by Eq. 3.114 holds rather well,[51] with $e'(q) \simeq (0.22 \pm 0.01)q$, that gives a Hurst exponent $H = 0.78 \pm 0.01$, quite close to the nominal value. This proves that the integrated process is, indeed, H-sssi.

Next, we apply the same method to one of the 10,000-point long realizations of fLm generated with $\alpha = 1.5$ and $H = 0.8$. Again, we construct the increments $\Delta Y^{(M)}$ and calculate the ratios in Eq. 3.114, but this time for a range of real q values between 1 and 3. The results are shown in Fig. 3.26. They prove that the integrated process is indeed H-sssi. As was the case with fLm (see Fig. 3.21), only the moments satisfying $q < \alpha$ scale adequately. In that range, we obtain an exponent $e'(q) = 0.24 \pm 0.03$, that corresponds to a Hurst exponent of $H = 0.76 \pm 0.03$, pretty close to the nominal one.

---

[51]The scaling however begins to deteriorate, in this case, for moments higher than 8, due to the lack of sufficient statistics of the Gaussian tail within the 10,000 points available.

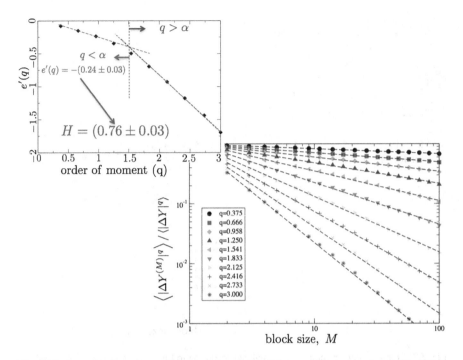

Fig. 3.26 Several real moments of orders between 1 and 3, normalized according to Eq. 3.114, for the increments of one single realization of fLm with 10,000 values generated with a tail-index $\alpha = 1.5$ and a nominal Hurst exponent $H = 0.8$. **Above, left:** The exponents obtained satisfy $e'(q) = (0.24 \pm 0.03) \cdot q$ only if restricted to $q < \alpha$. The resulting Hurst exponent is $H = 0.76 \pm 0.03$

3.4.2.3 Determination of the One-Point Pdf

The determination of the Φ function that defines the propagator of the integrated process (Eq. 3.67) becomes quite simple when using the increment series, since it coincides with the one-point pdf of the $M = 1$ increments (see Eq. 3.85). That is, the pdf of ΔY. The results of calculating it for one realization of the generated fBm and fLm series, using the CBC method (see Sect. 2.4), are shown in Figs. 3.27 and 3.28. A Gaussian is clearly identified for the fBm case, and a symmetric Lévy with $\alpha \simeq 1.4$ for the fLm case, that is very close to the nominal value ($\alpha = 1.5$).

3.4.3 Multifractal Analysis

It is sometimes observed that, while estimating the Hurst exponent H of a process suspected of being self-similar by means of its moments,[52] the exponent $e(q)$

---

[52]Or, when examining the behaviour of the moments of its increments when the spacing is varied, one finds that $e'(q) \neq qH$, contrary to what Eq. 3.114 predicts for a self-similar process.

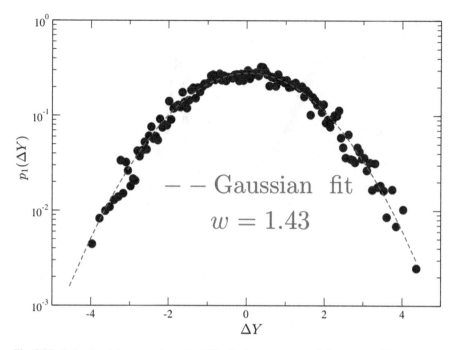

Fig. 3.27 Estimate of the one-point pdf of fGn for the increments of the generated fBm process (i.e., $h = 1$). A Gaussian with width $w \simeq 1.43$ is obtained, that should be compared with the theoretically generated, a Gaussian with $w = \sqrt{2}$

deviates from the linear relation, $e(q) = q(H - 1)$ that self-similarity dictates (Eq. 3.106). In some cases, this unexpected q-dependence can be attributed, as we did in the previous subsection, to poor statistics or the lack of finite moments above a certain value of q. In other cases, however, it may be a reflection of the fact that the underlying dynamics are governed by a combination of processes whose importance varies over time, thus producing a seemingly erratic succession of singularities that do not correspond to a single Holder exponent.

We illustrate how to quantify the degree to which this happens by means of the multifractal analysis (see Sect. 3.3.7) in this section. In particular, we will determine both the family of multifractal exponents $\xi(q)$ and the singularity spectrum, $f(h_s)$, of several time series. We proceed as follows. First, we divide the N-point signal in N_b non-overlapping blocks of size $M = N/N_b$. To each block we assign a weight, given by (Eq. 3.95):

$$w_i = \left| \frac{Y_{Mi+1} - Y_{M(i-1)+1}}{M} \right| = \left| \Delta Y_i^{(M)} \right|, \quad i = 1, \cdots N_b, \tag{3.115}$$

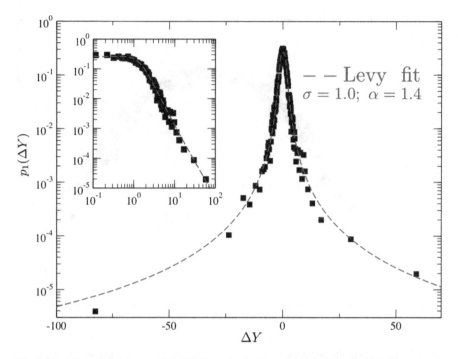

Fig. 3.28 Estimate of the one-point pdf of fLn for the increments of the generated fLm process. A symmetric Lévy pdf with tail-index $\alpha \simeq 1.4$ is obtained, that should be compared with the theoretically generated, a symmetric Lévy pdf with $\alpha = 1.5$. The pdf is now smoother than in the analysis for fLm done previously, in spite of the fact that we are using just one realization

where we have introduced the series of the block-averaged increments of the signal (Eq. 3.111). The partition function (Eq. 3.96) is then built as:

$$N(M_k, q_j) = \frac{1}{N_{b,k}} \sum_{i=1}^{N_{b,k}} |w_i|^{q_j} = \frac{1}{N_{b,k}} \sum_{i=1}^{N_{b,k}} \left| \Delta Y_i^{(M_k)} \right|^{q_j} \propto M_k^{\xi(q_j)}, \qquad (3.116)$$

where the indices k and j run to produce all the different choices for M and q. Similarly, $N_{b,k} = N/M_k$. The family of multifractal exponents $\xi(q)$ is estimated by performing a least-squares-fit for each q_j, concentrated on the smaller values of M. Then, we find the dominant singularity for each q_j with the help of Eq. 3.102, that in discretized form becomes:

$$h_{s,j} = 1 + \frac{\xi(q_{j+1}) - \xi(q_{j-1})}{q_{j+1} - q_{j-1}}. \qquad (3.117)$$

Finally, the fractal dimension of the space occupied by those instants in the temporal axis where the Holder exponent is $h_s = h_{s,j}$, is given by the discrete version of Eq. 3.103:

$$f(h_{s,j}) = 1 + q_j \left[h_{s,j} - 1 \right] - \xi(q_j). \qquad (3.118)$$

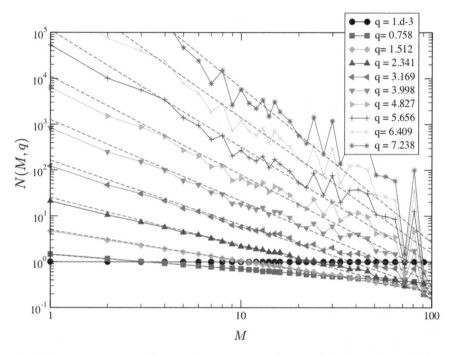

Fig. 3.29 Partition function $N(M, q > 0)$ (Eq. 3.116) estimated for a realization of a synthetic fBm process generated with a nominal $H = 0.8$. The fits used to estimate the exponents $\xi(q)$ are shown in dashed red lines

We illustrate the procedure by applying it to the 10,000-point realizations of synthetic fBm, with nominal exponent $H = 0.8$ that we used previously. The resulting partition function, for selected values of $q > 0$ and M, is shown in Fig. 3.29. It is clear from the figure that, the larger q is, the more visible the lack of statistics at the tail becomes. The multifractal exponents $\xi(q)$ are then obtained by fitting the resulting partition function, according to Eq. 3.116, for each value of q_j (see Fig. 3.30). The obtained multifractal exponents looks rather linear for $q > -1$, as expected for a monofractal process like fBm. The generalized Hurst exponent, calculated using Eq. 3.98, is almost constant in that region and equal to $H \simeq 0.8$. However, for $q < -1$ the q-scaling changes completely, a consequence of the divergence nature of those moments for the Gaussian distribution. In fact, to be on the safe side, it is always advisable to restrict the multifractal analysis for $q > 0$.

This idea is reinforced by calculating the singularity spectrum for the synthetic fBm (see Fig. 3.31). If all q values are included, one finds a very broad spectrum that does not look monofractal. However, if only $q > -1$ are used, the resulting spectrum is much more compact around $H \simeq 0.8$. In fact, the majority of the spectrum is concentrated below that value, with just a few values scattered to the right of it, that interestingly correspond to $-1 < q < 0$. Had we followed our own advice, with

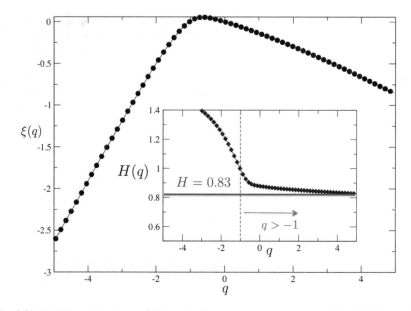

Fig. 3.30 Multifractal exponents $\xi(q)$ obtained by performing power-law fits on $N(M, q)$ (like those shown in dashed red in Fig. 3.29) for the synthetic fBm with nominal $H = 0.8$. As seen in the inset, the generalized Hurst exponent $H(q) = 1 + \xi(q)/q \sim 0.8$ for $q > 0$, as expected. However, the behaviour of $H(q)$ for $q < -1$ is rather different, being a consequence of the lack of finite negative moments of the normal distribution for $q \leq -1$

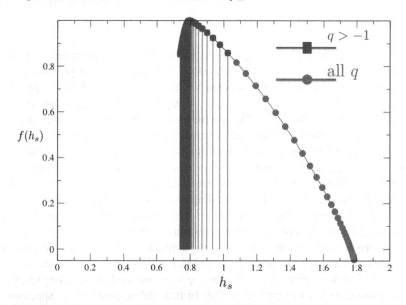

Fig. 3.31 Singularity spectrum of the synthetic fBm with $H = 0.8$. A broad spectrum is obtained (shown in red) if all q values are used. However, if only $q > -1$ are used, the spectrum (in blue) piles against $H = 0.8$ from below, with a width of about $\sigma_h \sim 0.03$–0.05. The few blue points that correspond to $-1 < q < 0$ are those few scattered to the right of $H \sim 0.8$

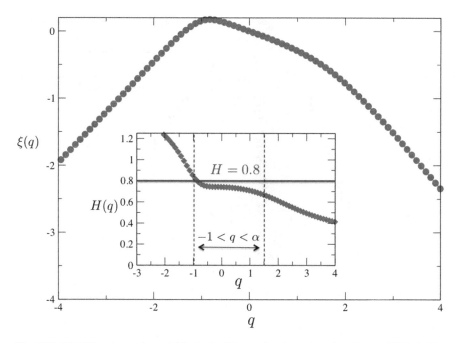

Fig. 3.32 Multifractal exponents $\xi(q)$ obtained by performing power-law fits on $N(M, q)$ (like those shown in dashed red in Fig. 3.29) for the synthetic fLm with nominal $H = 0.8$. As seen in the inset, the generalized Hurst exponent $H(q) = 1 + \xi(q)/q \sim 0.8$ for $0 < q > \alpha$, as expected. However, the behaviour of $H(q)$ for $q < -1$ or $q > \alpha$ is rather different, being a consequence of the lack of finite negative moments for $q \leq -1$, or moments of order $q \geq \alpha$ of any Lévy distribution

calculations restricted to $q > 0$, the resulting spectrum would have had a width of about (0.03–0.05) to the left of $H = 0.8$.

We have also performed the same analysis on one of the fLm synthetic realizations generated with $\alpha = 1.5$ and $H = 0.8$. The resulting family of multifractal exponents is shown, for positive and negative q's, in Fig. 3.32. It is apparent that $\xi(q)$ follows a linear scaling, within the range of q-values for which the symmetric Lévy has finite moments ($-1 < q < \alpha$), and yields a Hurst exponent $H \sim 0.8$ that is consistent with the nominal value. This is illustrated in the inset, that shows the generalized Hurst exponent. However, both for $q < -1$ and $q \geq \alpha$, the scaling of the multifractal exponents changes and the associated generalized Hurst exponent departs from the nominal value. As with the case of fBm, this misbehaviour is associated with the divergent moments of the Lévy for those values of q. This state of things is also apparent in the singularity spectrum for fLm, shown in Fig. 3.33. Again, a broad spectrum is obtained (in red) when all q-values are used. If the allowed values are restricted to $-1 < q < \alpha$, the spectrum is much narrower and reminiscent of what we found for fBm. In particular, it piles from below against $H \sim 0.8$, with a few points scattered above $H = 0.8$ that correspond to $-1 < q < 0$. The width of the spectrum is, in this case, larger than for fBm, about (0.1–0.2).

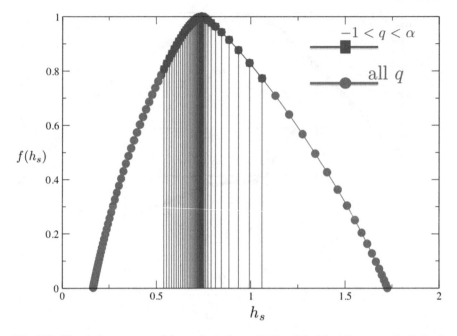

Fig. 3.33 Singularity spectrum of the synthetic fLm with $H = 0.8$. A broad spectrum is obtained (shown in red) if all q values are used. However, if only $-1 < q < \alpha$ are used, the spectrum (in blue) piles against $H = 0.76$ from below, with a width of about $\sigma_h \sim 0.03$–0.05. The few blue points that correspond to $-1 < q < 0$ are those few scattered to the right of $H \sim 0.76$

3.5   Case Study: The Running Sandpile

We continue with the characterization of the sandpile started in previous chapters, but focusing now on another popular diagnostic: the total amount of sand contained in the sandpile. We will refer to this quantity as its *global mass* or *total mass*. It is defined, at any given iteration k, as $M_k = \sum_{i=1}^{L} h_i^k$, where L is the number of sandpile cells and h_i^k the height of the i-th cell at iteration k. We will use a sandpile with $L = 10^3$, critical slope $Z_c = 200$, toppling size when unstable $N_F = 20$, rain probability $p_0 = 10^{-4}$ and rain size $N_b = 10$. It must be noted that the figure-of-merit $(p_0 L)^{-1} \sim 10 \ll L$, meaning that the probability of avalanche overlapping is larger than for the sandpile we examined in Sect. 2.6 (see Problem 3.7).

In what follows, we will be interested in determining the scale-invariant characteristics (if any) of the global mass M. Since we are in the sandpile steady state, the mean and other moments of M will be time invariant. In order to construct a time series that could be similar to a fractional noise, we first subtract the mean value from the mass time series, and then divide it by its standard deviation. The resulting time series, $\Delta M/\sigma$, is shown in Fig. 3.34 for 10^7 iterations. Clearly, it is stationary, symmetric around zero, and its time trace seems rather irregular, with a structure apparently rich in singularities.

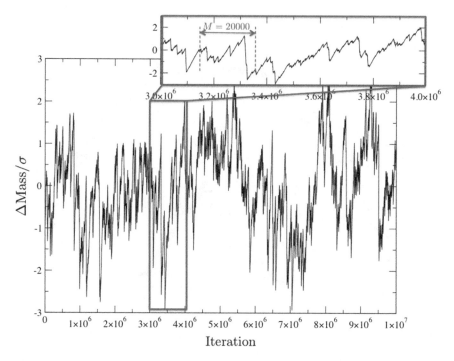

Fig. 3.34 Increments in the total amount of sand (normalized to its variance) contained in a sandpile ($L = 10^3$, $N_F = 20$, $Z_c = 200$, $p_0 = 10^{-4}$ and $N_b = 10$) as it is advanced for 10^7 iterations after saturation. **Inset:** mass increments for a period of 10^6 iterations. The extension of the smallest block size considered, $M = 20,000$ is marked for reference purposes

In order to investigate whether this record of mass increments shares the properties of either fGn or fLn, we will apply the methods discussed in Sect. 3.4.2. It is interesting to note, before starting, that a closer look shows that the singularities all appear to be very similar in shape, suggesting that the signal might be close to a monofractal (see the inset in Fig. 3.34).

Most interestingly, the time series only appears to experience significant variations over periods longer than a few thousand iterations. For smaller times, the signal seems pretty continuous. This observation provides a nice illustration of something that is often found in natural systems, in contrast to both the fGn and fLn mathematical models. Meaningful scales are limited from below and from above in a natural systems, while fGn and fLn exhibit self-similarity for all scales instead. In the sandpile the minimum scale is set by the cell size; the maximum, by the sandpile size. It is only in between these scales that self-similarity can be established. We often refer to this intermediate range of scales as the **self-similar mesorange** (see also the discussion in Sect. 1.3.2). A mesorange will also exist for time scales. In fact, *the boundaries of the mesorange will depend on the quantity we examine, even if all of them are measured in the same system.*

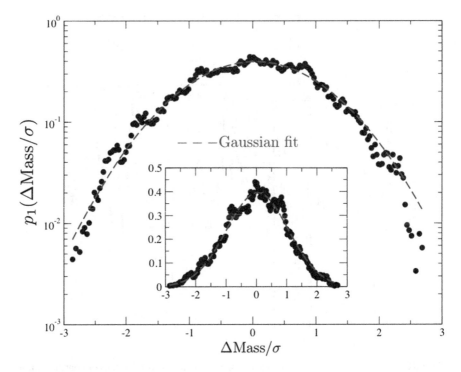

Fig. 3.35 Estimate of the one-point pdf of the normalized increments of sandpile global mass. A Gaussian with width $w \simeq 1.01$ is obtained. The value of the width is not surprising, since increments were normalized to their variance before starting the analysis

The existence of a mesorange becomes apparent after applying the methods described in Sect. 3.4.2 to the time series of the normalized, demeaned global mass increments. First, we have computed the Φ function (see Eq. 3.67) for the $\Delta M/\sigma$ time series. It turns out to be well approximated by a Gaussian (see Fig. 3.35). Since all positive moments of the Gaussian exist, we have then examined the rescaling properties of the moments of the block-averaged mass increments, using blocks of size M (Eq. 3.114). The first ten integer moments are shown in Fig. 3.36. The self-similar range is apparent in this figure, that does not show straight lines for all M (compare with the fBm case, shown in Fig. 3.26), but only within the range of M values that lies approximately between (30,000–800,000) iterations. Within this range, a power-law fit yields $e'(q) \sim (-0.17 \pm 0.02)q$, from which a value of the self-similarity exponent $H \sim 0.83 \pm 0.02$ can be inferred. Had this procedure been applied instead over the range of (1–30,000) iterations, we would have missed the mesorange and the results would have been rather different (see Problem 3.8).

It thus seems that, across the mesorange that extends in this case between (30,000–800,000) iterations, the increments of the global mass of the sandpile resemble fGn with a Hurst exponent $H \sim 0.83$. We can check whether such conclusion is robust by applying the techniques discussed in Sect. 3.4.3 to test for

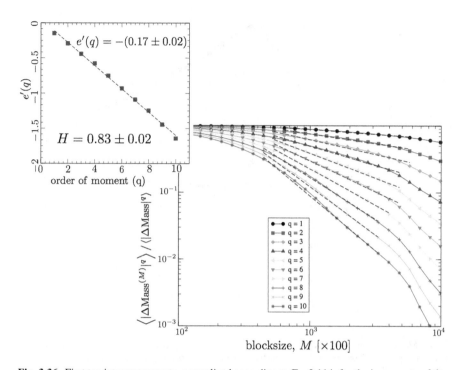

Fig. 3.36 First ten integer moments, normalized according to Eq. 3.114, for the increments of the sandpile total mass. **Above, left:** The exponents obtained from fitting the first ten moments over the range $M = 40,000–600,000$ satisfy $e'(q) \sim -(0.17 \pm 0.02) \cdot q$, which yields a Hurst exponent $H = 0.83 \pm 0.02$

multifractal behaviour. The structure function, $N(M, q)$ is shown in Fig. 3.37. It has been computed according to Eq. 3.116, although restricting the values of M to those that lie within the self-similar mesorange. It is apparent that the scaling behaviour of the structure function for $q > 0$ seems pretty well described by a power-law, similarly to what happened to fBm (see Fig. 3.32). The computation of the generalized Hurst exponent (Eq. 3.98) yields a rather uniform value for the exponent $H(q > 0) \sim 0.82–0.84$, that is also consistent with monofractal behaviour.

Finally, we have also computed the singularity spectrum of the sandpile mass increments by using Eqs. 3.117 and 3.118. Calculations are again restricted to the mesorange (i.e., $30,000 < M < 800,000$) and for $q > 0$. The result is that the singularity spectrum of the time series is concentrated between $H = 0.81–0.87$ (see Fig. 3.38). The width of the spectrum obtained is rather narrow and very similar to that found for fBm. These results reinforce our previous conclusion regarding the monofractal character of the time trace of the increments of the sandpile mass over the mesorange.

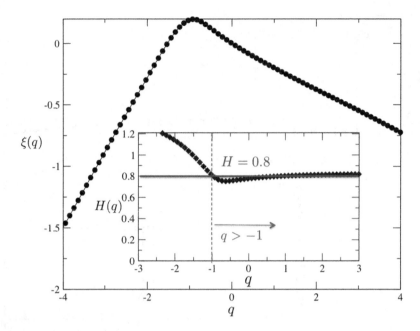

Fig. 3.37 Family of multifractal exponents $\xi(q)$ obtained by performing power-law fits on the partition function $N(M, q)$ of the global mass increments of the sandpile over the range $M = 40{,}000$–$600{,}000$. As seen in the inset, the generalized Hurst exponent $H(q) = 1 + \xi(q)/q \sim 0.82$–$0.84$ for $q > 0$. The behaviour of $H(q)$ for $q < -1$ is to be ignored, due to the lack of finite moments of the Gaussian for $q < -1$

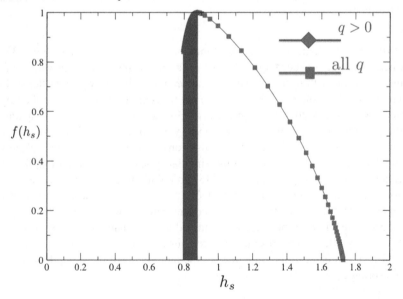

Fig. 3.38 Singularity spectrum of the increments of the total mass of the sandpile. By keeping only the Holder exponents for $q > -1$, the spectrum (in blue) piles against $H = 0.87$ from below, with a width of about $\sigma_h \sim 0.04$–0.07. The spectrum obtained when all q are included is shown for comparison

3.6 Final Considerations

In this chapter we have discussed the basic features of self-similarity and scale-invariance, both in spatial objects and in time random processes. Several methods have also been presented that can be used to estimate self-similarity exponents, all of them based on the comparison of the data of interest with exact monofractal mathematical models. Additional methods to estimate self-similarity exponents will be presented in Chap. 4, since self-similarity is also intrinsically connected to the idea of **memory** in complex systems. Finally, we have also shown that the "monofractality hypothesis" may be challenged in practice by means of several multifractal analysis techniques.

However, one of the most important concepts discussed in this chapter is that of the **mesorange**. In real systems, as illustrated by the running sandpile, self-similarity will always be limited by finite size effects. This will happen both at the smallest scales (by the properties of the individual components of the system) and at the largest scales (by the system size, lifespan, time to transverse the system, etc.). This fact should never be overlooked when examining data from real systems, specially while comparing them against mathematical models in which the scaling properties are not limited.

Appendix 1: Numerical Generation of Fractional Noises

Various algorithms have been proposed in the literature to generate numerical time series that approximate either fGn, or fLm with arbitrary tail-index α, for a prescribed Hurst exponent H [36]. Here, we will discuss one that is based on rewriting fBm (Eq. 3.71) and fLm (Eq. 3.89) in the form,

$$y^{H,\alpha}(t) = y^{H,\alpha}(t_0) + {_{t_0}}D_t^{-(H+1/\alpha-1)}\xi_\alpha. \tag{3.119}$$

This equation introduces a new type of operators known as *fractional operators* [47]. Equation 3.119 will be formally introduced in Chap. 5 under the name of the **fractional Langevin equation** (see Sect. 5.3.2). It is straightforward to show that it reduces to fBm for $\alpha = 2$, and yields all symmetric fLms if $\alpha < 2$.

The operator that appears in the fractional Langevin equation is known as a *Riemann-Liouville fractional operator*. They are integro-differential operators defined as [47],

$$_aD_t^p f(t) = \begin{cases} \dfrac{1}{\Gamma(k-p)} \dfrac{d^k}{dt^k} \left[\displaystyle\int_a^t dt'(t-t')^{k-p-1}f(t')dt' \right], & p > 0 \\[4mm] \dfrac{1}{\Gamma(-p)} \left[\displaystyle\int_a^t dt'(t-t')^{-(p+1)}f(t')dt' \right], & p < 0 \end{cases} \tag{3.120}$$

where k is the integer satisfying $k-1 \leq p < k$. They are called *fractional derivatives* if $p > 0$, and *fractional integrals* if $p < 0$. An introduction to the basic features of fractional operators[53] will be given in Appendix 2 of Chap. 5. For now, it suffices to say that fractional operators have many interesting properties. Among them, that for $p = \pm k$, they coincide with the usual k-th order derivatives (+) or integrals (−) and, for $p = 0$, with the identity. They provide interpolations between the usual integrals and derivatives of integer order. Fractional operators were first introduced by Gottfried von Leibniz, as early as 1697, although the form given in Eq. 3.120 was not introduced until the middle of the eighteenth century by Bernhard Riemann and Joseph Liouville.

We focus now on the properties of fractional operators that will allow us to generate synthetic fBm and fLm data, following a method introduced by A.V. Chechkin and V. Yu. Gonchar [48]. The first property is that, when acted on the right by a normal derivative, they satisfy that (see Appendix 2 of Chap. 5, Eq. 5.135):

$$\frac{d^m}{dt^m} \cdot {_a}D_t^p f(t) = {_a}D_t^{p+m} f(t), \tag{3.121}$$

for any positive integer m. Since fGn/fLn is essentially the derivative of fBm/fLm, fGn/fLn can be obtained by applying a normal derivative to Eq. 3.119 to get:

$$\Delta y^{H,\alpha} := \lim_{h \to 0} \Delta_h y^{H,\cdot} = \frac{dy^{H,\alpha}}{dt} = {_{t_0}}D_t^{-(H+1/\alpha-2)} \xi_\alpha. \tag{3.122}$$

The second property of interest has to do with the relation between the Fourier transforms (see Appendix 1 of Chap. 2 for an introduction to Fourier transforms) of a function and its fractional derivative/integral (see Appendix 2 of Chap. 5, Eq. 5.139)[54]:

$$\mathrm{F}\left[{_a}D_t^p f(t)\right] \simeq (-\imath\omega)^p \hat{f}(\omega), \tag{3.123}$$

where $\imath = \sqrt{-1}$ and ω stands for the frequency. Applying the Fourier transform to Eq. 3.122, one thus obtains:

$$\Delta\hat{y}^{H,\alpha}(\omega) = \frac{\hat{\xi}_\alpha(\omega)}{(-\imath\omega)^{H+1/\alpha-2}}. \tag{3.124}$$

---

[53] We will discuss these integro-differential operators and their physical meaning at length in Chap. 5, since they play an important role in the theory of **fractional transport**.

[54] Equation 3.123 is only exact if $a = -\infty$ [47]. However, we will 'abuse' the formula and assume it as valid when $a = t_0$, so that we can work out a reasonable algorithm to generate fGn/fLn series.

Thus, in order to generate a synthetic fGn/fLn series with arbitrary Hurst exponent H and tail-index α,[55] a possible procedure is simply to (see Problem 3.5):

- generate a random Gauss/Lévy noise sequence (see Appendix 2 of Chap. 2);
- carry out a discrete Fast Fourier transform (FFT) of the noise series;
- divide each Fourier harmonic by the corresponding factor $(-\imath\omega)^{H+1/\alpha-2}$;
- and Fourier invert the result to get the desired, approximated fGn/fLn series.

In addition, fBm/fLm synthetic series with arbitrary exponents H and α can easily be obtained by numerically integrating the fGn/fLn series for the same exponents that have been generated with this method. Or, by reusing the procedure just described but dividing instead the Gauss/Lévy noise series by $(-\imath\omega)^{H+1/\alpha-1}$.

Appendix 2: Detrended Fluctuation Analysis

Detrended fluctuation analysis (or DFA) [49, 50] is a popular method that can be applied to test for self-similarity *in both stationary and non-stationary signals.* In order to introduce the method, let's consider the series,

$$y = \{y_n, \quad n = 1, 2, \cdots, N\}, \tag{3.125}$$

that might be stationary or not. The method considers first the associated *integrated motion,*

$$Y_n = \sum_{i=1}^{n} y_i - n\bar{y}, \tag{3.126}$$

where the overall average of the motion, $\bar{y} = \sum y_i/N$ is removed.

Next, for every possible scale $l > 0$, the integrated motion is divided in (possibly overlapping) windows of size l. Inside each of these windows, a local least-squared-fit to a polynomial is done to capture the local trend. That is, if a possible linear local trend is assumed,[56] the fit would be against a straight line, $my + b$. The total squared error, χ^2, for each window k is then given by:

$$\chi_k^2 = \frac{1}{l} \sum_{i \in W_k} (Y_i - im_k - b_k)^2, \quad k = 1, N/l. \tag{3.127}$$

---

[55] All synthetic fGn/fLn and fBm/fLm series used in this chapter have been produced in this way.

[56] There are different orders of DFA, distinguished by the order of the polynomial used to remove the local trend. The one we just discussed is called DFA1, since it uses linear fits. DFAn, instead, uses polynomials of order n.

where W_k stands for the k-th window. The **fluctuation value at scale** l, $F(l)$, is given by the average of the squared-root of the error over all windows W_k of size l:

$$F^2(l) := \frac{l}{N} \sum_{k=1}^{N/l} \chi_k^2. \tag{3.128}$$

If the underlying process is self-similar once the trends have properly been removed,[57] it should happen that $F(l) \sim l^a$, for some exponent a. The interesting thing is that, if DFA is applied to fractional Gaussian noise, that is stationary, it is found that

$$F(l) \sim l^H, \tag{3.129}$$

for all scales $l >> \delta$, with H being the Hurst exponent that defines the associated fBm. But, more interestingly, DFA could also have been applied directly to the fractional Brownian motion signal itself, which is non-stationary. Then, one would find that,

$$F(l) \sim l^{H+1}, \tag{3.130}$$

so that the Hurst exponent can be obtained directly from fBm.

DFA can also be extended to probe for multifractality. The procedure is often referred to as M-DFA [51]. The twist here is to calculate, instead of the squared error per window, the quantity,

$$F_q(l) = \left[\frac{l}{N} \sum_{k=1}^{N/l} |\chi_k^2|^{q/2} \right]^{1/q}. \tag{3.131}$$

The generalized Hurst exponent is then defined by assuming the scaling,

$$F_q(l) \propto l^{H(q)}, \tag{3.132}$$

that, for a monofractal[58] (i.e., fGn), reduces to $H(q) = H$, $\forall q$. The same techniques that were discussed in Sect. 3.3.7 to the estimation of the multifractal spectrum can be used here on the $H(q)$ obtained from the application of MDFA.

---

[57]The removal of the hidden trends is where the subtlety of applying DFA correctly lies (see Problem 3.9). Since in most cases trends are unknown, one could remove more (or less) than the physically meaningful trends, thus affecting the actual process under examination. As it is the case of all the other methods in this book, DFA should be handled with care. Some knowledge of the underlying physics is always needed in order to be able to tell whether any trend that is removed is actually meaningful and not a feature of the process under study.

[58]Most DFA practitioners do not consider fLn to be a monofractal either, since M-DFA yields that $H(q) \neq q$ for $q \geq \alpha$. For instance, for fLm with $H = 1/\alpha$ it is found that $H(q) = 1/\alpha$ for $q < \alpha$ and $H(q) = 1/q$ for $q \geq \alpha$ [51].

Appendix 3: Multifractal Analysis Via Wavelets

Wavelets are another powerful technique to test for multifractal behaviour in time series. We will not discuss the basics of wavelets at any length, since there are some wonderful review papers and books available that discuss them much better than what we could do here [44, 45, 52]. Instead, we provide just a brief introduction to the topic, sufficiently long to clarify their relevance to the investigation of multifractal features (see also Problem 3.10).

Wavelets were introduced in the mid 1980s as an extension of Fourier analysis that permitted the examination of local properties in real time [53]. That is, while Fourier analysis expresses an arbitrary function as a linear combination of sines and cosines that are localized in frequency but not in time, the wavelet representation expresses the function as a linear combinations of rescaled versions of a prescribed basis function Φ, also called *wavelet*, that is localized *both in frequency and in time*.

The Φ-wavelet transform of a function, $f(t)$, at a scale r and time t is defined by the integral [45]:

$$\hat{f}_\Phi(r, t) = C^{-1/2} r^{-1/2} \int_{-\infty}^{\infty} \Phi\left(\frac{t - t'}{r}\right) f(t') dt', \tag{3.133}$$

where the constant C is defined by:

$$C := \int_{-\infty}^{\infty} \frac{|\hat{\Phi}(\omega)|^2}{|\omega|} d\omega, \tag{3.134}$$

where $\hat{\Phi}(\omega)$ is the Fourier transform (in time) of the wavelet basis function. Clearly, $C < \infty$, if Φ is to provide a valid wavelet, which requires that $\hat{\Phi}(0) = 0$ (or, in other words, that Φ has a zero mean). $\hat{f}_\Phi(r, t)$ can be interpreted as the part of $f(t)$ that contributes at time t to the scale r. A nice property of $\hat{f}_\Phi$ is that, if:

$$|f(t + r) - f(t)| \sim r^\alpha, \quad r \to 0, \tag{3.135}$$

then

$$\hat{f}_\Phi(r, t) \sim r^\alpha. \tag{3.136}$$

That is, the local Holder exponent α of $f(t)$ at any given time can be recovered by the scaling of *local wavelet spectrum* with r at the same time.

Therefore, the wavelet analysis of a time process allows in principle the *direct determination* of the local Holder exponent as a function of time, which opens up many new avenues of characterizing multifractality. In addition, wavelet multifractal analysis has been reported to be more robust (i.e., less sensitive to noise, for instance) that some of the methods that we have discussed in this chapter. The use of wavelets has also its own complications. A significant one is how to choose the more proper choice of the basis function, Φ, in order to best characterize singularities in time processes. Another one lies in the fact that the rescaled wavelets that enter in the continuous formulation given in Eq. 3.133 do not form an orthonormal set. Therefore, they provide a redundant representation whose interpretation can sometimes be confusing. This problem can be partially resolved through the introduction of discrete, orthonormal wavelet representations [44].

Problems

3.1 Scale Invariance: Power-Laws
Derive Eq. 3.4 from the Gutenberg-Richter law (Eq. 3.3).

3.2 Fractals: Cantor Set
Prove that the self-similar dimension of the Cantor Set is $D_{ss} = \log(2)/\log(3)$.

3.3 Fractals: The Mandelbrot Set
Consider the quadratic recurrence $z_{k+1} = z_k^2 + z_0$ in the complex plane. Mandelbrot set is composed of all the initial choices for z_0 for which the orbit predicted by the recurrence relation does not tend to infinity. Build a code to generate the Mandelbrot set. Determine its BC fractal dimension.

3.4 Time Processes: Brownian Motion
Write a numerical code to generate Brownian motion trajectories in two dimensions. Use the box-counting procedure and show that the BC fractal dimension of 2D Brownian motion is equal to 2. That is, if given enough time, Brownian motion would fill the whole plane.

3.5 Time Processes: Fractional Brownian Motion
Prove, by exploiting the self-similarity of fBm, that the fractal dimension of the time traces of fBm is given by $D = 2 - H$.

3.6 Generation of Synthetic fGn/fLn Series
Write a code that implements the algorithm described in Appendix 1 in order to generate synthetic series of fGn/fLn with arbitrary tail-index α and Hurst exponent, H. Implement the possibility of generating fBm/fLm as well.

3.7 Running Sandpile: Scaling Behaviour for Various Overlapping Regimes
Use the sandpile code (see Problem 1.5) to generate time series for the total mass of the sandpile over the SOC state using $L = 1000$, $N_f = 30$, $Z_c = 200$, $N_b = 10$, one for each the following values of $p_0 = 10^{-6}$, 10^{-5}, 10^{-3} and 10^{-2}. Repeat the scale-invariance analysis discussed in Sect. 3.5 for each cases. What is the mesorange for each case? How does the monofractal behaviour change as a function of the figure-of-merit that controls avalanche overlapping, $(p_0 L)^{-1}$?

3.8 Running Sandpile: Scaling Behaviour Below the Mesorange
Use the sandpile code (see Problem 1.5) and generate a time series for the total mass of the sandpile over the SOC state using $L = 1000$, $N_f = 30$, $Z_c = 200$, $N_b = 10$

and $p_0 = 10^{-4}$. Then, repeat the scale-invariance analysis discussed in Sect. 3.5, but for the range of block sizes $1 < M < 30{,}000$. How are the results different from what was obtained within the mesorange?

3.9 Advanced Problem: Detrended Fluctuation Analysis

Write a code that implements DFA1 (see Appendix 2). Then, use the code to estimate the Hurst exponent of fGn series generated with Hurst exponents $H = 0.25$, 0.45, 0.65 and 0.85 (see Problem 3.6). Compare the performance of DFA with the moment methods discussed in this chapter, both for the fGn series and their integrated fBm processes.

3.10 Advanced Problem: Wavelet Analysis

Write a code to perform the local determination of the Holder exponent using wavelets (see Appendix 3). Then, use the code on a synthetic fBm generated with nominal exponent $H = 0.76$ and show that the process is indeed monofractal, and the instantaneous local Holder exponent is constant and given by H. Refer to [45] to decide on the best option for the wavelet basis function.

References

1. Einstein, A.: Über die von der molekularkinetischen Theorie der Wärme geforderte Bewegung von in ruhenden Flüssigkeiten suspendierten Teilchen. Ann. Phys. 17, 549 (1905)
2. Gutenberg, B., Richter, C.F.: Seismicity of the Earth and Associated Phenomena. Princeton University Press, Princeton (1954)
3. Bak, P., Tang, C.: Earthquakes as a Self-organized Critical Phenomenon. J. Geophys. Res. 94, 15635 (1989)
4. Shaw, B.E., Carlson, J.M., Langer, J.S.: Patterns of Seismic Activity Preceding Large Earthquakes. J. Geophys. Res. 97, 478 (1992)
5. Hergarten, S.: Self-organized Criticality in Earth systems. Springer, Heidelberg (2002)
6. Crosby, N.B., Aschwanden, M.J., Dennis, B.R.: Frequency Distributions and Correlations of Solar X-ray Flare Parameters. Sol. Phys. 143, 275 (1993)
7. Frank, J., King, A., Raine, D.: Accretion Power in Astrophysics. Cambridge University Press, Cambridge (2002)
8. Richardson, L.F.: Variation of the Frequency of Fatal Quarrels with Magnitude. Am. Stat. Assoc. 43, 523 (1948)
9. Kleiber, M.: Body Size and Metabolism. Hilgardia 6, 315 (1932)
10. Frisch, U.: Turbulence: The Legacy of A. N. Kolmogorov. Cambridge University Press, Cambridge (1996)
11. Mandelbrot, B.B.: The Fractal Geometry of Nature. W.H. Freeman, New York (1982)
12. Mandelbrot, B.B., van Ness, J.W.: Fractional Brownian Motions, Fractional Noises and Applications. SIAM Rev. 10, 422 (1968)
13. Mandelbrot, B.B.: Fractals: Form, Chance, and Dimension. W.H. Freeman, New York (1977)
14. Mandelbrot, B.B.: How Long Is the Coast of Britain?: Statistical Self-similarity and Fractional Dimension. Science 156, 636 (1967)
15. Taylor, R.P.: The Art and Science of Foam Bubbles. Nonlinear Dynamics Psychol. Life Sci. 15, 129 (2011)
16. Li, J., Ostoja-Starzewski, M.: The Edges of Saturn are Fractal. SpringerPlus 4, 158 (2014)

17. Liu, H., Jezek, K.C.: A Complete High-Resolution Coastline of Antarctica Extracted from Orthorectified Radarsat SAR Imagery. Photogramm. Eng. Remote Sens. 70, 605 (2004)
18. Feder, J.: Fractals. Plenum Press, New York (1988)
19. Mandelbrot, B.B.: Intermittent Turbulence in Self-similar Cascades. J. Fluid Mech. 62, 331 (1977)
20. Martinez, V.J., Paredes, S., Borgani, S., Coles, P.: Multiscaling Properties of Large-Scale Structure in the Universe. Science 269, 1245 (1995)
21. Grassberger, P., Procaccia, I.: Measuring the Strangeness of Strange Attractors. Physica D 9, 189 (1983)
22. Kinsner, W.: A Unified Approach to Fractal Dimensions. Int. J. Cogn. Inform. Nat. Intell. 1, 26 (2007)
23. Hentschel, H.G.E., Procaccia, I.: The Infinite Number of Generalized Dimensions of Fractals and Strange Attractors. Physica D 8, 435 (1983)
24. Halsey, T.C., Jensen, M.H., Kadanoff, L.P., Procaccia, I., Shraiman, B.I.: Fractal Measures and Their Singularities: The Characterization of Strange Sets. Phys. Rev. A 33, 1141 (1986)
25. Wootton, A.J., Carreras, B.A., Matsumoto, H., McGuire, K., Peebles, W.A., Ritz, Ch.P., Terry, P.W., Zweben, S.J.: Fluctuations and Anomalous Transport in Tokamaks. Phys. Fluids B 2, 2879 (1990)
26. Carreras, B.A.: Progress in Anomalous Transport Research in Toroidal Magnetic Confinement Devices. IEEE Trans. Plasma Sci. 25, 1281 (1997)
27. van Milligen, B.P., Sanchez, R., Hidalgo, C.: Relevance of Uncorrelated Lorentzian Pulses for the Interpretation of Turbulence in the Edge of Magnetically Confined Toroidal Plasmas. Phys. Rev. Lett. 109, 105001 (2012)
28. Fernandez-Gomez, I., Martin-Solis, J.R., Sanchez, R.: Perpendicular Dynamics of Runaway Electrons in Tokamak Plasmas. Phys. Plasmas 19, 102504 (2012)
29. Feynman, R., Hibbs, A.R.: Quantum Mechanics and Path Integrals. McGraw-Hill, New York (1965)
30. Langevin, P.: Sur la theorie du mouvement brownien. C.R. Acad. Sci. (Paris) 146, 530 (1908)
31. Calvo, I., Sanchez, R.: The Path Integral Formulation of Fractional Brownian Motion for the General Hurst Exponent. J. Phys. A 41, 282002 (2008)
32. Huillet, T.: Fractional Lévy Motions and Related Processes. J. Phys. A 32, 7225 (1999)
33. Laskin, N., Lambadaris, I., Harmantzis, F.C., Devetsikiotis, M.: Fractional Lévy Motion and its Application to Network Traffic Modelling. Comput. Netw. 40, 363 (2002)
34. Calvo, I., Sanchez, R., Carreras, B.A.: Fractional Lévy Motion Through Path Integrals. J. Phys. A 42, 055003 (2009)
35. Lamperti, J.W.: Semi-stable Stochastic Processes. Trans. Am. Math. Soc. 104, 62 (1962)
36. Samorodnitsky, G., Taqqu, M.S.: Stable Non-Gaussian Processes. Chapman & Hall, New York (1994)
37. Mainardi, F., Luchko, Y., Pagnini, G.: The Fundamental Solutions for the Fractional Diffusion-Wave Equation. Appl. Math. Lett. 9, 23 (1996)
38. Sassi, R., Signorini, M.G., Cerutti, S.: Multifractality and Heart Rate Variability. Chaos 19, 028507 (2009)
39. Losa, G., Merlini, D., Nonnenmacher, T., Weiben, E.R.: Fractals in Biology and Medicine. Birkhauser, New York (2012)
40. Peltier, R.F., Lévy-Véhel, J.: Multifractional Brownian Motion: Definition and Preliminary Results. Rapport de recherche de INRIA, No 2645 (1995)
41. Lacaux, C.: Series Representation and Simulation of Multifractional Lévy Motions. Adv. Appl. Probab. 36, 171 (2004)
42. Ayache, A., Lévy-Véhel, J.: The Generalized Multifractional Brownian Motion. Stat. Infer. Stoch. Process. 3, 7 (2000)
43. Monin, A.S., Yaglom, A.M.: Statistical Fluid Mechanics. MIT Press, Boston (1985)
44. Meneveau, C.: Analysis of Turbulence in the Orthonormal Wavelet Representation. J. Fluid Mech. 232, 469 (1991)

45. Farge, M.: Wavelet Transforms and Their Applications to Turbulence. Annu. Rev. Fluid Mech. 24, 395 (1992)
46. Davidson, R., MacKinnon, J.G.: Econometric Theory and Methods. Oxford University Press, New York (2004)
47. Podlubny, I.: Fractional Differential Equations. Academic, New York (1998)
48. Chechkin, A.V., Gonchar, V.Y.: A Model for Persistent Lévy Motion. Physica A 277, 312 (2000)
49. Greene, M.T., Fielitz, B.D.: Long-Term Dependence in Common Stock Returns. J. Financ. Econ. 4, 339 (1977)
50. Peng, C.K., Buldyrev, S.V., Havlin, S., Simons, M., Stanley, H.E., Goldberger, A.L.: On the Mosaic Organization of DNA Sequences. Phys. Rev. E 49, 1685 (1994)
51. Kantelhardt, J.W., Zschiegner, S.A., Koscielny-Bunde, E., Havlin, S., Bunde, A., Stanley, H.E.: Multifractal Detrended Fluctuation Analysis of Nonstationary Time Series. Physica A 316, 87 (2002)
52. Mallat, S.: A Wavelet Tour of Signal Processing. Academic, New York (1998)
53. Grossmann, A., Morlet, J.: Decomposition of Hardy Functions into Square Integrable Wavelets of constant Shape. SIAM J. Appl. Anal. 15, 723 (1984)

Chapter 4
Memory

4.1 Introduction

Memory is a term that often appears in discussions about complex dynamics. What is it meant when a scientist says that a certain complex system exhibits '**memory**'? More often than not it is meant that *the past history of the system has some meaningful influence on its future evolution*. Let's consider, for instance, the case of earthquakes. It has often been claimed that the process of stress relaxation at a fault that can lead to the triggering of earthquakes exhibits long-term memory in the sense that when and where the next earthquake will happen, or how much stress and energy it will release, are somehow affected by how large past earthquakes were, as well as when and where they took place [1].

Although this imprecise explanation of the term '*memory*' grasps the essence of the idea, the concept must be specified much more if one is to apply it meaningfully to real systems. An element that needs to be considered is the concept of scale-invariance, that was discussed in Chap. 3, and that is a typical characteristic of many complex systems. One would expect that if memory is present in them, it should somehow inherit and exhibit some sort of scale-invariance. Or, in other words, that *memory in a complex system should lack any characteristic timescale*. Going back to the case of earthquakes previously mentioned, this would imply that it is not just the few earthquakes triggered during the last few weeks, months or even years that are important to determine when the next earthquake will happen, but that *all the past history of the fault system would have something to say as well*.

We have collected all these somewhat imprecise ideas within a working definition of **memory in a complex system** that we will profusely use throughout this chapter: *the ability of past events to act as an influence on future dynamics, that extends in time in a scale-invariant fashion, being only truncated by finite-size effects, such as the system size or lifetime*. We will try to make each of the terms used in this sentence more mathematically precise in the next sections.

© Springer Science+Business Media B.V. 2018
R. Sánchez, D. Newman, *A Primer on Complex Systems*,
Lecture Notes in Physics 943, https://doi.org/10.1007/978-94-024-1229-1_4

4.2 Memory and Correlation

We will start the discussion by describing first the mathematical tools that are traditionally used to detect and quantify the presence of memory in time processes. Or, more precisely, in *stationary random processes* (see Sect. 3.3.5). Readers should be aware that all these tools make an extensive use of averaging procedures. It is only fair to warn them at this point that, although the construction of these tools is often done using ensemble averages, we will shift the weight towards temporal averaging as soon as possible, thus implicitly assuming a large degree of ergodicity in the system under study (see discussion at the start of Chap. 2). The reason for this change is simply that temporal averaging is what most scientists can and do apply to real data.

The first tools that will be discussed are the popular **autocorrelation function** (Sect. 4.2.1) and its associated **power spectrum** (Sect. 4.2.2). Then, we will describe the less known **autodifference** (Sect. 4.2.3), that is one of the options to replace the autocorrelation function in cases in which the process exhibits divergent power-law (i.e., Lévy) statistics. Finally, we will discuss the use of **waiting-time statistics** (Sect. 4.4.5). We will always try to make as clear as possible the distinction between situations where a finite **memory timescale** may be found and those in which it is absent. It is only in the latter cases that one can actually speak of a **"complex memory"**.

4.2.1 The Autocorrelation Function

We will start introducing the autocorrelation function in the language of ensemble averages, as it is often done in theoretical texts. It is intimately related to the statistical **covariance** between two random processes.

4.2.1.1 Ensemble Average Definition of the Covariance Function

The **covariance** between two random processes, $y_1(t)$ and $y_2(t)$, is defined as the expected value[1] [2, 3]:

$$Co_{[y_1, y_2]}(t_1, t_2) := \langle y_1(t_1) y_2(t_2) \rangle - \langle y_1(t_1) \rangle \langle y_2(t_2) \rangle. \qquad (4.1)$$

The covariance is a rather interesting quantity in itself since, *whenever the processes* y_1 *and* y_2 *are* **independent of each other**, *its covariance vanishes.*[2] Indeed, one can

---

[1]Expected values were introduced in Sect. 2.2.5 (see Eq. 2.11).

[2]However, the reverse is not true in general, since the covariance may also vanish in cases in which the processes are not independent [3].

easily calculate the expected value in this case by using the fact that the probability of two independent events is then given by the product of the probabilities of each event[3]:

$$\langle y_1(t_1)y_2(t_2)\rangle = \int dy_1 \int dy_2\, p(y_1(t_1) \cup y_2(t_2))y_1(t_1)y_2(t_2)$$

$$= \int dy_1 p(y_1(t_1))y_1(t_1) \int dy_2\, p(y_2(t_2))y_2(t_2)$$

$$= \langle y_1(t_1)\rangle \langle y_2(t_2)\rangle \implies \mathrm{Co}_{[y_1,y_2]}(t_1,t_2) = 0,\ \forall t_1, t_2. \quad (4.2)$$

4.2.1.2   Ensemble Average Definition of the Autocorrelation Function

The **autocorrelation function** of a *stationary random process* is in essence the statistical *covariance of the process with a time-delayed copy of itself.* It is defined, for any time lag $\tau > 0$, as:

$$\mathrm{Ac}_{[y]}(\tau) := \frac{\mathrm{Co}_{[y,y]}(t, t+\tau)}{\mathrm{Co}_{[y,y]}(t, t)} = \frac{\langle y(t)y(t+\tau)\rangle - \langle y(t)\rangle\langle y(t+\tau)\rangle}{\langle y(t)^2\rangle - \langle y(t)\rangle^2}. \quad (4.3)$$

Here, the numerator is the covariance of the process at time t with a copy of itself delayed by a time lag τ. The denominator reduces to the variance of the process at time t. It should be noted that, thanks to the assumed stationarity of the process, the autocorrelation defined in Eq. 4.3 is independent of t.

4.2.1.3   Main Properties of the Autocorrelation Function

The autocorrelation function possesses some important properties. In particular [4]:

1. it satisfies that $|\mathrm{Ac}_{[y]}(\tau)| \leq 1$, with equality taking place for $\tau = 0$;
2. it is *symmetric* with respect to the lag value. That is, $\mathrm{Ac}_{[y]}(\tau) = \mathrm{Ac}_{[y]}(-\tau)$. That is the reason why we considered only $\tau > 0$;
3. it is *non-negative definite* in the sense that:

$$\lim_{T\to\infty} \int_0^T \mathrm{Ac}_{[y]}(\tau)d\tau \geq 0, \quad (4.4)$$

---

[3]That is, $p(A \cup B) = p(A)p(B)$. In fact, the degree of dependence between two events is measured through the probability [4]: $p(A \cap B) := p(A \cup B) - p(A)p(B)$, that somewhat resembles the covariances defined in Eq. 4.1. Naturally, $p(A \cap B) = 0$ if A and B are independent.

assuming that the limit of the integral is finite. This limit has an important physical interpretation in several contexts, as we will see in the next subsections. The existence of the integral requires, in the first place, that,

$$\lim_{\tau \to \infty} Ac_{[y]}(\tau) = 0, \tag{4.5}$$

which is usually the case for random processes. Secondly, it must also decay faster than τ^{-1} for $\tau \gg 0$, which not always happens.

4.2.1.4 Memory and Statistical Dependence

The autocorrelation is useful to investigate the presence of memory or, more precisely, **statistical dependence** in a process because it inherits from the covariance the following properties:

- if $Ac_{[y]}(\tau) = 0$ for all $\tau > 0$, it means that all the values of the process are independent of each other. That is, the process *lacks any kind of memory*.
- if $Ac_{[y]}(\tau) = 0$, $\forall \tau > \tau_c$, it implies that the process becomes independent of its past history only after a lapse of time τ_c. Or, in layman's terms, that the process has no memory about itself beyond a lapse τ_c, that provides a characteristic **memory timescale**.
- if $Ac_{[y]}(\tau) \sim \tau^{-a}, a > 0$, $\tau \gg 0$, one might suspect that *memory remains present in the system for all times in a self-similar manner*, as made apparent by the power-law dependence (although some restrictions on the valid values of the exponent a apply, as will be discussed soon), thus lacking any memory timescale.

In practice, things are never this clear-cut. For instance, the autocorrelation is almost never exactly zero. Therefore, one has to decide which threshold value is a reasonable one, below which one can safely consider that autocorrelation as negligible. Similarly, the presence of power-law decays is not easy to detect either, since the tails of the autocorrelation are typically very noisy due to lack of statistics. We will discuss how to deal with all these aspects in what follows.

4.2.1.5 Positive and Negative Correlation

Although the presence of memory (or dependence) can be inferred—between values of a process separated by certain time lag τ—from a non-zero autocorrelation value, much more can be said when considering the actual autocorrelation value, that could be anywhere within $[-1, 1]$. What does each of those values mean? Let's review each possibility:

- $Ac_{[y]}(\tau) = 0$, as already discussed, implies that *any two values of the process separated by the time lag τ* are **statistically independent of each other**. We often use the term **uncorrelation** to describe this situation.

- $Ac_{[y]}(\tau) = 1$ is possible only if $y(t + \tau) = y(t)$ for every t. That is, the process and its shifted copy are identical! In this case, we speak of **perfect correlation**. For random processes, this only happens for the trivial case of $\tau = 0$.[4]
- $Ac_{[y]}(\tau) = -1$ only happens if $y(t + \tau) = -y(t)$. In this case, one speaks of **perfect anticorrelation**. This negative extreme value is however never reached for random processes.[5]
- If $0 < Ac_{[y]}(\tau) < 1$, one usually speaks of **positive correlations** (or **persistence**). This word is used because the process $y(t)$ and its shifted version, $y(t+\tau)$, although not identical at every t, *share the same sign (positive or negative) more often than not*. Or, in other words, it is statistically more probable that any two values of the process separated by that time lag have the same sign.[6]
- Analogously, one uses the term **negative correlations** (or **anti-persistence**) when $-1 < Ac_{[y]}(\tau) < 0$. In this case, any two values of the process separated by a time lag τ *have opposite signs more frequently* than not.

4.2.1.6 Temporal Definition of the Autocorrelation Function

For stationary processes one is entitled to consider an alternative definition of the autocorrelation function based on temporal averages (see Sect. 2.1). It is given by the more familiar expression[7]:

$$Ac_{[y]}(\tau) = \lim_{T \to \infty} \frac{1}{2\sigma_y^2 T} \int_{-T}^{T} dt \, (y(t) - \bar{y})(y(t + \tau) - \bar{y}), \qquad (4.7)$$

---

[4]It may happen for $\tau \neq 0$ for other processes that are not stationary and random; for instance, for any periodic process when the lag $\tau = nP$, being n any integer and P the period of the process.

[5]Again, the extreme negative value can be reached for periodic processes when the lag is equal to a semiperiod, $\tau = \pm(n + 1/2)P$, with n integer.

[6]The autocorrelation function (as does the covariance) *quantifies statistical dependence, not causal dependence*. That is, it only tells us whether it is statistically more probable that two values share a sign, not if the sign of the first value causes that of the second value. Assuming causal instead of statistical dependence is a common misconception. It must always be kept in mind that although causal dependence often translates into statistical dependence, the opposite is not always true. Think, for instance, in the case in which two processes are caused by a third one. The first two processes are statistically dependent, but there is clearly no causal dependence between them.

[7]The equivalence of Eqs. 4.7 and 4.3 for stationary processes can be made apparent by replacing the ensemble averages in the latter by temporal averages:

$$Ac_{[y]}(\tau) = \frac{\left[\lim_{T \to \infty} \frac{1}{2T} \int_{-T}^{T} dt \, y(t)y(t + \tau) - \left(\frac{1}{2T} \int_{-T}^{T} dt \, y(t) \right)^2 \right]}{\left[\lim_{T \to \infty} \frac{1}{2T} \int_{-T}^{T} dt \, y(t)^2 - \left(\frac{1}{2T} \int_{-T}^{T} dt \, y(t) \right)^2 \right]}. \qquad (4.6)$$

Then, one can easily reorder the terms in order to convert this expression into Eq. 4.7.

where we have introduced the following notation for the temporal average and variance of the process,

$$\bar{y} := \lim_{T\to\infty} \frac{1}{2T} \int_{-T}^{T} dt\, y(t), \qquad \sigma_y^2 = \lim_{T\to\infty} \frac{1}{2T} \int_{-T}^{T} dt\, (y(t) - \bar{y})^2. \qquad (4.8)$$

It is also common to define the temporal definition of autocorrelation as,[8]

$$\mathrm{Ac}_{[y]}(\tau) = \lim_{T\to\infty} \frac{1}{2T} \int_{-T}^{T} dt\, y^*(t) y^*(\tau + t), \qquad (4.9)$$

by introducing the rescaled variable $y^*(t) := (y(t) - \bar{y})/\sigma_y$ that represents the **fluctuations** of the original process with respect to its mean value, normalized to its standard deviation.[9]

4.2.1.7   An Example: Plasma Turbulent Fluctuations

To illustrate the differences between the theoretical results just presented and reality, it is appropriate to discuss a real example. Figure 4.1 shows the values of the autocorrelation function for the first thousand time lags obtained[10] for a 10,000-long temporal record of plasma density fluctuations measured at the edge of a magnetically confined plasma (the TJ-II stellarator, operated in Madrid, SPAIN at the National Fusion Laboratory) using a Langmuir probe [5]. This is a stationary signal (see inset in Fig. 4.1) that is rather irregular and seemingly unpredictable, exhibiting many of the properties that one would expect of a (somewhat smoothed-out) random process.

Several things are clearly apparent in Fig. 4.1. First, we see that the autocorrelation is equal to 1 only at zero lag, as expected. Then, it quickly decays as the time lag is increased, although it never vanishes completely, wandering instead around zero. It maintains a significant positive value (above, say, 0.05 in magnitude) for lags $\tau < 80\,\mu$s. For larger lag values, it stays within the interval $(-0.05, 0.05)\,\mu$s or so, that is often interpreted as statistical noise.[11]

A common interpretation of this autocorrelation function would be to say that within the $\tau < 80\,\mu$s range local fluctuations are positively correlated, meaning that

---

[8]In practice, T will be finite. We will make this dependence explicit, when needed, by adding a T superscript to the symbol (i.e., $\mathrm{Ac}_{[y]}^T(\tau)$ or $\bar{y}^T$). To obtain meaningful results, T must be sufficiently large so that the dependence of the autocorrelation on T becomes negligible.

[9]Naturally, the process y^* has zero mean and unit variance.

[10]The precise way in which this autocorrelation function (or the power spectrum to be discussed later) is obtained will be discussed in Sect. 4.4. Here, to keep the discussion fluid, we just simply discuss the final results without dwelling too much on the details of how they are computed.

[11]We will argue soon that one has to be careful with such hasty interpretation of the tail of the autocorrelation function, though.

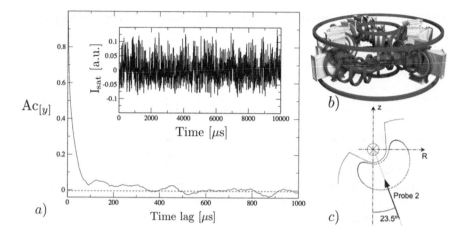

Fig. 4.1 Left: Typical autocorrelation function of turbulent data that corresponds, in this case, to plasma density fluctuations (shown in the inset) measured at the edge of a plasma magnetically confined within the TJ-II stellarator (**right, above**) device operated in the Laboratorio Nacional de Fusión at CIEMAT (Madrid, SPAIN). The data was recorded by means of a reciprocating Langmuir probe, that enters the toroidal plasma from the edge (**right, below**), and measures the ion saturation current, a surrogate for the plasma density [Credits: TJ-II model, courtesy of Víctor Tribaldos]

they maintain the same sign on average.[12] Thus, $\tau_c \sim 80\,\mu$s gives a first estimate for the memory timescale for this process. Most scientists would estimate this timescale in terms of the so-called **decorrelation time**, τ_d, for which several prescriptions can be found in the literature. The most popular one is probably the lag value at which $Ac_{[y]}(\tau_d) = 1/e \simeq 0.368$, yielding $\tau_c \sim 40\,\mu$s for the TJ-II data.[13]

The reader should be aware that the tail of the autocorrelation beyond $\tau \sim 80\,\mu$s is completely neglected in the analysis. It turns out that, in many complex systems, the tail becomes more important to characterize the presence of memory than the central part that defines the classical decorrelation time. Let's see why.

4.2.1.8   Typical Autocorrelation Tails: Exponentials and Power-Laws

The exact form of the tail of the autocorrelation function will depend on the specifics of the process at hand. However, there are a couple of shapes that illustrate well what

---

[12]Physically, this time would be interpreted as the average amount of time that turbulent structures take to pass by the Langmuir probe tip, if they are moving, or the average life of the local turbulent structures, if they remain relatively at rest with respect to the probe tip.

[13]In our case, we will prefer to use the value of the integral of the autocorrelation function over the extended temporal range (see Eq. 4.11, that will be discussed next), that yields $\tau_d \sim 60\,\mu$s.

is often found in many practical situations.[14] In particular, **exponential** and **power-law** decaying tails.

- **Exponential tails**

    Let's consider first the model autocorrelation function,

$$Ac_{[y]}(\tau) = \exp(-|\tau|/\tau_c). \tag{4.10}$$

This function is always positive, thus corresponding to a process that presents positive correlations for all lags. The important quantity here is τ_c, that provides an estimate for *how long the process remembers about itself.*[15] Why? Because, for lags $\tau \gg \tau_c$, the exponential becomes so small that the tail, although still above zero, is virtually indistinguishable from the case of absence of correlations.

This idea can be easily extended to other tail shapes that, although not strictly exponential, decay sufficiently fast. This can be done by defining the memory timescale τ_m by the limit,

$$\tau_m := \lim_{T \to \infty} \int_0^T Ac_{[y]}(\tau)d\tau, \tag{4.11}$$

that, for Eq. 4.10, yields

$$\tau_m = \lim_{T \to \infty} \int_0^T Ac_{[y]}(\tau)d\tau = \int_0^\infty \exp(-|\tau|/\tau_c)d\tau = \tau_c, \tag{4.12}$$

as demanded. This definition will however be also adequate for any other autocorrelation function for which the limit is finite. These could be functions with an asymptotically exponential tail (shown in Fig. 4.2, in blue), but also any other tail behaviours for which the limit converges.

- **Power-law tail (type I)**

    Another type of tail that is often encountered is the decaying power-law. It will be remembered that we said that power-laws are indicative of scale invariance (see Chap. 3). Of all possible power-law decays, we will focus on those of the form (shown in Fig. 4.2, in green),

$$Ac_{[y]}(\tau) \sim B|\tau|^{-a}, \quad 0 < a \le 1, \quad \tau \to +\infty, \tag{4.13}$$

---

[14]Things will never be as clear-cut as we discuss them in this section. The direct determination of the autocorrelation tail from a finite record is usually quite inaccurate. Tails usually exhibit irregular oscillations and are contaminated by statistical noise, making it rather difficult to tell exponential from power-law decays by direct inspection. Other methods must be usually called upon for this task. Be it as it may, the discussion of the tails that follows will still illustrate several important concepts quite clearly.

[15]In fact, it is the origin of the previously mentioned prescription, $Ac_{[y]}(\tau_d) = 1/e$.

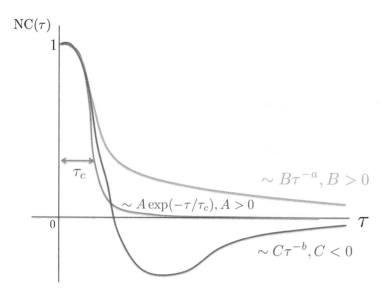

Fig. 4.2 Some meaningful autocorrelation function shapes: exponentially decaying (blue); positively correlated power law decaying (green); negatively correlated power-law decaying (red)

for some constant $B > 0$. The reason for restricting the exponent range is that, for $0 < a \leq 1$, the memory timescale τ_m defined by Eq. 4.11 fails to converge. Indeed, the dominant contribution to τ_m is then,[16]

$$\tau_m \sim \int^\infty Ac_{[y]}(\tau) \sim \left. \frac{B\tau^{1-a}}{1-a} \right|^\infty \to +\infty \tag{4.14}$$

Therefore, the case $a \leq 1$ corresponds to a very different class of processes, in which **long-term, scale-invariant, positive correlations** are present[17] for all lags.[18] It should thus come as no surprise that this kind of process is often encountered in the context of complex systems.

- **Power-law tail (type II)**

  Another interesting case that involves a power-law tail is (shown in Fig. 4.2, in red),

$$Ac_{[y]}(\tau) \sim C|\tau|^{-b}, \quad b > 1, \tag{4.15}$$

---

[16]For $a = 1$, the integral would diverge instead logarithmically.

[17]Note that B must be positive in the case of the divergent tail; otherwise, Eq. 4.13 would not exhibit the non-negativeness property we discussed earlier.

[18]It must be remembered that exact scale-invariance does not happen in nature, as we discussed in Chap. 3, being always limited to a mesorange set by finite-size effects. Thus, one should never expect an autocorrelation function with a perfect power-law scaling extending to infinitely long times, even if long-term memory is present in the system. The scaling should be eventually overcome by noise.

with $C < 0$ taking the precise value that makes *the limit that defines τ_m vanish*:

$$\lim_{T\to\infty} \int_0^T Ac_{[y]}(\tau) = 0. \tag{4.16}$$

Note that $b > 1$ is required here to ensure the convergence of the integral, whilst $C < 0$ so that the negative tail can compensate the always positive contribution of the central region of the autocorrelation around $\tau = 0$.

It is important to note that this case is rather different from the completely uncorrelated case (i.e., the $\tau_c \to 0$ limit of Eq. 4.10), that also has $\tau_m = 0$. Here, the whole extent of the negative asymptotic tail is needed to achieve a total cancellation of the integral in the limit $T \to \infty$. Thus, there is not a finite lag value beyond which the tail can be neglected, and still have a good estimate of τ_m. As a result, τ_m loses its meaningfulness as a memory timescale, and the process is said to exhibit **long-term, self-similar, negative correlations**. We will see examples of complex processes with this type of autocorrelation function soon.[19]

4.2.1.9 An Example: Plasma Turbulent Fluctuations (Continued)

It is illustrative to re-examine now the tail of the autocorrelation function of the turbulent fluctuations measured at the edge of the TJ-II stellarator (previously shown in Fig. 4.1) in the light of the discussion we just had. Is the tail close to an exponential? Or does it scale like a power-law? In the latter case, is the tail fat enough to reveal the presence of any kind of scale-free, long-term memory?

In order to address these questions, we have magnified the tail of the autocorrelation function by plotting it using both log-log and log-lin scales (see Fig. 4.3). Power-law tails appear as straight lines in log-log plots; exponential tails appear linear in log-lin plots. The result of this exercise illustrates some of the typical

---

[19]The definition of τ_m given by Eq. 4.14 is intimately related to the so-called *Green-Kubo relation* that appears in stochastic transport theory [6]. The Green-Kubo relation states that, if $v(t)$ is a random variable that represents the instantaneous velocity of a particle, any population of these particles will **diffuse**, at long times, according to the famous diffusive transport equation,

$$\frac{\partial n}{\partial t} = D\frac{\partial^2 n}{\partial x^2} \tag{4.17}$$

where n is the particle density, and with the **diffusivity** D given by the integral:

$$D = \sigma^2 \left(\lim_{T\to\infty} \int_0^T Ac_{[v]}(\tau)d\tau \right). \tag{4.18}$$

σ^2 is the velocity variance. There are cases, however, when the velocity fluctuations are such that this integral may vanish or diverge. We will discuss the type of transport that appears in these cases in Chap. 5 under the respective names of **subdiffusive** and **superdiffusive** transport. Both correspond to typical behaviours often found in the context of complex dynamics.

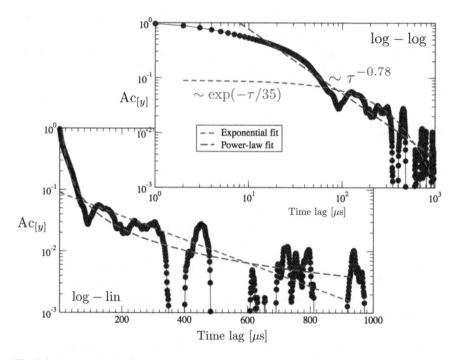

Fig. 4.3 Autocorrelation function of the same TJ-II turbulent data shown in Fig. 4.1 displayed in log-log (**right, above**) and log-lin (**left, below**) representations. In dashed lines, exponential (red) and power law (blue) fits to the tail are shown

problems of trying to make these estimations directly on the autocorrelation function. Large fluctuations are seen at the larger lag values, that are least resolved due to the shortness of the record. As a result, it is difficult to tell the behaviour of the tail, particularly beyond $\tau \sim (80–100)\,\mu$s. However, a power-law scaling as $\sim \tau^{-0.8}$ seems much more probable than an exponential one, as shown by the dashed curves included in the figure to help guide the eye. The value of the tail exponent also appears to be within the interval $0 < a \leq 1$, thus suggesting that long-term, self-similar, positive correlations might indeed be present in the signal for all scales above $\tau > 100\,\mu$s for which data is available.

A final (and important!) observation to make is that, even when a power-law tail seems probable, the finite length of our time record implies that the integral of the autocorrelation function over the available data set always remains finite. In this case, such integral yields $\tau_m \simeq 60\,\mu$s. The observation of the divergent power-law tail suggests however that this value is rather meaningless as a memory diagnostic.[20]

---

[20]Indeed, one should expect that, if a longer record was available, one could calculate the autocorrelation function for larger values of τ. The divergent power-law scaling, if maintained for these larger lags, would then make the value of τ_m to increase towards infinity as the record

At any rate, it is clear that the large fluctuations that are apparent over the tail region make this analysis rather imprecise. Better estimation methods are thus needed. One such method is based on the **power spectrum**, that we will discuss next. An even better method is provided by the R/S **analysis**, that will be discussed in Sect. 4.4.4.

4.2.2 The Power Spectrum

The **power spectrum** of an stationary random process $y(t)$ is defined as the Fourier transform of its autocorrelation function [2],

$$\hat{S}_{[y]}(\omega) =: \int_{-\infty}^{\infty} d\tau \, e^{\imath \omega \tau} \mathrm{Ac}_{[y]}(\tau). \tag{4.19}$$

The power spectrum is well defined as long as this Fourier transform can be computed. As discussed in Appendix 1 of Chap. 2, a sufficient condition for the convergence of the integral is that the autocorrelation function is Lebesgue integrable (i.e., that its tail decays faster than $|\tau|^{-1}$ as $\tau \to \pm\infty$). However, the integral may also exist in other cases.

The main reason to use the power spectrum comes from the fact that it can be estimated directly from the process, *without having to calculate the autocorrelation function first*, which is a very nice feature in practice due to the inaccuracies involved. This can be done by considering that,[21]

$$
\begin{aligned}
\hat{S}_{[y]}(\omega) &= \int_{-\infty}^{\infty} d\tau \, e^{\imath \omega \tau} \mathrm{Ac}_{[y]}(\tau) \\
&= \int_{-\infty}^{\infty} d\tau \, e^{\imath \omega \tau} \left(\lim_{T \to \infty} \frac{1}{2T} \int_{-T}^{T} dt \, y^*(t) y^*(\tau + t) \right) \\
&= \lim_{T \to \infty} \frac{1}{2T} \left[\int_{-\infty}^{\infty} d\tau \, e^{\imath \omega \tau} \left(\int_{-T}^{T} dt \, y^*(t) y^*(\tau + t) \right) \right] \\
&= \lim_{T \to \infty} \frac{|\hat{y}_T^*(\omega)|^2}{2T}.
\end{aligned}
\tag{4.20}
$$

Here, $\hat{y}_T^*(\omega)$ is the Fourier transform of the process $y^*(t)$ *after it is truncated to the finite interval* $[-T, T]$.[22]

---

length is increased. This is in contrast to what would happen for an exponential decay, for which τ_m will eventually become independent of the record length.

[21] We consider here that the process is such that the interchange of limits, averages and integrals is allowed.

[22] It is only by introducing the truncated process that we can extend the limits of the last integral in Eq. 4.20 to infinity, allowing us to invoke the convolution theorem (see Appendix 1 of Chap. 2) to get the final result for the power spectrum in terms of $y_T^*(\omega)$.

It is also nice to note that one can estimate the memory time τ_m (defined in Eq. 4.11) directly from the power spectrum:

$$\tau_m = \int_{-\infty}^{\infty} d\tau \, \text{Ac}_{[y]}(\tau) = \frac{1}{2} \lim_{\omega \to 0} \hat{S}_{[y]}(\omega). \tag{4.21}$$

4.2.2.1 Typical Power Spectra Shapes

Some further insight, that will be useful to interpret power spectra from real data, can be gained by analyzing the power spectra associated to the model autocorrelation tails discussed at the end of Sect. 4.2.

1. **Exponential tail** In the case of the exact exponential (Eq. 4.10), its power spectrum can be computed analytically (see Table 2.7). The result is:

$$\hat{S}_{[y]}(\omega) = \int_{-\infty}^{\infty} d\tau \, e^{i\omega\tau} \exp(-|\tau|/\tau_c) = \frac{2\tau_c}{1 + \tau_c^2 \omega^2}, \tag{4.22}$$

that decays as ω^{-2} for large frequencies (i.e., $\omega \gg \tau_c^{-1}$), and becomes flat for smaller frequencies (i.e., $\omega \ll \tau_c^{-1}$). The break-point between the two regions happens at $\omega_{bp} \sim \tau_c^{-1}$ (see Fig. 4.4, in blue), that coincides with the inverse of the memory timescale of the process (see Eq. 4.21),

$$\tau_m = \lim_{\omega \to 0} \frac{\hat{S}_{[y]}(w)}{2} = \tau_c. \tag{4.23}$$

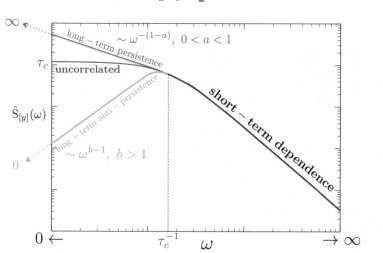

Fig. 4.4 Sketch showing the three meaningful types of power spectra discussed in text. They share the same short-term dynamics (for $\omega \gg \tau_c^{-1}$), but the long-term dynamics are uncorrelated (blue), self-similarly persistent (red) and self-similarly antipersistent (green). In reality, spectra can be much more complicated, often including more than two scaling regions

This **flat region** is particularly interesting. In the limit $\tau_c \to 0$ in Eq. 4.22, that represents a process that is uncorrelated for all lags, the flat region naturally extends from $\omega \to 0$ towards $\omega \to \infty$, thus filling the whole spectrum. Therefore, we can use the presence of a flat region for $\omega \ll \tau_m^{-1}$ as an indication that τ_m gives the timescale above which correlations disappear. Even more interestingly, the power spectrum will also become eventually flat for frequencies $\omega \ll \tau_m^{-1} = 2/\hat{S}_{[y]}(0)$ for any other autocorrelation function that is not exactly exponential but that decays sufficiently fast at long times. Furthermore, *the association of flatness and lack of correlations also holds for any other flat regions in the power spectrum even if they do not extend for arbitrarily low frequencies, but are instead restricted to a limited frequency range.* Physically, the flatness means that the available energy is equally distributed over all those frequencies, without any distinguishing feature. This is in fact how noise is usually characterized.

On the other hand, the behaviour of the power spectrum for $\omega \gg \tau_m^{-1}$, however, will depend on the details of the autocorrelation function for $\tau \ll \tau_m$. If the behaviour is not exactly exponential, the power spectrum could decay either faster or slower than ω^{-2}. Sometimes, it may even exhibit several scaling regions, depending on how complicated the autocorrelation function structure is.

2. **Power-law tail (type I)** In the case of an autocorrelation function that decays with a divergent power law tail (Eq. 4.13), one can invoke a nice property of Fourier transforms (see Appendix 1 of Chap. 2) to find out the asymptotic behaviour of the power spectrum. Namely, that if $f(t) \sim t^{-s}$ with $0 < s < 1$ when $t \to \pm\infty$, it then follows that $\hat{f}(\omega) \sim \omega^{-(1-s)}$, for $\omega \to 0$. Therefore, the power spectrum we are looking for asymptotically decays according to the **decreasing power law,**

$$\hat{S}_{[y]}(\omega) \sim \omega^{-(1-a)}, \text{ for } \omega \to 0. \tag{4.24}$$

The important thing here is that the $\omega \to 0$ limit of the spectrum now diverges. As a result, a finite τ_m does not exist (see Eq. 4.21), as is expected for any process with long-term, self-similar persistence. Furthermore, similarly to what happened with finite flat regions, *any region in the power spectrum that behaves like a decreasing power-law with exponent between 0 and -1 can be related to the action of positive correlations over those frequencies.* The decreasing behaviour implies that more energy exists at the lower than at the higher frequencies, a reflection of the process tendency to vary slowly, thus keeping the same sign for extended periods of time.

On the other hand, the behaviour of the power spectrum for $\omega \to \infty$ will depend again on the structure of the autocorrelation function around $\tau = 0$, that may be rather different from the power-law scaling at the other end (see Fig. 4.4, in red).

3. **Power-law tail (type II)** In the case in which the autocorrelation function decays according to Eq. 4.15, the fact that $C < 0$ has such value as to make τ_m vanish implies that (see Eq. 4.21),

$$\lim_{\omega \to 0} \hat{S}_{[y]}(\omega) = 0. \tag{4.25}$$

As a result, the asymptotic behaviour of the power spectrum in this case is given by the **increasing power-law**:

$$\hat{S}_{[y]}(\omega) \sim \omega^{b-1}, \text{ for } \omega \to 0. \tag{4.26}$$

This is the behaviour expected from processes that exhibit long-term, self-similar antipersistence. Furthermore, *any region in the power spectrum that behaves like an increasing power-law is indicative of the presence of negative correlations over those frequencies.* The increasing behaviour now implies that energy tends to accumulate at the higher frequencies, that speaks of a process that tends to vary quickly over time, thus changing sign very often.

As in the other cases, the behaviour of the power spectrum as $\omega \to \infty$ will be determined, by the structure of the autocorrelation function around $\tau = 0$ (see Fig. 4.4, in green).

4.2.2.2   An Example: Plasma Turbulent Fluctuations (Continued)

We continue now the analysis of the TJ-II turbulent edge data already discussed in Sect. 4.2 in the light of what we have said about the power spectrum. Figure 4.5 shows the power spectrum of the data as computed using Eq. 4.20. It has been represented using a double logarithmic scale, in order to make any power-law regions that might be present to appear as straight segments.

The first thing one notes from looking at the figure is that, although some fluctuations are still present, they are definitely of much lower amplitude that in the autocorrelation function plots (see Fig. 4.3). The spectrum seems to be composed of three distinct regions, a flat region at the smallest frequencies ($\omega \ll 10^{-3}$ MHz), a power-law region scaling as $\sim \omega^{-0.3}$ for intermediate frequencies (i.e., 10^{-3} MHz $\ll \omega \ll 10^{-2}$ MHz), and a second power-law region that extends for all larger frequencies ($\omega \gg 10^{-2}$ MHz), that scales as $\sim \omega^{-0.8}$.

The presence of a flat region at the lowest frequencies would suggest that the process becomes eventually uncorrelated. However, it must be noted that Eq. 4.21 gives a memory timescale $\tau_m = \hat{S}_{[y]}(0)/2 \simeq 60\,\mu\text{s}$, much smaller than the timescale at which the flat region scaling starts, $\tau \sim 10^3\,\mu\text{s}$. Such a large discrepancy is not expected for any process for which τ_m provides the memory timescale, since both timescales should coincide as discussed earlier (see Fig. 4.4). For that reason, it seems more plausible that the flat region at the lowest frequencies is caused by the truncation of the divergent power law region above $\tau \simeq 10^3$. In fact, if longer

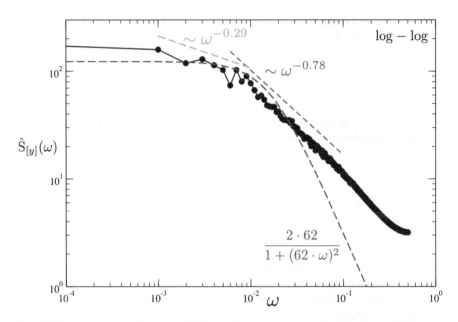

Fig. 4.5 Power spectrum of the same TJ-II turbulent data shown in Fig. 4.1 displayed in log-log scale. Also included are a best fit to Eq. 4.22, that yields a "memory time" $\tau_c \sim 60\,\mu$s (in blue) and a power-law fit to the flat part of the spectrum (in red)

time records of the same data were available, we should expect the $\sim \omega^{-0.3}$ scaling region to extend to smaller frequencies, eventually flattening at values of $\hat{S}_{[y]}(0)$ that steadily increase (towards infinity) with increasing record lengths. In this case, the actual value of τ_m obtained from our data set would become pretty meaningless.

Let's focus now on the intermediate region of the power spectrum that scales as $\sim \omega^{-0.3}$. This exponent value is consistent with what is expected for a process exhibiting long-term, self-similar persistence (see Eq. 4.24). This result suggests that some kind of complex dynamics might perhaps be active in the plasma. It is also interesting to note that the value $a \simeq 0.3$ is consistent with the power-law fits we did over the autocorrelation tail region, that yielded $\sim \tau^{-0.8}$ (see Fig. 4.3). From this value, one would expect a power spectrum with a decreasing power-law region at low frequencies with exponent $a = 1 - 0.8 \sim 0.2$, that is not far from the one we have obtained now.

We conclude by discussing the last scaling region, that goes like $\sim \omega^{-0.8}$ for the largest frequencies (i.e., the shortest timescales). This region is determined by the shape of the autocorrelation function around $\tau = 0$, for timescales $\tau < 100\,\mu$s. As we mentioned previously, an exponential decay of the autocorrelation function would appear here as an $\sim \omega^{-2}$ region at the largest frequencies. Here, the decay is much slower. It is difficult to say what it means without knowing more about the physics in place. In fact, the decay is sufficiently slow that it might reflect some other persistent process taking place at the shortest timescales in the local turbulent dynamics, maybe related to the dynamics that govern a single eddy. Or it could be related to the inner workings of the fast response of the Langmuir probe!

4.2.3   The Autodifference Function

The autocorrelation function and the power spectrum have been very popular approaches to study the presence of correlations and memory in time series for many years. However, these quantities do not exist for all random processes. In order to guide the discussion, we repeat here the ensemble average definition of the autocorrelation function (Eq. 4.3):

$$Ac_{[y]}(\tau) := \frac{\langle y(t)y(t+\tau)\rangle - \langle y(t)\rangle\langle y(t+\tau)\rangle}{\langle y(t)^2\rangle - \langle y(t)\rangle^2}. \tag{4.27}$$

Clearly, this definition ceases to be meaningful if the process y lacks finite second moments.[23] This will be the case, for instance, if the process exhibits any kind of Lévy statistics (see Sect. 2.3).

   This fact has lead to the introduction of other expected values that could replace the autocorrelation function in order to investigate and quantify the presence of memory. One such expected value is the so-called s-**autodifference function** [7], that is defined for any *stationary process* as[24]:

$$Ad_{[y]}^{s}(\tau) = \left\langle \left| y^{\#}(t+\tau)\right|^{s}\right\rangle + \left\langle \left| y^{\#}(t)\right|^{s}\right\rangle - \left\langle \left| y^{\#}(t+\tau) - y^{\#}(t)\right|^{s}\right\rangle, \tag{4.29}$$

for any $0 < s \le 2$, and with,

$$y^{\#}(t) = \frac{y(t) - \langle y(t)\rangle}{(\langle | y(t) - \langle y(t)\rangle |^{s}\rangle)^{1/s}}. \tag{4.30}$$

It is trivial to check that, for $s = 2$, the s-autodifference reduces to the autocorrelation function. Similarly to the autocorrelation function, the s-autodifference also

---

[23]If one uses the temporal-average formulation instead, the signature of the non-existence of the finite second moment is that $Ac_{[y]}^{T}(\tau)$, the autocorrelation computed with a record of finite length T, will never become independent of T. Instead, the autocorrelation will diverge as $T \to \infty$.

[24]The autodifference is a particular case of more general function known as the s-**codifference**, that is defined for two arbitrary time processes **with zero mean**, x and y, as [7]:

$$Cd_{[x,y]}^{s}(\tau) = \langle |x(t+\tau)|^{s}\rangle + \langle | y(t)|^{s}\rangle - \langle |x(t+\tau) - y(t)|^{s}\rangle. \tag{4.28}$$

The s-codifference reduces to (twice) the standard covariance for $s = 2$. In contrast to the cross-correlation, it can be calculated for processes x and y with divergent variance, as long as the moments of order s, $\langle | \cdot |^{s}\rangle$, do exist. It vanishes for any s for which it is finite when x and y are independent processes.

vanishes at time lag τ, for any s for which it is finite, *if $y(t)$ is independent of $y(t + \tau)$*.[25] Therefore, the s-autodifference can be used in a similar way to the autocorrelation, if the statistics of the process of interest lack finite second-order moments. It is sufficient to choose s sufficiently low so that all s-moments are finite.

In practical applications with time series, however, it is rare to use the ensemble average version of the s-autodifference. Instead, thanks to the assumed stationarity of the processes of interest, one can use the expression:

$$\langle |y(t)|^s \rangle = \lim_{T \to \infty} \frac{1}{2T} \int_{-T}^{T} |y(t)|^s dt, \qquad (4.31)$$

that allows us, to rewrite the s-autodifference as[26]:

$$\mathrm{Ad}_{[y]}^s(\tau) = \lim_{T \to \infty} \frac{1}{2T} \int_{-T}^{T} dt \left(\left| y^{\#}(t + \tau) \right|^s + \left| y^{\#}(t) \right|^s - \left| y^{\#}(t + \tau) - y^{\#}(t) \right|^s \right). \qquad (4.32)$$

The interpretation of the autodifference is however somewhat more involved. In contrast to the autocorrelation, that could take values only within the interval $[-1, 1]$, the autodifference satisfies that [7]:

$$1 \geq \mathrm{Ad}_{[y]}^s(\tau) \geq \begin{cases} 0, & 0 < s \leq 1 \\ 1 - 2^{s-1}, & 1 \leq s \leq 2. \end{cases} \qquad (4.33)$$

The upper bound (i.e., 1) is reached, for any s, for perfect correlation (i.e., when $y(t) = y(t + \tau)$). However, the lowest possible value for $0 < s \leq 1$ (i.e., 0) is reached whenever $y(t)$ is completely independent of $y(t + \tau)$; on the other hand, the lowest possible value for $1 \leq s \leq 2$ (i.e., $1 - 2^{s-1} < 0$) is reached for perfect anti-correlation (i.e., $y(t) = -y(t + \tau)$). This requires to interpret the s-autodifference values with some care (see Problem 4.1). The guiding principle rests on the mathematical interpretation of the autodifference: *the larger it is, the greater "the dependence" will be between $y(t)$ and $y(t + \tau)$, since a larger autodifference implies that it is less likely that $y(t)$ and $y(t + \Delta t)$ be different from each other* [7]. This principle permits us, when $1 < s \leq 2$, to differentiate between regions of **positive dependence** (i.e., $\mathrm{Ad}_{[y]}^s > 0$) and **negative dependence** (i.e., $\mathrm{Ad}_{[y]}^s < 0$). Regretfully, the same cannot be done when $0 < s < 1$, where the minimum value corresponds to absence of dependence, but one cannot separate positive and negative dependences.

---

[25] As it happened in the case of the autocorrelation, the inverse is not true in general.

[26] As it also happened with the autocorrelation, T should be chosen large enough as to guarantee that the result becomes independent of it.

4.3 Memory in Self-Similar Time Random Processes

In Chap. 3 we introduced fractional Brownian and Lévy motions as the poster children of self-similar random processes. In contrast to natural processes, fBm and fLm are self-similar over all scales. In spite of that, real processes are often compared against them to reveal their scale-invariant properties. Thus, it is quite illuminating to discuss how the various tools discussed in the previous sections perform on them. Or more precisely, on their increments, since fBm and fLm are not stationary processes (see Sect. 3.3.6).

Not surprisingly, it will be shown that all memory properties of fBm and fLm are completely determined by their self-similarity exponent H. This is a direct consequence of their monofractal character. As a result, any method that quantifies the self-similarity exponent (some of them discussed in Chap. 3) could in principle be used to characterize the type of memory present in fBm and fLm as well. This equivalence does however not hold if self-similarity is only valid over a finite range of scales or when the process is multifractal, as often happens with real data. The analysis of these cases must then be done with some care, always guided by our knowledge of the physics at work.

4.3.1 Fractional Brownian Motion

It will be remembered that fractional Brownian motions (Eq. 3.71) are H-sssi processes, with a self-similarity exponent $H \in (0, 1]$. fBm is not a stationary process. Its increments, however, are stationary for any value of the spacing h (see Eq. 3.83). They are often known as **fractional Gaussian noise** (fGn). Their autocorrelation function and power spectrum have some interesting properties.

4.3.1.1 Autocorrelation Function of fGn

The autocorrelation function for fGn with self-similarity exponent H behaves asymptotically as [8]:

$$\text{Ac}_{[\Delta_h y^{\text{fGn}}]}(\tau) \sim AH(2H - 1)|\tau|^{-2(1-H)}, \quad A > 0, \quad \tau \gg h. \tag{4.34}$$

That is, it has a power law tail. It is worth analyzing the meaning of the possible values of its exponent in the light of the discussions had in Sect. 4.2. In the case $H = 1/2$ it is apparent that the autocorrelation vanishes for every possible lag value except $\tau = 0$, meaning that the increments are all uncorrelated and that the related fBm process lacks memory on any scale.

For $H > 1/2$, however, the fGn autocorrelation tail exhibits a type I power law scaling (see Eq. 4.13). Indeed, the tail remains positive and the tail exponent is $a = 2(1-H) \in (0, 1]$. Therefore, fGn with $H > 1/2$ exhibits long-term, self-similar persistence for arbitrarily long lags. Clearly, the degree of persistence will be more pronounced the closer H is to unity.

The case $H < 1/2$, on the other hand, exhibits a tail that scales as a type II power law (see Eq. 4.15). The constant in front becomes negative, and the tail exponent is now $b = 2(1 - H) > 1$. We will soon argue, when discussing its power spectrum, that the integral of the autocorrelation function actually vanishes for this type of fGn in the limit $h \rightarrow 0$. As a result, fGn with $H < 1/2$ also exhibits long-term, self-similar memory over all scales, but with an antipersistent character instead. Antipersistence will be more pronounced as H approaches zero.

4.3.1.2 Power Spectrum of fGn

The power spectrum[27] of fGn with self-similarity exponent H has been shown to be [8]:

$$\hat{f}_h^{\text{fGn}}(\omega) \sim \frac{K(H)}{|\omega|^{2H-1}}, \quad \omega \ll h^{-1}, \tag{4.35}$$

with the constant $K(H) > 0, \quad \forall H$. Let's discuss this spectrum in the light of our previous discussions. The first thing one notes is that the exponent of the fGn autocorrelation function (i.e, $a = 2H - 2$) and the fGn power spectrum (i.e., $c = 1 - 2H$) satisfy the relation $c = a - 1$, as predicted by Eqs. 4.24 and 4.26. Secondly, we find that the power spectrum becomes flat for $H = 1/2$; for $H > 1/2$, the spectrum is a decreasing power-law over all scales, decreasing more steeply as H approaches unity, as expected from a persistent process; finally, for $H < 1/2$, the power spectrum is an increasing power-law, increasing faster as H approaches zero. These are precisely the scalings that were predicted, respectively, for uncorrelated, persistent and antipersistent long-term, self-similar memories (see Sect. 4.2).

4.3.2 Fractional Lévy Motion

Fractional Lévy motion (Eq. 3.76) was the other H-sssi process discussed in Chap. 3 (with $H \in (0, \max(1, \alpha^{-1})]$). In contrast to fBm, fLm lacked any finite moments for orders $r \geq \alpha$, where $\alpha \in (0, 2)$ is the tail-index that characterizes the symmetric noise, ξ_α that drives it. In particular, all second order moments diverged for any fLm. They also diverged for the series of their increments, known as fractional Lévy noise (fLn). As a result, neither the autocorrelation function nor the power spectrum

---

[27]The derivation of Eq. 4.35 can be done either by considering the limit of truncated versions of fGn, along the lines of what we did to derive Eq. 4.20, or by moving to a generalization of the Fourier representation known as a *Fourier-Stieltjes representation* [9], that exists for stationary random processes [10].

can be used with fLn. Instead, one needs to turn to other diagnostics such as the **autodifference** (Eq. 4.29) of the fLn series in order to characterize memory in fLm.

The autodifference of fLn has been studied in detailed in the literature [7]. Its asymptotic behaviour is given *almost everywhere* by [7]:

$$\mathrm{Ad}^{\alpha,H}_{[\Delta_h y^{\mathrm{fLn}}]}(\tau) \sim A_1(H,\alpha)\mathrm{sgn}(\alpha H - 1)\,|\alpha H - 1|^\alpha\, \tau^{-\alpha(1-H)}, \tag{4.36}$$

with the constant $A_1(H,\alpha) > 0$. The exception for this scaling is a small region in $\alpha - H$ space (shown in blue in Fig. 4.6), that is defined by the parameter ranges $\alpha_c < \alpha < 2$ and $0 < H < H_c(\alpha)$. The critical values are given by,

$$\alpha_c = \frac{(1 + \sqrt{5})}{2} \simeq 1.618, \tag{4.37}$$

$$H_c(\alpha) = 1 - \frac{1}{\alpha(\alpha - 1)}. \tag{4.38}$$

Within the excluded region, the autodifference scales instead as:

$$\mathrm{Ad}^{\alpha,H}_{[\Delta_h y^{\mathrm{fLn}}]}(\tau) \sim A_2(H,\alpha)\,(H\alpha - 1)\,\tau^{-(1+\frac{1}{\alpha}-H)}, \tag{4.39}$$

with the constant $A_2(H,\alpha) > 0$.

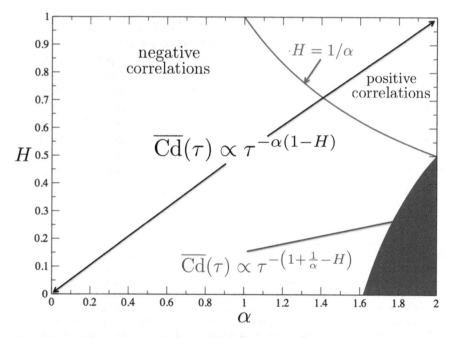

Fig. 4.6 Distribution of scaling regions for the autodifference of fLn in the $H-\alpha$ plane. The region where the scaling $\overline{\mathrm{Cd}}^\alpha(\tau) \sim \tau^{-(1+\alpha^{-1}-H)}$ holds is shown in blue. In the rest, $\overline{\mathrm{Cd}}^\alpha(\tau) \sim \tau^{-\alpha(1-H)}$. The boundary $H = 1/\alpha$ is shown in red. Above it, correlation is positive. Beneath it, negative

Let's analyze these scalings. The first thing to note is that the fLn autodifference vanishes if $H = 1/\alpha$, for any $\alpha \in (0, 2)$. Since $H \in (0, 1]$, *uncorrelated fLm is only possible for* $1 \leq \alpha \leq 2$ (see Fig. 4.6). Secondly, inspired by the fBm terminology, it is customary to say that fLm exhibits long-term persistence (i.e., positive correlations) for $H > 1/\alpha$, and long-term antipersistence (negative correlations) for $H < 1/\alpha$. This naming convention is consistent with the sign of the asymptotic tail of the autodifference. Indeed, the tail remains positive for $H > 1/\alpha$ and negative otherwise. However, note that the fact that $H \in (0, 1]$ implies that *no fLm with long-term persistence exists for* $\alpha < 1$ (see Fig. 4.6). For $1 < \alpha < 2$, on the other hand, persistent and antipersistent cases do exist depending on the specific value of H. It should also be noted that all fLm within the excluded region are antipersistent.

Finally, it is worth mentioning that the integral of the autodifference (over all lag values) diverges for any fLm with positive correlations, since the exponent of the tail is $\alpha(1 - H) < 1$ when $H > 1/\alpha$ and $\alpha > 1$. This fact is very reminiscent of what happened for the autocorrelation function in cases of long-term dependence, such as any fBm with $H > 1/2$.

4.4 Techniques for Detecting Memory in Stationary Time Series

We describe now several methods that can be used to characterize the presence of memory in stationary time series. These methods are reliable, easy to implement and have a straightforward interpretation.[28] We will start by presenting discrete formulations of the classical autocorrelation function, of the autodifference function and of the power spectrum that can be readily applied to any stationary time series. Then, we will proceed to introduce other popular methods such as the study of the **statistics of waiting-times** or the **R/S analysis**, introduced in the 1950s by Harold E. Hurst to look for correlations in hydrologic data. To illustrate all these methods, we will apply them to various synthetic series for fGn and fLn with some prescribed Hurst exponent (generated using the algorithms already discussed in Appendix 1 of Chap. 3).

---

[28]It is worth mentioning at this point that some authors often use many of the methods discussed in Chap. 3 to characterize scale-invariance (i.e., moment methods, rescaling methods, DFA, etc) to characterize the presence of memory. The rationale for their use is that, for the monofractal fBm and fLm processes, the self-similarity exponent H also characterizes completely the memory properties of their associated noises. However, one must be careful when adhering to this philosophy since monofractal behaviour does not usually extend over all scales in real data, being instead restricted to within the mesorange (either because of finite-size effects, or by having different physics governing the dynamics at different scales). As a result, the self-similarity exponent H obtained say, from DFA, and the exponent c of the power spectrum of a real set of data, will often will not verify the relation $c = 1 - 2H$ that we found for fBm in Sect. 4.3.1.

4.4.1 Methods Based on the Autocorrelation Function

Let's assume the discrete temporal record, $\{y_j, j = 1, \cdots, N\}$. In order to calculate its autocorrelation function, one could use the discrete version of Eq. 4.7, that is given by (for $k \geq 0$):

$$Ac^N_{[y]}(k) = \frac{1}{(N-k)\sigma_y^2} \sum_{j=1}^{N-k}(y_j - \bar{y})(y_{j+k} - \bar{y}), \quad k \geq 0 \qquad (4.40)$$

where discrete means and variances are given by:

$$\bar{y} = \frac{1}{N}\sum_{j=1}^{N} y_j, \quad \sigma_y^2 = \frac{1}{N}\sum_{j=1}^{N}(y_j - \bar{y})^2. \qquad (4.41)$$

These formulas have some limitations, though. First, N should be sufficiently large so that the results are independent of it. In addition, the reader must be aware of the fact that the resulting autocorrelation values become increasingly more imprecise as k gets larger. The reason is that the number of instances over which the autocorrelation value at lag k is averaged is $n(k) = N - k$, which decreases as k gets larger. To account for this deterioration, one typically disregards the values corresponding the largest k's. But how many of them? We advise to neglect at least all $k \geq N/10$.[29]

We have illustrated the autocorrelation method by applying it to three independent realizations of fGn with nominal self-similarity exponent $H = 0.8$ (see Fig. 4.7). The only difference between the realizations is their length: one is 10,000 points long, the second one is five times longer, and the last one is ten times longer. The fGn autocorrelation decays as a power-law with exponent given by Eq. 4.34. That is, $a = 2(1 - H) = 0.4$. This scaling is also included in the figure with a blue dashed line to guide the eye. Clearly, the tails of the three autocorrelation functions obtained from Eq. 4.40 follow the theoretical scaling. However, fluctuations (with an amplitude that increases with k) are apparent starting at relatively small lag values. These fluctuations appear, on the one hand, because of the lack of good statistics as k increases, but also because the monofractal character is lost in the fGn series at the shortest scales, due to its discreteness.

In fact, the fluctuations become so bad that the tail sometimes become negative, which should not happen for fGn with $H = 0.8$ for which antipersistence is absent. The fluctuations also make more difficult to determine the tail exponent precisely, which is particularly apparent for the shortest series. One possible technique to facilitate the extraction of this information, specially for short series, is to consider

---

[29]That is the reason why, in the autocorrelation of the TJ-II data we discussed at the beginning of this chapter, the values only run up to 10^3 lags, although the data set contained 10^4 points.

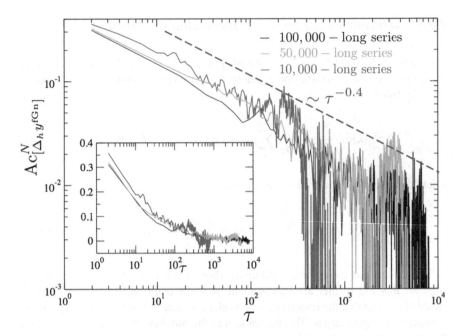

Fig. 4.7 Autocorrelation function of three independent realizations of synthetic fGn data generated with nominal $H = 0.8$, shown in log-log coordinates to better appreciate any power-law decay. Inset: the same functions, in cartesian coordinates

the *integral of the autocorrelation function up to time lag τ* instead (see Fig. 4.8):

$$\text{Ic}_{[y]}^{N}(k) = \sum_{j=0}^{k} \text{Ac}_{[y]}^{N}(k), \quad k \geq 0 \tag{4.42}$$

Since integration acts as a low-pass pseudo-filter, its application to the auto-correlation function reduces high-frequency fluctuations. In the case of fGn, the integrated autocorrelation function should scale as $\tau^{(2H-1)}$ for large τ, which should yield an exponent $+0.6$ for the cases under examination. The values obtained from the three series examined lie within 0.49–0.64, yielding a self-similarity exponent $H \simeq (0.75$–$0.82)$, that is not a bad estimate.

4.4.2   Methods Based on the Power Spectrum

To estimate the power spectrum of the time series $\{y_k, \ k = 0, 1, \cdots, N-1\}$, one can use a discretized version of Eq. 4.20. The nice thing about this approach is that

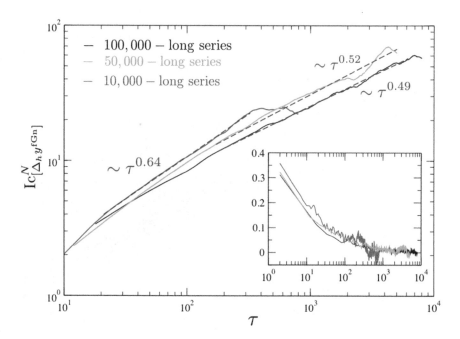

Fig. 4.8 Accumulated autocorrelation function of three independent realizations of synthetic fGn data generated with nominal $H = 0.8$, shown in log-log coordinates to better appreciate any power-law decay. Inset: corresponding autocorrelation functions, in lin-log scales

one does not need to evaluate the autocorrelation.[30] The discrete version we will use is:

$$\hat{S}_y^N(\omega_n) = \frac{|\hat{y}_N^*(\omega_n)|^2}{N}. \tag{4.43}$$

where the Fourier transform (see Appendix 1 of Chap. 2) is discretized as[31]:

$$\hat{y}_N^*\left(\omega_n = \frac{n}{N}\right) = \sum_{k=0}^{N-1} y_k e^{ink/N}, \quad n = 0, 1, \cdots, N-1. \tag{4.45}$$

---

[30]In fact, another way of estimating the autocorrelation of any time series is by computing its discrete power spectrum first, and then Fourier-inverting it to get the discrete autocorrelation.

[31]One could also use one of the many fast Fourier transform (FFT) canned routines available [11] to evaluate Eq. 4.43. In that case, subroutines perform the sum,

$$\sum_{k=0}^{N-1} y_k e^{2\pi ink/N}, \quad n = 0, 1, \cdots, N-1. \tag{4.44}$$

that provides the Fourier transform at frequencies $\omega_n' = 2\pi(n/N)$ instead of $\omega_n = n/N$.

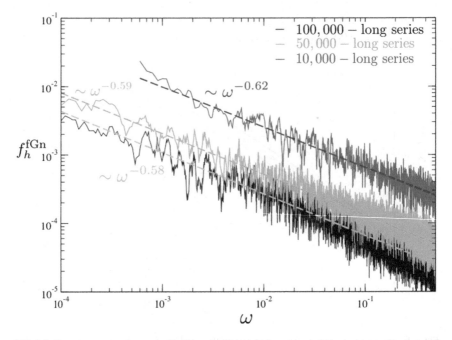

Fig. 4.9 Power spectra, shown in log-log coordinates, of two same independent realizations of synthetic fGn data generated with nominal $H = 0.8$ and whose autocorrelation functions were shown in Fig. 4.7

We have also calculated the discrete power spectrum (see Fig. 4.9) for the three fGn signals that were used to illustrate the calculation of the autocorrelation function in Sect. 4.4.1. The resulting power spectra (see Fig. 4.7) all exhibit fluctuations whose amplitude grow larger as the frequency increases. In spite of that, all spectra clearly show a power-law decay, ω^{-c}, with the exponent c easily obtained from a simple power-law fit. The value obtained is $c \simeq (0.60 \pm 0.02)$ is obtained, which is very close to the theoretical value $c = 2H - 1 = 0.6$. It is apparent that the power spectrum method is much less sensitive to the length of the available signal than the direct estimate of autocorrelation function, at least for fBm.[32]

4.4.3   Methods Based on the Autodifference

In the case in which the dataset of interest follows an statistical distribution that lacks finite second moments (for instance, any Lévy pdf), both its truncated

---

[32]However, the interpretation of the power spectrum becomes much more involved when dealing with non-monofractal signals. We will see an example when applying the technique to the analysis of the running sandpile, carried out as always at the end of the chapter (see Sect. 4.5).

autocorrelation function, $Ac^N(\tau)$ and power spectrum $\hat{S}^N(\omega)$, albeit finite, would scale non-trivially with N. This makes their analysis rather unreliable.[33] A possible way to go around this problem could be to estimate instead their s-autodifference (Eq. 4.29), using a value of $s < \alpha$, being α the tail index of the signal statistics. One might be tempted to use a discrete version of the s-autodifference in the same spirit to what we did for the autocorrelation. Namely,

$$
Ad_{[y]}^{s,N}(k) = \frac{\displaystyle\sum_{j=1}^{N-k}\left(|y_j - \bar{y}|^s + |y_{j+k} - \bar{y}|^s - \left|(y_{j+k} - y_k|^s\right)\right)}{\displaystyle\sum_{j=1}^{N-k}\left(|y_j - \bar{y}|^s + |y_{j+k} - \bar{y}|^s\right)}.
\tag{4.46}
$$

However, this discretization does not work very well in practice. Among other things, because as s becomes smaller, errors at the tail region become larger, making the determination of the tail exponent very difficult.[34] For that reason, other estimators for the autodifference have been proposed in the literature [12, 13]. We encourage the interested readers to study those references, but we will not discuss them in this book. Instead, we will discuss another method to deal with Lévy signals using the rescaled range analysis presented in the next section.

4.4.4 Hurst's Rescaled Range (R/S) Method

There are methods to characterize memory that do not rely on either the autocorrelation or the power spectrum. One of the most popular ones is the **rescaled range (or R/S) method**, introduced by Harold E. Hurst in the 1950s [14] while studying the long-term dependencies in the water content of the river Nile over time.

The R/S method must be applied on a *stationary time series*,

$$
\{y_k, \quad k = 1, 2, \cdots N\},
\tag{4.47}
$$

that is first demeaned,

$$
\tilde{y}_k = y_k - \frac{1}{N}\sum_{i=1}^{N} y_i,
\tag{4.48}
$$

---

[33]Indeed, second moments are infinite for an infinite series; they are finite, for a finite record, but their values diverge with the length of the series.

[34]These difficulties were also apparent when computing the discrete autocorrelation function, but are made more apparent as s becomes smaller.

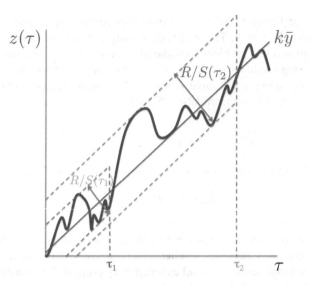

Fig. 4.10 Illustration of the R/S method. The range explored by the integrated process Z is shown at two different times, τ_1 and τ_2

and then integrated to form the cumulative variable,[35]

$$z_k = \sum_{i=1}^{k} \tilde{y}_i, \quad k = 1, 2, \ldots, N. \tag{4.49}$$

Then, the so-called *rescaled range* is formed ($k = 1, 2, \cdots, N$):

$$(R/S)_k = \sigma_k^{-1} \left(\max\{z_i, \ i = 1, \cdots k\} - \min\{z_i, \ i = 1, \cdots k\} \right). \tag{4.50}$$

with the variance up to time k, σ_k^2, is defined as:

$$\sigma_k^2 = \frac{1}{k} \sum_{i=1}^{k} \tilde{y}_i^2 - \left(\frac{1}{k} \sum_{i=1}^{k} \tilde{y}_i \right)^2. \tag{4.51}$$

In essence, the rescaled range measures *the size of the range of values explored by the process during the first k iterations* (see Fig. 4.10), normalized to its running variance up to iteration k. Hurst's idea was that, if the original stationary series (i.e., Y) lacked any long-term dependence, its demeaned version should essentially

---

[35]The $\tilde{y}$-series is thus formed by the increments of the z-series.

be a random noise.[36] Therefore, the resulting integrated process (i.e., Z) should have the properties of a random walk, whose range is known to grow as $\sqrt{t}$ (see Chap. 5). Therefore, Hurst expected that, if long-term correlations were present:

$$(R/S)_k \sim k^{H^{R/S}}, \qquad (4.52)$$

with $H^{R/S} \neq 1/2$ at least over an extended range of scales. More precisely, if positive correlations were present in $\tilde{Y}$, the increments should tend to maintain their sign more often than not, and the range explored should increase faster with k. That is, $H^{R/S} > 1/2$ in that case. The limiting value of the R/S exponent should be $H^{R/S} = 1$, corresponding to the case in which the process $\tilde{Y}$ is constant, what would make the range to increase linearly with k. On the other hand, the presence of long-term negative correlations would favor a frequent sign reversal for the increments. The explored range would then grow slower than in the uncorrelated case. Therefore, $H^{R/S} < 1/2$ would follow. Clearly, the minimum value would be $H^{R/S} = 0$.

Hurst first applied his method to several water reservoirs in the Nile river in the 50s. He found that the temporal series of the maximum level of water in the reservoirs yielded rescaled ranges that scaled with a Hurst exponent $H^{R/S} \sim 0.7$, that he interpreted as a signature of long-term positive dependence [14]. The R/S method was made popular by B.B. Mandelbrot in the late 60s, when he showed that the rescaled range of fractional Brownian motion satisfied Eq. 4.52 for all scales, with a value of $H^{R/S} = H$, being H the self-similarity exponent of fBm (Eq. 3.71) [8]. Since then, the method has remained popular.[37]

We will illustrate the R/S method on the same three realizations of fGn with nominal $H = 0.8$ previously discussed in Sect. 4.4.1. The resulting rescaled ranges are shown in Fig. 4.11. In order to improve the statistics, we repeat here a trick that was previously used in Sect. 3.4 when discussing moment methods. Namely, to take advantage of the fact that, to compute the rescaled range up to iteration k using Eq. 4.50, only the first k values in the signal are needed. Thus, we can break the signal in non-overlapping blocks of size k, calculate the rescaled range associated to each block, and then average the range over the blocks. The result is a much better resolved rescaled range.[38] Fitting the obtained ranges to a power law yields a value of $H^{R/S} = 0.79 \pm 0.01$, which is pretty close to the nominal self-similarity exponent.

---

[36]Hurst always assumed the statistics of the increment process (i.e., $\tilde{Y}$), to be near-Gaussian. As a result, his method only applies, as originally formulated, to processes with statistics with finite second moments.

[37]In fact, Mandelbrot showed that this scaling property of the range requires only that the random process be self-similar with exponent H and with stationary increments. As such, the same range scaling also applies to fLm.

[38]Although the average becomes less effective as k grows larger, since the number of blocks in the signal, roughly N/k, becomes smaller. This is apparent in the increasing perturbations that appear at the end of each of the curves in Fig. 4.11.

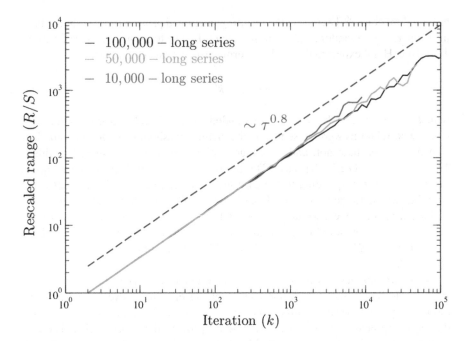

Fig. 4.11 Rescaled range as a function of time for the same three realizations of fGn with nominal $H = 0.8$ that were discussed in Figs. 4.7, 4.8, and 4.9. The loss of effectiveness of the block-averaging procedure becomes apparent in the form of oscillations at the largest iteration values

Although R/S analysis is very popular tool to look for memory in stationary time series, many authors prefer other schemes in order to estimate their self-similarity exponent, such as detrended fluctuation analysis [15] discussed in Appendix 2 of Chap. 3. The main reason is that R/S gives $H^{R/S} \simeq H$ for finite realizations of fGn with H exponents in the range (0.4–0.8), but tends to overestimate H for $H < 0.4$ and to underestimate it for $H > 0.8$, particularly when applied to short fGn series [16] (see Problem 4.2). Some prescriptions exist that manage to improve the method, making it to perform much better when the series length goes down to even just a few hundred data points [17].

In spite of these limitations, we will stick to the R/S method in large parts of this book because of two of its most outstanding properties. Namely, its *robustness against noise contamination* (see Problem 4.3) and its *predictable behaviour in the presence of periodic perturbations*[39] (see Problem 4.4).

---

[39]These two properties are quite important in the context of plasmas [18], due to the fact that signals are always contaminated by all kinds of noise and, quite often, also by mid- to low-frequency MHD modes. The R/S analysis provides a robust tool with which self-similarity and memory can be looked for in these contexts. On the other hand, plasma signals typically have tens (or even hundreds) of thousands of values, which makes the imprecisions of the R/S technique at small record lengths less worrisome.

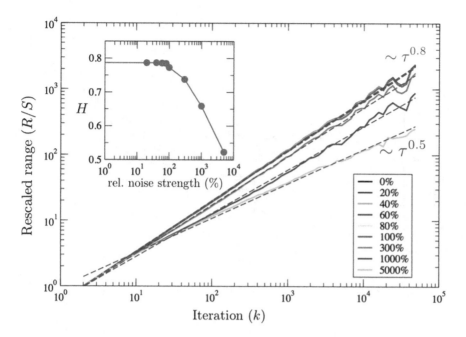

Fig. 4.12 Distortion of the rescaled for a 50,000 point long realization of fGn with nominal $H = 0.8$ when random uniform noise is added with a various relative amplitudes (with respect to the fGn standard deviation). It remains unchanged until the values of the noise relative amplitude is closer to or larger than unity

To illustrate the performance of R/S in the presence of noise, we have added a random perturbation to a fGn series generated with nominal $H = 0.8$:

$$\tilde{y}_k = y_k^{\text{fGn}} + \epsilon \sigma^{\text{fGn}} u_k, \tag{4.53}$$

where u is a random, uniform noise in $[-1, 1]$, $\epsilon > 0$ and σ^{fGn} is the standard deviation of the fGn signal. Therefore, for $\epsilon = 0$, the original signal is recovered, whilst for $\epsilon = 1$, the relative strength of the added noise is 100%. We have applied the R/S technique to several time series obtained by varying ϵ between 0 and 50. The results are shown in Fig. 4.12. The good performance of R/S is apparent, since the scaling remains basically unchanged from the original one until $\epsilon \sim 1$. That is, *up to the case in which the relative strength of the noise is almost a 100% of the original fGn signal.*

To illustrate the performance of R/S in the presence of a periodic perturbation, we have constructed the signal,

$$\tilde{y}_k = y_k^{\text{fGn}} + \epsilon \sigma^{\text{fGn}} \sin(2\pi k/T), \tag{4.54}$$

where T is the periodicity of the perturbation. The resulting rescaled ranges, using a synthetic fGn with nominal $H = 0.8$ perturbed with different periodic signals

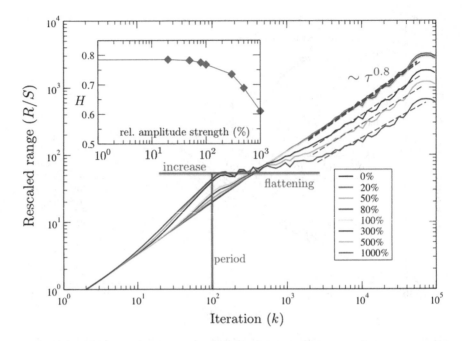

Fig. 4.13 Distortion of the rescaled range for a 100,000-long fGn series with $H = 0.8$ when perturbed by a periodic signal with period $T = 100$ and various relative strengths. The rescale range increases for iterations prior to the external periodicity and flattens behind it. If the estimation of H is restricted to iterations much larger than the external periodicity, it remains accurate until the relative amplitude of the external oscillation is close to or larger than unity

with $T = 100$, are shown in Fig. 4.13. The first thing that becomes apparent is that the periodicity has the largest effect for the rescaled range at time lags close to T: it increases for lags just below T, and decreases for lags just above the period. This signature is in fact extremely useful to detect hidden periodicities in the signal under examination.[40] Once periodicities have been spotted, one can restrict the determination of H to the timescales much longer than their periods, where the scaling behaviour of the rescaled range remains pretty much unchanged until the relative amplitude of the periodic perturbation is of similar to that of the original signal, as shown in the inset of Fig. 4.13.

4.4.4.1 R/S Analysis for Lévy Distributed Time Series

As we already mentioned in passing, Hurst developed his method with near-Gaussian fluctuations in mind. If the statistics of the process are however not

---

[40]This comes extremely handy in some contexts, such as plasma turbulence, where magnetohydrodynamic modes may sometime coexist with the background turbulence.

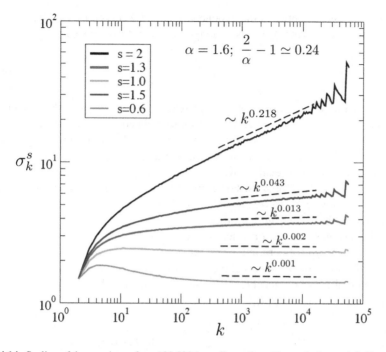

Fig. 4.14 Scaling of the s-variance for a 100,000-long fLn series with nominal $\alpha = 1.5$. For $s > \alpha$ ($s = 2$ is shown in black) the s-variance diverges with k. It should be noted, however, that although the s-variance is finite for $s < \alpha$, the closer to α the longer it takes for the s-variance to become independent of k

near-Gaussian, one could still apply Eq. 4.50, but the resulting exponent $H^{R/S}$ may no longer coincide with the self-similarity exponent of the process (assuming it has one). This is, for instance, what happens in the case of fractional Lévy noise. All instances of fLn lack a finite variance (see Chap. 2). As a result, σ_k^2 (Eq. 4.51) will scale with the fLn series length (see Fig. 4.14). This scaling can be estimated as,

$$\sigma_k^2 \simeq 2 \int_0^{\tilde{y}_k^{max}} \tilde{y}^2 L_{\alpha,0,\sigma}(\tilde{y}) d\tilde{y} \sim \left(\tilde{y}_k^{max}\right)^{2-\alpha} \propto k^{\left(\frac{2}{\alpha}-1\right)}, \quad k \gg 1, \tag{4.55}$$

where $\tilde{y}_k^{max} \sim k^{1/\alpha}$ is the estimated maximum value that the process can take up to iteration k, if distributed according to a symmetric Lévy pdf of tail-index α [19].

Since fLm has stationary increments (as fBm), it follows that its range should scale with its self-similarity exponent H [8]. Thus, when normalized to the divergent variance, the fLn rescaled-range will scale with k as:

$$(R/S)_k^{fLn} \sim \frac{k^H}{k^{\frac{1}{\alpha}-\frac{1}{2}}} \sim k^{H-\frac{1}{\alpha}+\frac{1}{2}} \sim k^{H^{R/S}}, \quad k \gg 1. \tag{4.56}$$

That is, the relation between the exponent obtained form the R/S analysis and the self-similarity exponent of fLm is no longer the identity, but instead:

$$H \simeq H^{R/S} - \frac{1}{2} + \frac{1}{\alpha}. \qquad (4.57)$$

This is an interesting result for several reasons. First, because the reader will remember that $H = 1/\alpha$ corresponds to uncorrelated Lévy motion, and that we previously associated $H > 1/\alpha$ with persistence, and $H < 1/\alpha$ with antipersistence (see Sect. 4.3.2). Therefore, if the standard R/S analysis is to be applied to a Lévy-distributed series, these relations would translate into $H^{R/S} = 1/2$ for uncorrelated motion, $H^{R/S} > 1/2$ for persistence and $H^{R/S} < 1/2$ for antipersistence. That is, *the same general interpretation already given for near-Gaussian statistics!* The conclusion is thus, that although the R/S method may not give the correct self-similarity exponent of the time series under examination, it can still be used to characterize the type of correlations present in it.

Secondly, the possible values of the self-similarity exponent of fLm are restricted to the interval $H \in \left(0, \max\left(1, \alpha^{-1}\right)\right]$ (see Sect. 3.3.4). This restriction becomes, when expressed in terms of $H^{R/S}$,

$$H^{R/S} \in \left(\frac{1}{2} - \frac{1}{\alpha}, \frac{3}{2} - \frac{1}{\alpha} \right], \quad \text{if } 1 < \alpha < 2, \qquad (4.58)$$

and

$$H^{R/S} \in \left(\frac{1}{2} - \frac{1}{\alpha}, \frac{1}{2} \right], \quad \text{if } 0 < \alpha < 1. \qquad (4.59)$$

This means that, although the R/S method can never yield values of $H^{R/S} > 1$ for any fLn, it could yield negative values for any $\alpha < 2$.[41] This would happen whenever the square of the variance grows faster than the range with time, that is possible for sufficiently antipersistent cases.

A possibility to make $H^{R/S}$ and H coincide again for fLm is to modify the standard R/S procedure as follows. One simply needs to replace the variance used to normalize the rescaled range by any other finite moment of the process,[42]

$$(R/S)_k^s := \frac{\max\{Z_i, \ i = 1, \cdots k\} - \min\{Z_i, \ i = 1, \cdots k\}}{\left[\frac{1}{k} \sum_{i=1}^{k} |\tilde{x}_i|^s \right]^{1/s}}, \quad s < \alpha, \qquad (4.60)$$

---

[41]For instance, for fLn with $\alpha = 1$, $H^{R/S}$ varies within $(-1/2, 1/2]$; for $\alpha = 0.5$, $H^{R/S}$ varies within $(-3/2, 1/2]$.

[42]That is, one that does not scale with the record length!

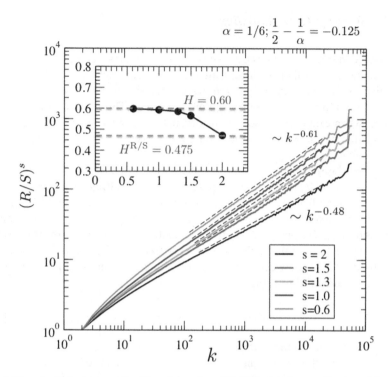

Fig. 4.15 *s*-rescaled range as a function of time for a 100,000 points long realization of fLn with $\alpha = 1.6$ and nominal self-similarity exponent $H = 0.6$. Values of $s < \alpha$ used to normalize the range are shown in the legend

which ensures that $(R/S)_k^s \propto k^H$, $k \gg 1$, for any fLm with self-similarity exponent H. In practice however, it is advisable to choose s much smaller than α since, although all these moments (that we will refer to as *s*-variances) are finite in theory, the value of k above which the moment becomes independent of k increases as k approaches α (see Fig. 4.14). As a result, $H^{R/S}$ may still differ significantly from the self-similarity exponent H as s approaches α from below. The process is illustrated in Fig. 4.15 for a synthetic series of fLn with $\alpha = 1.6$ and nominal exponent $H = 0.6$. As can be seen, the standard R/S analysis yields an exponent $H^{RS} \simeq 0.48$.[43] If one applies the $(R/S)^s$ procedure, the exponent obtained is virtually identical to H for $s < 1$, whilst there is a small difference for $s > 1$ that increases towards $s = 1.6$, in spite of the fact that all these *s*-variances are finite in theory. The reason for the discrepancy, as explained previously, is the increasingly longer time series needed for the *s*-variance to settle to an stationary value for $s > 1$ (see Fig. 4.14).

---

[43]Thus implying slight antipersistence, since $H^{R/S} < 0.5$, what is consistent with the self-similarity exponent $(H = 0.6 < 1/\alpha = 0.625)$.

4.4.5   Waiting Time Statistics

The last technique we will discuss to characterize the presence of memory in time series is the study of their **waiting-time statistics**. This technique is often applied to stationary processes in which *relatively large events can be easily made out from a background of smaller, irregular variations.* This is often the case, for instance, when examining time records of solar activity (see Chap. 7), radioactive disintegration of a substance, requests on a web server or beats in the human heart, to name a few.

The procedure consists on quantifying the statistics of *the lapses of time in which no significant activity takes place*, often referred to as **waiting-times**. Memory can be made apparent from the shape of their pdf. The reason is that the collection of all the points along the temporal axis at which the triggerings take place can be considered as a **point process** [20]. If no correlation exists between the triggerings, the point process becomes a **Poisson process** (see Sect. 2.3.4), and the statistics of the time intervals between successive triggerings (i.e., the waiting times) must follow an **exponential pdf** (Eq. 2.59):

$$E_{\tau_0}(w) = \tau_0^{-1} \exp(-w/\tau_0), \qquad (4.61)$$

with τ_0^{-1} providing the **mean triggering rate** of the process.

If the waiting-time pdf turns out to be not exponential, one should naturally suspect of the presence of some kind of correlation between the triggerings of the meaningful events in the stationary process.[44] However, our interest is on detecting the kind of long-term, self-similar memory that might appear in complex systems. In this case, observing a non-exponential waiting-time pdf is not sufficient. We should then expect that the waiting-time pdf has the asymptotic form,

$$p(w) \sim w^{-\lambda}, \qquad (4.62)$$

with $\lambda \in (1, 2]$ to ensure that no finite triggering rate could be defined (since the first moment of $p(w)$, that gives the triggering rate, would then be infinite), while keeping the pdf integrable.

It must be noted, however, that there is always a certain degree of arbitrariness when estimating waiting-time statistics. This arbitrariness is introduced while making the decision of what is to be considered a meaningful event in the time

---

[44]In some cases, non-exponential waiting-time pdfs have been interpreted as evidence of non-stationarity, instead of memory. That is, of processes in which the triggering rate varies with time, generally known under the name of **inhomogeneous Poisson processes** [20]. When the triggering rate is itself a random variable, the process is known as a **Cox process**. As usual, it is important to know the physics of the problem at hand, and to contrast any result with other analysis tools, so that one can distinguish between any of these non-stationary possibilities and truly self-similar behaviour.

series. Usually, this is done in practice by choosing a **threshold**, above which events are considered meaningful. Luckily, Poisson processes have the interesting property that, *if any subprocess is formed by randomly selecting elements with a certain probability* $0 < p < 1$, *the resulting subprocess is again a Poisson process but with the reduced triggering rate* $p\tau_0^{-1}$ [20]. Therefore, the process of thresholding does not introduce artefacts that could be mistaken by some kind of memory if memory was absent to begin with! Regretfully, this is not always the case when correlations are present. A widely prescription to choose a reasonable threshold is based on estimating the standard deviation (or, a finite *s*-variance, in case the variance diverges) of the series, σ, and setting the threshold to be a few times σ [21], but many other options are also useful. We will show an example of thresholding in the next section, when we revisit the analysis of the dynamics of the running sandpile. Be it as it may, thresholding must always be done carefully and, if possible, guided by what is known about the physics of the problem at hand.

4.5 Case Study: The Running Sandpile

We continue to characterize the dynamics of the running sandpile, this time by means of several of the diagnostics discussed in this chapter. In particular, we will make use of the R/S analysis, the power spectrum and waiting-time statistics. The signal that will be analyzed is the one usually known as the *instantaneous activity* [22], $I(t)$. It is defined (see inset of Fig. 4.16) as the *total number of unstable cells present in the sandpile at any given time*. $I(t)$ is a positive quantity by definition, being zero only when the sandpile is inactive. It is also bounded from above by the total length of the sandpile, L. In this case, we will consider a sandpile with size $L = 200$, critical slope $Z_c = 200$, toppling size $N_F = 20$, rain probability $p_0 = 10^{-4}$ and rain size $N_b = 10$. As a result, the overlapping figure-of-merit $(p_0 L)^{-1} \sim 50 < L$, meaning that the sandpile is effectively driven harder than the one considered in Sect. 3.5.

The sandpile intensity is rather different from most of the signals that we have examined in this chapter, since it is not symmetrically distributed. Indeed, Fig. 4.16 shows that its pdf is quite close to an exponential (except at $I = 0$, that corresponds to the moments in which the sandpile remains inactive). Most of the tools that we have discussed in previous sections do not require symmetric statistics to be applicable, since they only test whether the values of the process are statistically independent (or not) of each other. However, the lack of symmetry makes drawing comparisons with symmetric mathematical models, such as fGn or fLn, not just of little use, but sometimes even misleading.

We start by discussing the power spectrum of the intensity (see Fig. 4.17), that has been computed following the methodology discussed in Sect. 4.4.2. The obtained power spectrum has three distinct power-law scaling regions, $f(\omega) \sim \omega^{-c}$. This is in contrast to what one expects from a monofractal signal (such as fGn, for instance), but is typical of real signals where monofractality can be limited by either finite-size

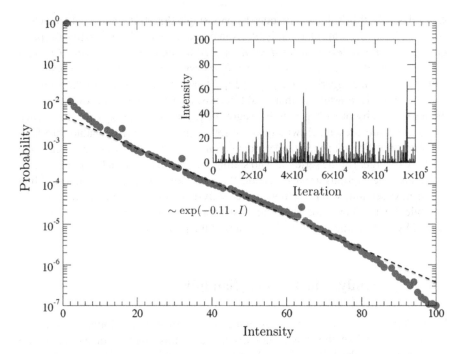

Fig. 4.16 Probability density function of the instantaneous intensity of the SOC state of a running sandpile ($L = 200, N_F = 20, Z_c = 200, p_0 = 10^{-4}$ and $N_b = 10$). An exponential fit to it (excluding its value at $I = 0$) is also shown as a black dashed line. **Inset:** intensity time trace for the first 100,000 iterations showing its irregular nature in time

effects or by different physics governing dynamics at different scale ranges. The region in the spectrum at the smallest frequencies ($\omega < 10^{-5}$) seems to be associated to some kind of negative correlations, as the increasing power-law behaviour (i.e., $c \sim -1.58$) reveals. The other two regions exhibit decaying power law behaviours, possibly associated with some kind of persistence. The first one scales as $c \sim 0.44$ for almost three decades, extending from $\omega \sim 10^{-5}$ to $\omega \sim 10^{-2}$; the second one scales as $c \sim 3.32$ for $\omega > 10^{-2}$, that dominates at the shortest timescales.

As was already pointed out, the fact that multiple scaling regions are present points to a global multifractal character for the sandpile intensity, although it might be dominated by a monofractal component over certain regions. The same conclusion is reached from the analysis of the results obtained with the R/S method. Again, three regions are found with different power-law scalings $R/s(\tau) \sim \tau^{H_{R/S}}$ (see Fig. 4.18). The first one, at the shortest timescales (i.e., $\tau < 10^2$), scales as $H_{R/S} \sim 1$; a second one scaling as $H_{R/S} \sim 0.78$ extends for almost three decades, between $\tau \sim 5 \times 10^2$ and $\tau \sim 5 \times 10^6$; the last one, scaling as $H_{R/S} \sim 0.21$, is found for $\tau > 10^6$. The correlation nature of these regions, as well as the timescales over which they extend, coincides with those found in the analysis of the power spectrum. The first two regions are dominated by persistence (i.e., $H^{R/S} > 1/2$), while the one at the longest timescales exhibits a strong antipersistent character (i.e., $H^{R/S} < 1/2$).

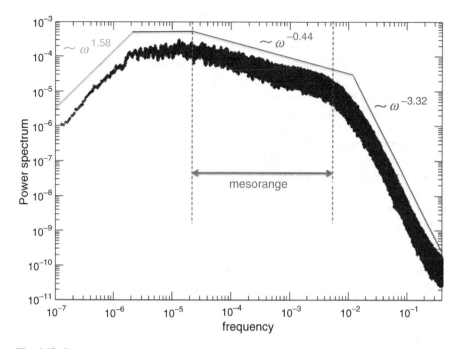

Fig. 4.17 Power spectrum of the sandpile intensity, exhibiting three different power-law scaling regions

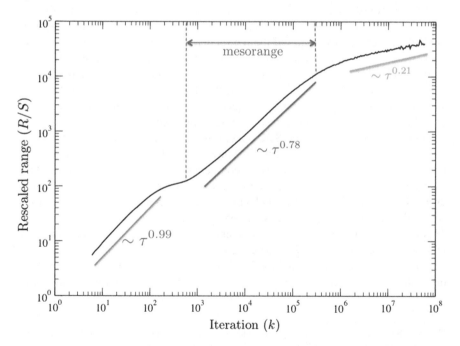

Fig. 4.18 R/S analysis of the intensity signal of the running sandpile. It exhibits three different power-law scaling regions, consistently with what was found in the power spectrum

An important point to make here is that the exponents of the R/S analysis and the power spectrum do not satisfy the relation $c = 2H_{R/S} - 1$ that is however satisfied by fGn. This was to be expected, since our signal is clearly neither monofractal nor symmetrically distributed. In fact, a one-to-one correspondence between the exponents c and $H^{R/S}$ should not be expected either in this case. Furthermore, it has been shown with the running sandpile that, by changing the sandpile drive intensity, p_0, one can significantly change the value of the exponent c in the mesorange whilst keeping $H_{R/S}$ basically unchanged [23]. It is observed, however, that regions that appear dominated by persistence (or antipersistence) in the power spectrum, do exhibit the same type of correlations when looked at using the R/S analysis. Or, in other words, that extended regions in the power spectrum scaling with exponents $0 < c < 1$ are always associated to extended regions, over the same timescales, that scale with exponents in $1/2 < H^{R/S} < 1$. The same happens for those regions dominated by antipersistence.

The physical interpretation of each of the three distinct regions previously identified[45] is well understood [22, 23]. The only region that is a direct consequence of the properties characteristic of the SOC state is the one that appears at the intermediate scales. We have previously referred to it as the **mesorange** (see Sect. 1.3). This region is a reflection of the *existence of long-term positive correlations* (thus the values of $H_{R/S} \sim 0.78 > 0.5$ and $c \sim 0.44 < 1$) *between the many different avalanches happening in the sandpile, that is transmitted through the footprints left in the sand slope profile by previous avalanching activity*. It is interesting to note that the mesorange extends, for this particular case, up to timescales of the order of 10^6 iterations, much longer than the duration of typical activity periods, that are limited for this sandpile finite size to a few hundred iterations. This estimate is obtained from the extension of the region that appears at the shortest timescales (largest frequencies), whose scaling is one of almost perfect correlation (as reflected by the obtained $H_{R/S} \simeq 0.99$). This quasi-perfect correlation is related to the average *correlation of each intensity period with itself*. The other scaling region that appears at the longest timescales (or shortest frequencies) is antipersistent, and is a reflection of finite-size effects. To understand its origin, one needs to consider that, as the sandpile is driven while in the SOC state, avalanches tend to be initiated at those locations that are closer to the local slope threshold, and to stop at those cells where the slope is flatter. As a result of this concomitant process, the slope profile will tend over time to alternate intermittently between periods where the profile is, almost everywhere, respectively closer and farther from marginal. When at the former states, the probability of global, system-size discharges will become much larger. This situation is maintained until a sufficient number of large events take place that manage to push the profile away from marginal over large fractions of the sandpile. The overall avalanching probability then becomes smaller until, under the external drive, the process starts all over again. This charge/discharge intermittent cycle is a

---

[45]Although these three regions are the most relevant regions, it can be shown that other regions can also appear in the sandpile, particularly as p_0 is varied [23] (see also Problem 4.6).

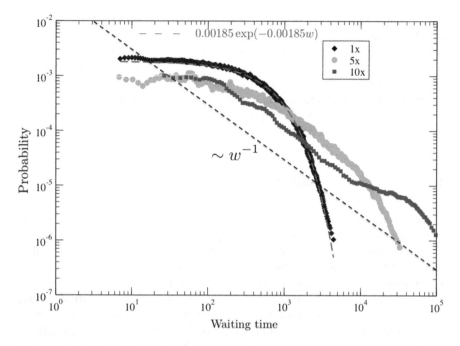

Fig. 4.19 Probability of the waiting-times between intensity events larger than 1σ, 5σ and 10σ, where σ is the standard deviation of the intensity signal. When most events are included, the pdf is very close to an exponential. However, for larger thresholds, a clear power-law behaviour appears, signalling to the existence of long-range dependences between sufficiently large events

direct consequence of the finite size of the sandpile. Clearly, if $L \to \infty$, the sandpile could never approach global marginality and this scaling region would not exist.[46]

We conclude by carrying out the analysis of the waiting-time statistics of the sandpile intensity. In order to identify the meaningful events, we have chosen the *threshold* equal to a prescribed number of times the standard deviation of the intensity signal. The results for one, five and ten times are shown in Fig. 4.19. As can be seen in the figure, the behaviour of the waiting-time pdf for the smallest threshold is exponential, a direct reflection of the random character of the sandpile drive. Indeed, since sand is dropped randomly on the sandpile, and since avalanches can only start at locations that have just received some sand, it should be expected that the triggering process be completely random. The exponential pdf obtained for one standard deviation, clearly confirms this fact.[47] However, the waiting-time pdf develops an extended power-law region as the threshold value is further increased.

---

[46]In fact, as the sandpile size increases, the timescale at which the anticorrelated region appears is pushed to larger timescales (or smaller frequencies)[23].

[47]In fact, it was at first believed that such exponential pdf was a fundamental feature of SOC models [25]. It was later shown that this was not the case (see Sect. 1.3), and that very similar SOC dynamics could also be obtained in a sandpile driven with coloured noises, which yielded a power-law waiting-time pdf at small thresholds that reflected the correlation in the drive [24], or by choosing the location where a random drive is applied in a correlated manner [26].

For ten times the standard deviation, such power law scales as $\sim w^{-1}$ for more than two decades, for this sandpile size. It turns out that this change in scaling behaviour coincides with choosing a sufficiently large threshold so that the events considered as avalanches lie within the mesoscale of avalanche sizes [24]. The appearance of the power-law scaling thus warns of the fact that avalanche triggering is strongly correlated if sufficiently large avalanches are considered. This correlation is a reflection of the presence of long-term memory present in the sandpile dynamics (see discussion in Sect. 1.3.2).

4.6   Final Considerations

In this chapter we have discussed the concept of memory, focusing on the type of self-similar, long-term memory (both persistent and antipersistent) that one often encounters in complex systems. Several methods have been presented to search for memory and to characterize it, including popular tools such as the autocorrelation function, the power spectrum, the R/S analysis or the study of waiting-time statistics. All these methods can be applied to any stationary time series. However, there is a tendency in the literature to interpret the results in the light of what is expected from monofractal, self-similar processes such as fractional Brownian noises. There is a hidden danger in abusing these associations since finite-size effects in some cases, and the coexistence of different type of dynamics over different range scales in others, can distort this simple picture and lead to erroneous conclusions. The lack of a one-to-one correspondence between exponents in the power spectrum and the R/S analysis, although valid for fGn, is just one example. In other cases, it is because of the presence of divergent statistics that the various methods need to be modified to yield meaningful results.

In addition, the important concept of the mesorange, the intermediate range of scales over which a real system scales (approximately) as a monofractal, has appeared again. The analysis of the sandpile activity has shown that the range of scales over which the mesorange becomes apparent depends on the quantity under examination, as can easily verified by comparing the relevant scaling regions in Fig. 4.18 (that shows the power spectrum of the sandpile activity) and in Fig. 3.19 (showing the moments of the sandpile global mass). The neat thing is, however, that all of them are reflections of the same underlying dynamics. In this case, self-organized criticality.

Problems

4.1 Autodifference
Prove that the relations given in Eq. 4.33 hold, with the upper bound obtained for perfect correlation, and the lower bound reached for perfect anticorrelation, for $1 < \alpha < 2$.

4.2 DFA vs R/S: First Round

Generate two fGn series (use the code written for Problem 3.6) with 50,000 points and nominal self-similarity exponents $H = 0.25$, $H = 0.4$, $H = 0.75$ and $H = 0.9$. Then, apply DFA1 and R/S and estimate the Hurst exponent of the series. Which method performs better for each value of H?

4.3 DFA vs R/S: Second Round

Use a uniform random number generator to add a random perturbation to the two fGn series with 50,000 points and self-similarity exponents $H = 0.25$ and $H = 0.75$. Set the amplitude of the noise to various values (say, $\sigma/5$, $\sigma/2$, σ and 3σ) relative to the standard deviation of the fGn series. Then, apply DFA1 and R/S to try to estimate the Hurst exponent of the series. Which method performs better as a function of the amplitude of the noise?

4.4 DFA vs R/S: Third Round

Generate two fGn series with 50,000 points and nominal self-similarity exponents $H = 0.25$ and $H = 0.75$. Add a periodic perturbation with period $T = 300$ for various amplitudes (say, $\sigma/5$, $\sigma/2$, σ and 3σ) relative to the standard deviation of the fGn series. Then, apply DFA1 and R/S and estimate the Hurst exponent of the series. Which method performs better as a function of the amplitude?

4.5 Running Sandpile: Global Mass Analysis

Use the sandpile code previously built (see Problem 1.5) and produce a time (at least 10^8 iterations!) record of the global mass evolution for a sandpile with $L = 200$, $Z_c = 200$, $N_f = 20$, $N_b = 10$ and $p_0 = 10^{-3}$. Carry out the R/S, power spectrum and waiting-time analysis done in Sect. 4.5 but for the global mass. Do the results compare better with what could be expected from fGn than what we found for the sandpile activity in this chapter?

4.6 Running Sandpile: Effect of Varying Drive

Use the sandpile code previously built (see Problem 1.5) and produce a time record (at least 10^8 iterations!) of the global mass evolution for a sandpile with $L = 200$, $Z_c = 200$, $N_f = 20$, $N_b = 10$ and rain probabilities $p_0 = 10^{-5}$, 10^{-4} and 10^{-2}. Carry out the R/S, power spectrum and waiting-time analysis for each of their activity time records. How do the results change with respect to the case analysed in Sect. 4.5? Are there any new regions? How do the transition points between regions change? How about the values of the exponents in each region?

References

1. Hergarten, S.: Self-Organized Criticality in Earth Systems. Springer, Heidelberg (2002)
2. Box, G.E.P., Jenkins, G.M., Reinsel, G.C.: Time Series Analysis: Forecasting and Control. Prentice-Hall, New York (1994)
3. Feller, W.: Probability Theory and Its Applications. Wiley, New York (1950)
4. Fuller, W.A.: Introduction to Statistical Time Series Analysis. Wiley, New York (1996)

5. Hidalgo, C., Balbin, R., Pedrosa, M.A., Garcia-Cortes, I., Anabitarte, E., Sentíes, J.M., San Jose, M., Bustamante, E.G., Giannone, L., Niedermeyer, H.: Edge Plasma Turbulence Diagnosis by Langmuir Probes. Contrib. Plasma Phys. 36, 139 (1996)
6. Kubo, R.: Fluctuation-Dissipation Theorem. Rep. Prog. Phys. 29, 255 (1966)
7. Samorodnitsky, G., Taqqu, M.S.: Stable Non-Gaussian Processes. Chapman and Hall, New York (1994)
8. Mandelbrot, B.B., van Ness, J.W.: Fractional Brownian Motions, Fractional Noises and Applications. SIAM Rev. 10, 422 (1968)
9. Katznelson, Y.: An Introduction to Harmonic Analysis. Dover, New York (1976)
10. Yaglom, A.M.: Correlation Theory of Stationary and Related Random Functions: Basic Results. Springer, New York (1987)
11. Press, W.H., Teukolsky, S.A., Vetterling, W.T., Flannery, B.P.: Numerical Recipes. Cambridge University Press, Cambridge (2007)
12. Rosadi, D., Deistler, M.: Estimating the Codifference Function of Linear Time Series Models with Infinite Variance. Metrika 73, 395 (2011)
13. Yu, J.: Empirical Characteristic Function Estimators and Its Applications. Econ. Rev. 23, 93 (2004)
14. Hurst, H.E.: Long Term Storage Capacity of Reservoirs. Trans. Am. Soc. Civ. Eng. 116, 770 (1951)
15. Kantelhardt, J.W., Zschiegner, S.A., Koscielny-Bunde, E., Havlin, S., Bunde, A., Stanley, H.E.: Multifractal Detrended Fluctuation Analysis of Nonstationary Time Series. Physica A 316, 87 (2002)
16. Mercik, S., Weron, R., Burnecki, K.: Enigma of Self-Similarity of Fractional Lévy Stable Motions. Acta Phys. Pol. B 34, 3773 (2003)
17. Weron, R.: Estimating Long Range Dependence: Finite Sample Properties and Confidence Intervals. Physica A 312, 285 (2002)
18. Carreras, B.A., van Milligen, B.P., Pedrosa, M.A., Balbin, R., Hidalgo, C., Newman, D.E., Sanchez, E., Frances, M., Garcia-Cortes, I., Bleuel, J., Endler, M., Riccardi, C., Davies, S., Matthews, G.F., Martines, E., Antoni, V., Latten, A., Klinger, T.: Self-Similarity of the Plasma Edge Fluctuations. Phys. Plasmas 5, 3632 (1998)
19. Chechnik, A.V., Gonchar, V.Y.: Self and Spurious Multi-Affinity of Ordinary Lévy Motion, and Pseudo-Gaussian Relations. Chaos Solitons Fractals 11, 2379 (2000)
20. Cox, D.R., Isham, V.: Point Processes. Chapman and Hall, New York (1980)
21. Spada, E., Carbone, V., Cavazzana, R., Fattorini, L., Regnoli, G., Vianello, N., Antoni, V., Martines, E., Serianni, G., Spolaore, M., Tramontin, L.: Search of Self-Organized Criticality Processes in Magnetically Confined Plasmas: Hints from the Reversed Field Pinch Configuration. Phys. Rev. Lett. 86, 3032 (2001)
22. Hwa, T., Kardar, M.: Avalanches, Hydrodynamics and Discharge Events in Models of Sand Piles. Phys. Rev. A 45, 7002 (1992)
23. Woodard, R., Newman, D.E., Sanchez, R., Carreras, B.A.: Persistent Dynamic Correlations in Self-Organized Critical Systems Away from Their Critical Point. Physica A 373, 215 (2007)
24. Sanchez, R., Newman, D.E., Carreras, B.A.: Waiting-Time Statistics of Self-Organized-Criticality Systems. Phys. Rev. Lett. 88, 068302 (2002)
25. Boffetta, G., Carbone, V., Giuliani, P., Veltri, P., Vulpiani, A.: Power Laws in Solar Flares: Self-Organized Criticality or Turbulence? Phys. Rev. Lett. 83, 4662 (1999)
26. Sattin, F., Baiesi, M.: Self-Organized-Criticality Model Consistent with Statistical Properties of Edge Turbulence in a Fusion Plasma. Phys. Rev. Lett. 96, 105005 (2006)

Chapter 5
Fundamentals of Fractional Transport

5.1 Introduction

In many physical systems, *transport* plays a central role. By transport we mean *any macroscopic process that moves some physical quantity of interest across the system*. For instance, transport processes are responsible for transporting mass and heat, including pollutants, throughout the atmosphere. Or for transporting water and energy, but also debris, across the ocean. Or for transporting plasma density and energy, but also impurities, out of a tokamak. Or for transporting angular momentum out of an accretion disk, but also mass and energy into the black hole at its center. In all these examples, the understanding of how transport behaves and its quantification are extremely important from a theoretical and practical point of view.[1]

The most popular *phenomenological* description of macroscopic transport is probably **Fick's law**, introduced by A. Fick in the mid 1800s [1]. In particular, Fick proposed that *the local flux of any physical quantity of interest was proportional to the local value of its gradient*.[2] A linear dependence is indeed the simplest hypothesis one could make, but it turns out that it can also be often justified (although not always!) by the dynamics of the microscopic processes responsible, as we will soon discuss in Sect. 5.2.

If, for the sake of simplicity, one focuses on particle transport only and restricts the discussion to one dimension, Fick's law expresses the particle flux at position x and time t as,

$$\Gamma(x, t) = -D \frac{\partial n}{\partial x}(x, t). \tag{5.1}$$

---

[1] Note also that, in all of these examples, transport is mainly carried out by turbulence!

[2] Other famous examples that share the same spirit that Fick's law are **Fourier's law**, that establishes the *flux of heat* to be proportional to the *temperature gradient*, $q = -\kappa \partial T / \partial x$, where κ is known as the **thermal conductivity**; and **Newton's law**, that relates the *kinematic stress* and the *velocity shear in a perpendicular direction*, $\tau_x = \nu \partial u_x / \partial y$. ν is known as the **kinematic viscosity**.

© Springer Science+Business Media B.V. 2018
R. Sánchez, D. Newman, *A Primer on Complex Systems*,
Lecture Notes in Physics 943, https://doi.org/10.1007/978-94-024-1229-1_5

The proportionality coefficient, $D > 0$, is known as the **particle diffusivity**. It has dimensions of (length)^2/time. The physical meaning of the minus sign in Fick's law is made apparent after introducing the flux into the **continuity equation**,[3]

$$\frac{\partial n}{\partial t} + \frac{\partial \Gamma}{\partial x} = S(x, t). \tag{5.2}$$

that becomes the famous **classical diffusion equation**,

$$\frac{\partial n}{\partial t} = D\frac{\partial^2 n}{\partial x^2} + S(x, t). \tag{5.3}$$

The diffusion equation is very familiar to all physics and science students since their high school days. The minus sign thus ensures that the transport driven by the gradient will tend to flatten it out. Once $\partial n/\partial x$ vanishes, the flux becomes zero.

In order to use Fick's law in practical cases, one needs to estimate the constant that relates flux and gradient. Sometimes this can be done from the knowledge of the microscopic processes taking place in the system; but more often than not, these coefficients are directly obtained from experiments. In both cases, however, the soundness of the linearity behind Fick's approach is implicitly assumed. This assumption may not always be justified, though. For instance, this is the situation in some complex systems for reasons that will be discussed at length throughout this chapter. Turbulent plasmas seem to be particularly good at providing situations in which Fick's law fails. For instance, this seems to be the case of fusion plasmas magnetically confined in tokamaks, for which multiple observations suggest that the radial transport of particles and energy does not behave in a Fickian way [2, 3]. Examples can however be found in many other physical systems. For instance, in the transport of magnetic vortices across Type-II superconductors in the so-called Bean state [4]. Or in the transport through fractal landscapes and random media [5], such as the percolation of water through porous rock or the propagation of cracks across rocks and ice [6].

In all these cases and many others other phenomenological formalisms must be found to characterize transport. A relationship between flux and gradient that appears to be sufficiently general to capture transport in many complex systems is[4]:

$$\Gamma(x, t) = \int_0^t dt' \int_{-\infty}^{\infty} dx' K(x - x', t - t') \left[\frac{dn}{dx}(x', t')\right]. \tag{5.4}$$

---

[3]The continuity equation simply expresses mathematically the fact that the number of particles must be conserved.

[4]This relation could be made even more general since, as it stands, it assumes translational invariance in time and space. Otherwise, the kernel would depend on either time or space, either explicitly or through other fields. In this chapter, however, we will always assume, for simplicity's sake, that spatial and temporal translational invariance holds.

Here, a new player has been introduced: the **spatio-temporal kernel** $K(\Delta x, \Delta t)$, that makes it possible to endow transport dynamics with a **non-local** character in space, and **non-Markovian** nature in time. By non-local it is meant that the value of the gradient at other positions ($x' \neq x$) influences the flux at the current position x; by non-Markovian, that the value of the gradient at past times ($t' < t$) influences the flux at the current time t.[5] Clearly, Fick's law is recovered from Eq. 5.4 if the kernel is chosen to be:

$$K(\Delta x, \Delta t) = -D\delta(\Delta x)\delta(\Delta t), \qquad (5.5)$$

that collapses the integral so that only the local value of the density gradient (in space and time) is relevant to determine the local flux.

We could say that the main objective of this chapter is to discuss and clarify what constitutes a reasonable choice for the kernel in order to describe transport across complex systems that exhibit spatial scale invariance (as discussed in Chap. 3) and long-term, scale-invariant memory (as discussed in Chap. 4). In order to reach these goals, we will need to revisit first the physical basis of Fick's law (Sect. 5.2) and to identify the reasons why it fails in the case of scale-invariant complex systems. Once this is well established, we will be able to discuss various generalizations that resolve the identified shortcomings (Sect. 5.3). The final product of this lengthy tour will be a generalized phenomenological transport framework, based on the use of **fractional derivatives**, that could prove more suitable to model transport in the context of complex systems.

5.2 Diffusive Transport: Fundamentals

In order to gain some insight on the physical basis of Fick's law, it is worth to review two *"microscopic"* mathematical models[6] from which it can be *"derived"*: the **continuous-time random walk** (CTRW) and the **Langevin equation**.[7]

5.2.1 The Continuous-Time Random Walk

The **continuous-time random walk** (or CTRW) is a mathematical model, introduced in the 1960s by E.W. Montroll [7], that generalizes the classical random walk

---

[5]This last statement establishes a direct connection between transport and the concept of memory that was discussed in Chap. 4.

[6]The quotes are used to stress the fact that these two models are just that: mathematical models that provide with a naive, although rich, idealization of the processes that might be taking place at the microscopic level.

[7]It might appear to some readers that we discuss them in much deeper detail than what would be needed for the purposes of this subsection. The reasons for doing this will become clearer soon, since these models will provide the basis of the generalizations leading to fractional transport.

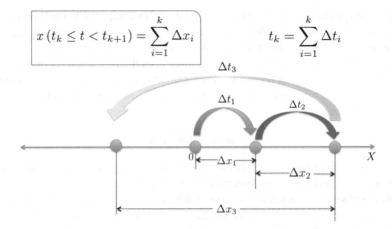

Fig. 5.1 Sketch of the CTRW. Each walker moves along the real axis by taking successive jumps of length Δx_i, $i = 1, \cdots, N$. In between jumps, it sits at rest for a lapse of time Δt_i. The position of the walker as a function of time is thus given by $x(t_k \leq t < t_{k+1}) = \sum_{i=1}^{k} \Delta x_i$, where $t_k = \sum_{i=1}^{k} \Delta t_i$. The choice for the probability density functions for step lengths, $p(\Delta x)$ and waiting times, $\psi(\Delta t)$, defines the various CTRW

[8]. The model considers a population of particles (or *walkers*) whose motion is defined probabilistically. In its one-dimensional version (see Fig. 5.1) each walker stays at its initial position, x_0, for a **waiting time**, Δt. Then, it carries out a **jump** of length Δx, that takes it to its new position, $x_0 + \Delta x$. The process is repeated infinitely by each walker. The CTRW is defined[8] by prescribing the pdfs of the jump sizes, $p(\Delta x)$, and the waiting-times, $\psi(\Delta t)$.

How is the CTRW related to the classical diffusion equation (Eq. 5.3)? To illustrate this point, we have followed numerically (see Prob. 5.1) the evolution of $N = 10^6$ walkers in the case in which $p(x)$ is a Gaussian law (Eq. 2.30) and $\psi(\Delta t)$ is an exponential (Eq. 2.59). That is,

$$p(\Delta x) = N_{0,2\sigma^2}(\Delta x), \quad \psi(\Delta t) = E_{\tau_0}(\Delta t). \tag{5.6}$$

Figure 5.2 shows the evolution of the walker density when all walkers are initiated at $x = 0$. The figure also includes the numerical solution of the classical diffusive equation using a diffusivity $D = \sigma^2/\tau_0$, starting from $n(x, 0) = N\delta(x)$.[9] It is

---

[8]It is possible to define more general CTRWs. For instance, one could define non-separable CTRWs by specifying instead a joint probability $\xi(\Delta x, \Delta t)$ of taking a jump of size Δx after having waited for a lapse of time Δt. It is also possible to define CTRWs that are not invariant under spatial and/or temporal translations, and that depend on time, space or both either explicitly [9, 10] or through dependencies on other fields [11, 12].

[9]More precisely, the initial condition distributes the N walkers uniformly in the interval $(-dx/2, dx/2)$, where dx is the spacing used to discretize the problem.

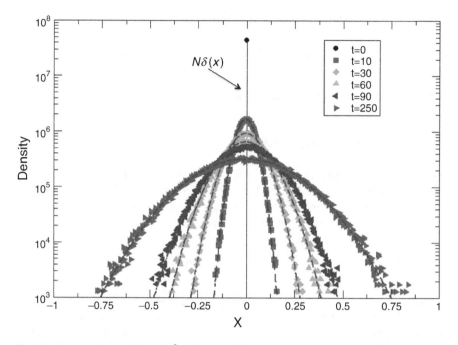

Fig. 5.2 Time evolution of $N = 10^6$ walkers, initially located at $x = 0$, advanced according to a Gaussian jump size distribution with $\sigma^2 = 5 \times 10^{-5}$, and an Exponential waiting time distribution with $\tau_0 = 5$. In blue, the analytical solution of the classical diffusion equation with $D = 10^{-5}$ is shown, at the same times

apparent that the two solutions are very close to each other, which suggests that there must be a way to derive the classical diffusion equation from the CTRW defined by Eq. 5.6, at least in some limiting form.[10]

In order to make this connection we need to solve the CTRW first. That is, we need to find its **propagator**, $G(x, t|x_0, t_0)$, as a function of the choices made for $p(\Delta x)$ and $\psi(\Delta t)$.[11] Propagators were already introduced in Sect. 3.3. It will be remembered that $G(x, t|x_0, t_0)$ provides the *probability of finding a walker at position x and time t , assuming that they started at position x_0 at time t_0.* Once known, one can write the general solution of the CTRW, for arbitrary initial

---

[10]From our discussions in Chap. 2 about the central limit theorem, the fact that Gaussians seem to be related to diffusion also suggests that the underlying physical processes behind diffusive transport are of an additive, and not multiplicative nature. This idea will also play a role when extending the description of transport beyond diffusion, as will be seen soon.

[11]The impatient reader may ignore this relatively lengthy calculation and jump directly to the final result (Eq. 5.18), and continue reading without any loss of coherence.

condition $n_0(x)$ and external source $S(x, t)$, as [13]:

$$n(x, t) = \int_0^t dt' \int_{-\infty}^{\infty} dx' G(x, t | x', t') S^{\text{aug}}(x', t'),\qquad(5.7)$$

where the augmented source is defined as $S^{\text{aug}}(x, t) = S(x, t) + n_0(x)\delta(t)$.

5.2.1.1 Propagator of the CTRW

The propagator of the unbounded CTRW defined by $p(\Delta x)$ and $\psi(\Delta t)$ is more easily found in terms of its Fourier-Laplace transform,[12,13] $\bar{G}^{\text{CTRW}}(k, s)$. Following *Montroll* [7], and assuming that $t_0 = 0$ without loss of generality, we first write the probability of a walker being found at position x at time t as[14]:

$$G^{\text{CTRW}}(x, t | x_0, 0) = G^{\text{CTRW}}(x - x_0, t) = \int_0^t \eta(t - t') Q(x - x_0, t') dt',\qquad(5.8)$$

where $Q(x - x_0, t')$ is the total probability of the walker arriving at x (from x_0) at time t' after executing any number of jumps, and $\eta(t - t')$ is the probability of **staying** at site x for an interval of time $\Delta t = t - t'$. Note that $Q(x - x_0, t')$ differs from $G^{\text{CTRW}}(x - x_0, t')$ in the fact that the latter includes the possibility of having arrived at x (from x_0) at time $t' < t$ and having stayed there until time t. Temporal convolutions, as the one appearing in Eq. 5.8, always become much simpler after transforming them to Laplace-Fourier space (see Appendix 1). Therefore, Eq. 5.8 becomes[15]:

$$\bar{G}^{\text{CTRW}}(k, s) = \tilde{\eta}(s)\bar{Q}(k, s).\qquad(5.9)$$

Next, we compute $\tilde{\eta}(s)$ and $\bar{Q}(k, s)$. Obtaining $\tilde{\eta}(s)$ is quite straightforward, since the probability of staying quiet during the time interval $[0, t]$, for any t, is one minus

---

[12]A primer on the Laplace transform has been included in Appendix 1 for those readers unfamiliar with it. The Laplace transform is very useful to simplify the solution of initial-value problems. When used with partial differential equations, they are often combined with Fourier transforms, that were already introduced in Appendix 1 of Chap. 2.

[13]We will use the notation $\bar{f}(k, s)$ to denote the double Laplace-Fourier transform. It will be remembered that $\hat{f}(k, t)$ is used for the Fourier transform, whilst $\tilde{f}(x, s)$ is used for the Laplace transform.

[14]Since the CTRW, as it has been defined here, is translationally invariant in both time and space, its propagator can only depend on the difference $x - x_0$ and not separately on x and x_0.

[15]The spatial and temporal variables associated to the Fourier (k) and Laplace (s) variables are, respectively, $\Delta x = x - x_0$ and τ.

the probability of leaving x between 0 and t (see Appendix 1, Eq. 5.116):

$$\eta(t) = 1 - \int_0^t \psi(t')dt' \implies \tilde{\eta}(s) = \frac{1 - \tilde{\psi}(s)}{s}. \tag{5.10}$$

Finding $\bar{Q}(k, s)$ is a more demanding task. Indeed, since $Q(x, t)$ gives the total probability of the walker making it to position x by time t **by any possible mean, all paths** to x contribute to it *independently of the any number of jumps taken.* By denoting the probability of arriving at x at time t in j steps by $Q^j(x, t)$, we can calculate this quantity is:

(*probability of the walker executing its j-th jump at time t*)

$\times$

(*probability of the walker reaching x after j jumps*)

and respectively referring to each factor by $\psi_j(t)$ and $p_{n,j}(x)$, one has that:

$$Q^j(x, t) = \psi_j(t)p_{n,j}(x). \tag{5.11}$$

Thus,

$$Q(x, t) = \sum_{j=0}^{\infty} Q^j(x, t) = \sum_{j=0}^{\infty} \psi_j(t)p_{n,j}(x). \tag{5.12}$$

Since $\psi_j(t)$ gives the probability of the j-th jump occurring at time t is related to the probability of the $(j-1)$-th jump through the following recurrence relation:

$$\psi_j(t) = \int_0^t \psi(t - t')\psi_{j-1}(t')dt', \tag{5.13}$$

where $\psi_1(t) = \psi(t)$, is the waiting-time pdf. Taking again advantage of the convolution theorem for the Laplace transform (see Appendix 1), we Laplace-transform Eq. 5.13 to get:

$$\tilde{\psi}_j(s) = [\tilde{\psi}(s)]^j. \tag{5.14}$$

Similarly, the probability of finding our walker t at a given location x after j iterations obeys the recurrence formula:

$$p_{n,j+1}(x) = \int_{-\infty}^{\infty} p(x - x')p_{n,j}(x')dx', \tag{5.15}$$

where $p_{n,1}(x) = p(x)$, the jump size pdf. Taking now advantage of the convolution theorem for the Fourier transform (see Appendix 1 of Chap. 2), we Fourier-transform this relation to get to,

$$\hat{p}_{n,j}(k) = [\hat{p}(k)]^j. \tag{5.16}$$

We can now use Eqs. 5.16 and 5.14 into the Fourier-Laplace transform of Eq. 5.11:

$$\bar{Q}(k,s) = \sum_{j=0}^{\infty} \left[\tilde{\psi}(s)\hat{p}(k) \right]^j = \frac{1}{1 - \tilde{\psi}(s)\hat{p}(k)}, \tag{5.17}$$

after adding the geometric series.

By collecting all these results (namely, Eqs. 5.14 and 5.17), we find that the Laplace transform of the propagator of the CTRW is given by:

$$\bar{G}^{\text{CTRW}}(k,s) = \frac{[1 - \tilde{\psi}(s)]/s}{1 - \tilde{\psi}(s)\hat{p}(k)}, \tag{5.18}$$

that could be Fourier-Laplace inverted to get the propagator in real space, if so desired. We will write instead the Fourier-Laplace transform of the general solution for the CTRW (Eq. 5.7), taking advantage again of the convolution theorems for the Laplace and Fourier transforms:

$$\bar{n}(k,s) = \bar{G}^{\text{CTRW}}(k,s) \cdot \bar{S}^{\text{aug}}(k,s) = \frac{1 - \tilde{\psi}(s)}{1 - \tilde{\psi}(s)\hat{p}(k)} \frac{\bar{S}^{\text{aug}}(k,s)}{s}. \tag{5.19}$$

This equation gives the final solution to the CTRW. In order to find the time-space solution, one needs to Fourier-Laplace invert it to the real time and space variables. However, we will see that Eq. 5.19 is sufficient for our purposes.

5.2.1.2   Connecting the CTRW to Classical Diffusion

The classical diffusion equation can be derived from the "microscopic" CTRW model in the so-called **fluid limit**, in which one only retains *the long-distance, long-time features of the CTRW*. This limit is particularly easy to take in Laplace-Fourier space, since the limit of long distances is tantamount to considering $k \to 0$, whilst the limit of long times is equivalent to assuming $s \to 0$.

In the case of $p(\Delta x) = N_{[0,2\sigma^2]}(\Delta x)$ and $\psi(\Delta t) = E_{\tau_0}(\Delta t)$, their Fourier and Laplace transforms are respectively given by (see Appendix 1 of Chap. 2):

$$\hat{p}(k) = \exp(-\sigma^2 k^2), \quad \tilde{\psi}(s) = \frac{1}{\tau_0 s + 1}. \tag{5.20}$$

In the fluid limit ($k \to 0$, $s \to 0$), these transforms behave as:

$$\hat{p}(k) \simeq 1 - \sigma^2 k^2, \quad \tilde{\psi}(s) \simeq 1 - \tau_0 s, \tag{5.21}$$

that inserted into Eq. 5.18 gives the fluid limit of the CTRW propagator,

$$\lim_{s\to 0, k\to 0} \bar{G}^{\text{CTRW}}(k, s) \simeq \frac{1}{s + (\sigma^2/\tau_0)k^2}. \tag{5.22}$$

Inserting the asymptotic propagator in Eq. 5.19, that provides the solution of the CTRW, and after working out some straightforward algebra, one obtains:

$$\bar{n}(k, s) \simeq \frac{\bar{S}^{\text{aug}}(k, s)}{s + (\sigma^2/\tau_0)k^2}. \tag{5.23}$$

Next, this expression can be easily reordered as,

$$(s\bar{n}(k, s) - \hat{n}_0(k)) = -\frac{\sigma^2}{\tau_0}k^2 \bar{n}(k, s) + \bar{S}(k, s), \tag{5.24}$$

that is trivially Fourier-Laplace inverted. The result is:

$$\frac{\partial n}{\partial t} = \frac{\sigma^2}{\tau_0}\frac{\partial^2 n}{\partial x^2} + S(x, t), \quad \breve{a}n(x, 0) = n_0(x). \tag{5.25}$$

That is, we obtain precisely the classical diffusive equation (Eq. 5.3) with a diffusive coefficient $D = \sigma^2/\tau_0$. It is no wonder that the CTRW and the numerical solution of the classical diffusive equation that were compared in Fig. 5.2 matched so well!

5.2.1.3   Underlying Assumptions: Locality in Time and Space

We have just shown that the classical diffusive equation (and Fick's law) can be obtained as the fluid limit of any CTRWs with a Gaussian jump-length pdf and an exponential waiting time pdf. Or more precisely, as the fluid limit of any CTRW whose jump-length pdf has a **finite second moment** so that $\hat{p}(k) \sim 1 - \sigma^2 k^2$, and whose waiting-time pdf has a **finite mean**, so that $\tilde{\psi}(s) \sim 1 - \tau_0 s$. Is this result sufficient to explain why Fick's law seems to provide such an adequate description of transport in so many practical situations? Of course not. The CTRW is just a simple mathematical model, and the microscopic reality is clearly much more complicated and system dependent. However, it is apparent that the (Gaussian/Exponential) CTRW model somehow manages to capture the basic features of transport in many of these systems. Which are these basic features? Clearly, one appears to be that the nature of the underlying transport is rather additive. The second one is *the existence of **characteristic scales** (in time and* *space) for the transport process.* That is precisely what σ^2 and τ_0 represent. $\sqrt{\sigma^2}$ sets a characteristic length, and τ_0 a characteristic time for the transport process. It is the existence of these scales that seems to matter in order for the basics of transport to be well captured by Fick's law, at least for scales much longer than τ_0 and much larger than $\sqrt{\sigma^2}$. It is only then that the strong locality assumptions hidden within Fick's law—that the flux at one location at a given time only depends on the present value of the gradient at the same location—become justified.

Characteristic transport scales can be easily identified in many natural systems. Take, for instance, the case of particles moving throughout a gas. Particles will move on average for a mean free path, λ, in between successive collisions. The motion can happen in any direction, thus giving a zero mean displacement. The variance of the motion, however does not vanish. Instead, it satisfies $\sigma^2 \propto \lambda^2$. Therefore, the typical scale length for transport is given by λ. The characteristic timescale, on the other hand, is given by the (inverse of the) *collision frequency*, ν_c. As a result, one expects that the effective transport through the gas be well described by considering a (Gaussian/Exponential) CTRW that ultimately leads to a diffusivity $D \sim \lambda^2 \nu_c^{-1}$. In a turbulent fluid or plasma, on the other hand, one would think that the typical turbulent eddy size, l_e and the typical eddy lifetime τ_e might play similar roles, and an effective eddy diffusivity, given by $D \simeq l_e^2 / \tau_e$, should do a good job describing the overall transport for $x \gg l_e$ and $t \gg \tau_e$. One must be careful here, though, since there are regimes in turbulent fluids and plasmas in which characteristic transport scales do not exist in spite of the fact that one can still estimate both l_e and τ_e! As a result, Fick's law is no longer a good model to model transport in these regimes.[16]

5.2.2 The Langevin Equation

The second "microscopic" model that we will discuss is the *Langevin formulation of Brownian motion*, introduced by P. Langevin in the early 1900s [14]. We already introduced the Langevin equation in Chap. 3, when discussing self-similar random processes.[17] The **Langevin equation** gives the position of a single particle on the real line at time t as (see Fig. 5.3):

$$x(t) = x_0 + \int_0^t dt' \xi_2(t'), \tag{5.26}$$

---

[16]We will discuss several examples of systems and regimes in which this is indeed the case, in the second part of this book. Particularly, in Chaps. 6 and 9.

[17]At that time, we referred to it as the only uncorrelated member of the fractional Brownian motion (fBm) family (see Eq. 3.64).

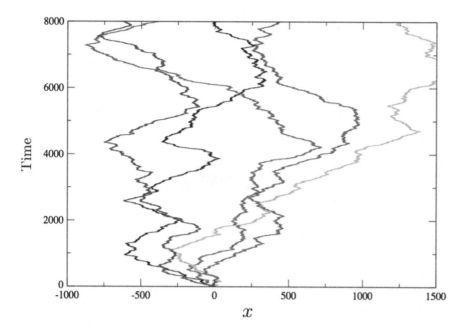

Fig. 5.3 Examples of particle trajectories, all started from $x_0 = 0$ and advanced according to the Langevin equation Eq. 5.26

where $\xi_2(t)$ is a Gaussian random variable (or noise) with zero mean and covariance function (see Chap. 4, Eq. 4.1) given by:

$$Co_{[\xi,\xi]}(t,t') = \langle \xi(t)\xi(t') \rangle = 2D\delta(t-t'). \tag{5.27}$$

Similarly to the case of the CTRW with jumps distributed according to the Gaussian $p(\Delta x)$ and waiting-times following the exponential $\psi(\Delta t)$, the Langevin equation is also closely related to the classical diffusive equation.[18] To illustrate this fact, Fig. 5.4 shows the evolution of the particle density according to the Langevin equation, for a collection of $N = 10^6$ particles all initialized at $x = 0$ (see Prob. 5.2). This evolution is compared against that of the numerical solution of the classical diffusive equation with diffusivity D. It is apparent that the two solutions are indeed very close. However, in order to establish a more formal connection, we will need to consider the propagator of the Langevin equation, similarly to what was already done for the CTRW.

---

[18]The additive nature of the physical processes behind diffusion are also apparent in the formulation of Langevin equation, where the noise acts as a surrogate for the instantaneous displacement.

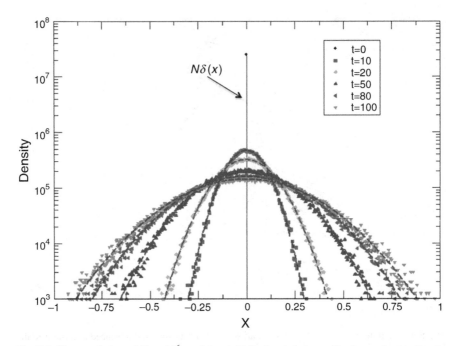

Fig. 5.4 Time evolution of $N = 10^6$ particles, initially located at $x = 0$, advanced according to the Langevin equation (Eq. 5.26) using Gaussian noise with a variance $\sigma^2 = 2 \cdot 10^{-4}$. In blue, the analytical solution of the classical diffusion equation with $D = 10^{-4}$ is shown, at the same times, for comparison

5.2.2.1   Propagator of the Langevin Equation

We will calculate the propagator of the Langevin equation through its moments [15]. That is, by estimating the expected values,

$$\langle (x - x_0)^n \rangle := \int (x - x_0)^n G^{\text{Lang}}(x - x_0, t) dx. \tag{5.28}$$

The moments can be calculated using the equation that defines the particle path (Eq. 5.26), even if the propagator is not known. For instance, the first moment is obtained as:

$$\langle x - x_0 \rangle = \left\langle \int_0^t dt' \xi(t') \right\rangle = \int_0^t dt' \, \langle \xi(t') \rangle = 0. \tag{5.29}$$

Naturally, we have assumed that the integral and the ensemble averages commute, and that the noise has a zero mean.

Regarding the second moment, the calculation goes like this:

$$\left\langle (x - x_0)^2 \right\rangle = \left\langle \int_0^t dt' \xi(t') \int_0^t ds' \xi(s') \right\rangle = \frac{1}{2} \int_0^t dt' \int_0^{t'} ds' \left\langle \xi(t') \xi(s') \right\rangle =$$

$$= D \int_0^t dt' \int_0^{t'} ds' \delta(t' - s') = Dt, \tag{5.30}$$

where the value of the covariance of the noise (Eq. 5.27) has been used. The remaining moments can be computed in a similar way (see Prob. 5.3). It turns out that all odd moments vanish, whilst the even ones are equal to:

$$\left\langle (x - x_0)^n \right\rangle = (Dt)^n (n - 1)!!, \tag{5.31}$$

where $n!! = n(n - 2)(n - 4) \cdots$ represents the *double factorial*.

The attentive reader will recognize the moments of a Gaussian distribution in these formulas (see Prob. 2.4). Therefore, we can conclude that the propagator for the Langevin equation is given by:

$$G^{\text{Lang}}(x, t; |x_0, 0) = \frac{1}{\sqrt{2\pi Dt}} \exp\left(-\frac{(x - x_0)^2}{2Dt} \right), \tag{5.32}$$

a Gaussian that spreads out in time. As a result, the center of the propagator, given by its mean value, remains unmoved and centered at $x = x_0$ throughout the motion. Its width, on the other hand, grows with time as $\sigma \propto t^{1/2}$.

5.2.2.2 Connecting the Langevin Equation to Classical Diffusion

The connection, in this case, is even more direct that in the case of the CTRW with jumps distributed according to the Gaussian $p(\Delta x)$ and waiting-times following the exponential $\psi(\Delta t)$. It turns out that the propagator of the classical diffusion equation, $G^{\text{CDE}}(x, t|x_0, t_0)$, is also given by Eq. 5.32. Indeed, the propagator is the solution of the equation,

$$\frac{\partial}{\partial t} G^{\text{CDE}}(x, t|x_0, 0) = D \frac{\partial^2}{\partial x^2} G^{\text{CDE}}(x, t|x_0, 0), \tag{5.33}$$

with initial condition given by $G^{\text{CDE}}(x, 0|x_0, 0) = \delta(x - x_0)$. We proceed by Laplace-Fourier transforming this equation to find,

$$s \bar{G}^{\text{CDE}}(s, k) - 1 = -Dk^2 \bar{G}^{\text{CDE}}(s, k) \implies \bar{G}^{\text{CDE}}(s, k) = \frac{1}{s + Dk^2}, \tag{5.34}$$

that can be easily Laplace-Fourier inverted to yield (see Appendix 1),

$$G^{CDE}(x, t; x_0, 0) = \frac{1}{\sqrt{2\pi Dt}} \exp\left(-\frac{(x - x_0)^2}{2Dt}\right) = G^{Lang}(x, t; |x_0, 0). \qquad (5.35)$$

In passing, one should also note that Eq. 5.34 coincides with the fluid limit of the CTRW propagator (Eq. 5.22).

5.2.2.3 Underlying Assumptions: Locality in Time and Space

Ultimately, the reason why the ensemble average of the Langevin equation behaves like the classical diffusion equation is because of the *locality assumptions, both in time and space, implicitly hidden in its formulation*. These assumptions are however hidden in a different way that in the case of the CTRWs with Gaussian $p(\Delta x)$ and exponential $\psi(\Delta t)$, when spatial and time locality is enforced separately. In the case of the Langevin equation they are enforced jointly when assuming the dependence of the noise covariance function (Eq. 5.27) and the fact that D, that has dimensions of length^2/time, has a finite value.

5.3   Scale Invariant Formulations of Transport

After having shown that the validity of Fick's law (and consequently, of classical diffusion) ultimately relies on the existence of finite characteristic scales associated to an underlying transport process of additive nature, we are in a position of exploring what happens when these characteristic scales are absent. This situation is of relevance for complex systems since they often exhibit scale-invariance, that impedes the existence of any characteristic scale (see Chap. 3). Therefore, a question arises regarding which is the proper way of characterizing "macroscopic transport" in these cases, given that Fick's law is not suitable for the job. Clearly, a framework that is devoid of any characteristic scale would appear to be more adequate. Here, we will discuss two such frameworks. The first one is based on the use of a particular kind of **scale invariant CTRWs**. The second framework employs a self-similar generalization of the Langevin equation that we have already seen elsewhere (see Sect. 3.3.4): **fractional Levy motion** (fLm).

5.3.1   Scale Invariant Continuous-Time Random Walks

CTRWs can be made scale invariant simply by choosing pdfs for the jump size, $p(\Delta x)$, and waiting-time, $\psi(\Delta t)$, that are themselves self-similar [16–19].

Furthermore, to ensure that the resulting CTRW also lacks characteristic scales, one just needs to choose $p(\Delta x)$ and $\psi(\Delta t)$ so that they respectively lack a finite variance and a finite mean. In fact, the most natural choice should probably be any member of the family of Lévy distributions (see Sect. 2.3.1) since they, as the Gaussian law, are also stable attractors favoured by the Central Limit Theorem in the case of underlying additive processes.

5.3.1.1   Removing Characteristic Scales from the CTRW

We will thus build a CTRW with no spatial characteristic length by choosing the jump size pdf, $p(\Delta x)$ from within the family of symmetric, Lévy distributions[19]:

$$p(\Delta x) = L_{[\alpha,0,0,\sigma]}(\Delta x), \quad \alpha \in (0,2). \tag{5.36}$$

All of these pdfs lack a finite variance since, asymptotically, they scale as (Eq. 2.33),

$$L_{[\alpha,0,0,\sigma]}(\Delta x) \sim |\Delta x|^{-(1+\alpha)}, \tag{5.37}$$

for large (absolute) values of their argument Δx. Of particular interest for our discussion is their Fourier transform, which is given by (Eq. 2.32):

$$\hat{L}_{[\alpha,0,0,\sigma]}(k) = \exp\left(-\sigma_L^\alpha |k|^\alpha\right) \sim 1 - \sigma^\alpha |k|^\alpha, \tag{5.38}$$

in the limit of large distances (i.e., $k \to 0$).

To avoid any characteristic timescale, we will pick the waiting-time pdf $\psi(\Delta t)$ from within the family of *positive extremal Lévy distributions* (see Sect. 2.3.1). These pdfs all have their asymmetry parameter equal to $\lambda = 1$, and a tail index $\beta < 1$. As a result, they all lack a finite mean[20]:

$$\psi(\Delta t) = L_{[\beta,1,0,\tau]}(\Delta t), \quad \beta \in (0,1), \tag{5.39}$$

The reason for this choice is that, as discussed in Sect. 2.3.1, extremal distributions with $\lambda = 1$ are *only defined for positive values of their argument*, making them perfect candidates for a waiting-time pdf [12]. Although extremal pdfs lack an analytical form as a function of Δt, it turns out that their Laplace transform is

---

[19]In the sake of simplicity we have assumed symmetry for the underlying microscopic motion. Naturally, asymmetric CTRWs could also be considered and might be important in some contexts (see, for instance Sect. 5.5).

[20]We will refer to the tail-index of these extremal Lévy pdfs as β (instead of α) in order to avoid any confusion with the index of the step-size pdf. We will also use τ (instead of σ) for the scale parameter.

known [20]:

$$\tilde{L}_{[\beta,1,0,\tau]}(s) = \exp\left(-\frac{\tau^\beta}{\cos(\pi\gamma/2)}s^\beta\right) \sim 1 - \frac{\tau^\beta s^\beta}{\cos(\pi\beta/2)}, \qquad (5.40)$$

in the limit of long times (i.e., $s \to 0$).

As an illustration of some of the features of these scale invariant CTRWs, Fig. 5.5 shows the evolution of the particle density for a set of $N = 10^6$ particles initiated at $x = 0$ that execute jumps distributed according to Eq. 5.36 (using $\alpha = 1.25$ and $\sigma = 0.01$) after having waited for lapses of time distributed according to Eq. 5.39 (using $\beta = 0.75$ and $\tau = 5$). Each snapshot is compared with its best Gaussian fit. It is apparent that the CTRW evolution behaves rather differently from an spreading Gaussian. In particular, note the different tails of the solution at large values of x, that seem to decay as $n(x) \sim |x|^{-2.25}$, a divergent power law. Also, the CTRW

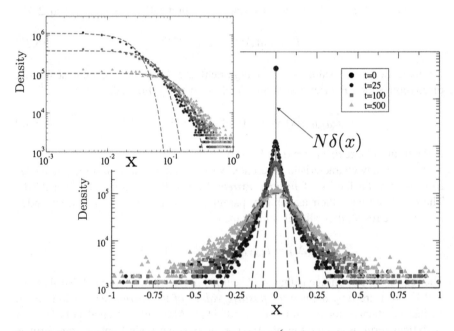

Fig. 5.5 Time evolution of the particle density for $N = 10^6$ walkers, initially located at $x = 0$, that are advanced according to a symmetric Lévy jump size distribution with tail-index $\alpha = 1.25$ and $\sigma = \times 10^{-2}$, and an extremal Lévy waiting time distribution with $\beta = 0.75$ and $\tau_0 = 5$. In blue, Gaussian fits for the particle density are shown, at the same instants, in order to illustrate their difference. **Inset:** the same evolution is shown in log-log scale, for $x > 0$

density seems to spread away from the initial location faster than the diffusive one.[21] Clearly, the diffusive equation would do a very poor job at modelling transport for this CTRW, no matter the value of D chosen. How can this be done better?

5.3.1.2   Fluid Limit of the Scale Invariant CTRW

In order to obtain a "macroscopic" model for transport, in the spirit of the classical diffusive equation, but able to capture the scale invariant features of these CTRWs, we will proceed by taking the same fluid limit that we used with the CTRW with Gaussian $p(\Delta x)$ and exponential $\psi(\Delta t)$ [16]. This only requires inserting the asymptotic behaviours of $p(\Delta x)$ for $k \to 0$ (see Eq. 5.36), and $\psi(\Delta t)$ for $s \to 0$ (see Eq. 5.39), into the general solution of the CTRW (Eq. 5.19). One quickly obtains that,[22]

$$\bar{n}(k, s) \simeq s^{\beta-1} \left(s^{\beta} + \cos(\pi\beta/2) \frac{\sigma^{\alpha}}{\tau^{\beta}} |k|^{\alpha} \right)^{-1} \bar{S}^{\text{aug}}(k, s). \tag{5.41}$$

After some reordering, this equation can be easily rewritten as [17]:

$$s\bar{n}(k, s) - \hat{n}_0(k) = -s^{1-\beta} \left[D_{(\alpha,\beta)} |k|^{\alpha} \bar{n}(k, s) \right] + \bar{S}(k, s). \tag{5.42}$$

Here, we have defined the coefficient $D_{(\alpha,\beta)} := (\sigma^{\alpha}/\tau^{\beta}) \cos(\pi\beta/2)$, Also, the augmented source $S^{\text{aug}}(k, s) = n_0(k) + S(k, s)$ has been made explicit to reveal the initial condition $n_0(k)$ and the external source $S(k, s)$. Therefore, Eq. 5.42 is the scale invariant version of Eq. 5.24, that represented the Fourier-Laplace transform of the classical diffusion equation.[23]

The scale-invariant macroscopic equation we have set ourselves to find is then obtained by Fourier-Laplace inverting Eq. 5.42. The inverse of the left hand side is simple, since it is equal to the temporal partial derivative of the density (as it was in the case of the classical diffusive equation). The inverse of the right hand side is more involved, though. The source is pretty easy, but the other term is not. In fact, its inverse cannot be written in terms of ordinary differential operators. One must use instead **fractional derivatives**.[24]

---

[21] Or, *superdiffusively*, as this behaviour is often referred to.

[22] It should be remarked that the same fluid limit would be obtained if one defines the CTRW in terms of non-Lévy pdfs, as long as they have the same asymptotic behaviours. This is completely analogous to what happened in the case of the CTRW with jumps following the Gaussian $p(\Delta x)$ and waiting times distributed according to the exponential $\psi(\Delta t)$.

[23] In fact, note that the classical diffusion equation is recovered if $\beta \to 1$ and $\alpha \to 2$.

[24] The date in which fractional derivatives were born is actually known exactly. In a letter to L'Hôpital in 1695, Leibniz wrote: *Can the meaning of derivatives with integer order be generalized to derivatives with non-integer orders?"* Apparently, L'Hôpital was curious about this, and replied himself writing back to Leibniz: *What if the order will be 1/2?* Leibniz replied that—with a

Fractional derivatives are *integro-differential operators* whose most interesting property is that they provide *interpolations in between integer derivatives* [21]. The easiest way to introduce them is to stay in Fourier-Laplace space. For instance, let's consider first the spatial derivative of integer order $n > 0$ of an arbitrary function $f(x)$. The Fourier transform of this derivative is given by (Appendix 1 of Chap. 2):

$$\mathbf{F}\left[\frac{d^n f}{dx^n}\right] = (-\imath k)^n \hat{f}(k). \tag{5.43}$$

The *fractional derivative of real order* $r > 0$ is then introduced as the operator whose Fourier transform is[25]:

$$\mathbf{F}\left[-_\infty D_x^r f\right] = (-\imath k)^r \hat{f}(k). \tag{5.44}$$

It turns out that the Fourier inverse of the right hand side of Eq. 5.44 can be calculated. The result is [21]:

$$-_\infty D_x^r f = \frac{1}{\Gamma(k-r)} \frac{d^k}{dx^k}\left[\int_{-\infty}^{x}(x-x')^{k-r-1}f(x')dx'\right], \tag{5.45}$$

where k is the integer satisfying $k-1 \leq r < k$ and $\Gamma(x)$ is the usual Euler's Gamma function. Similarly, we can introduce another fractional derivative of order r, but whose Fourier transform is instead,

$$\mathbf{F}\left[^\infty D_x^r f\right] = (+\imath k)^r \hat{f}(k). \tag{5.46}$$

The Fourier inverse of Eq. 5.46 is given by [21]:

$$^\infty D_x^r f = \frac{1}{\Gamma(k-r)} \frac{d^k}{dx^k}\left[\int_{x}^{\infty}(x'-x)^{k-r-1}f(x')dx'\right], \tag{5.47}$$

where k is again the integer satisfying $k-1 \leq r < k$. As advertised, both fractional derivatives are integro-differential operators that only reduce to the local usual derivatives when $r = n$, being n and integer.

If we go back now to Eq. 5.42, a quick inspection reveals that, in order to Fourier-Laplace it, we will need to deal first with a term of the form,

$$- |k|^\alpha \hat{f}(k). \tag{5.48}$$

---

letter dated September 30, 1695—*(The idea) will lead to a paradox, from which one day useful consequences will be drawn.* Appendix 2 contains a crash course on fractional derivatives, that might be worth reading before the reader continues to read this section.

[25]More specifically, $-_\infty D_x^q$ is a *left-sided Riemann-Liouville fractional derivative of order* r, with starting point at $-\infty$ (see Appendix 1).

Which fractional derivative is this? It turns out it is a combination of the two derivatives just introduced. Indeed, thanks to the complex identity,

$$(\imath k)^r + (-\imath k)^r = 2\cos(r\pi/2)|k|^r, \tag{5.49}$$

we can rewrite that term as:

$$-|k|^r \hat{f}(k) = -\frac{1}{2\cos(r\pi/2)}\left(\mathbf{F}\left[^{\infty}D_x^r f\right] + \mathbf{F}\left[_{-\infty}D_x^r f\right]\right) \tag{5.50}$$

and therefore, the fractional derivative we will need to Fourier-Laplace invert Eq. 5.42 is:

$$\mathbf{F}^{-1}\left[-|k|^r \hat{f}(k)\right] = -\frac{1}{2\cos(r\pi/2)}\left(^{\infty}D_x^r f + _{-\infty}D_x^r f\right) := \frac{\partial^r f}{\partial|x|^r}. \tag{5.51}$$

This symmetric sum defines a famous fractional derivative: the **Riesz symmetric fractional derivative of order** r (see Appendix 3). Its form, in real space, is given by the integral operator:

$$\frac{\partial^r f}{\partial|x|^r} := \frac{1}{2\Gamma(r)\cos(r\pi/2)}\int_{-\infty}^{\infty}\frac{f(x')dx'}{|x-x'|^{r-1}}. \tag{5.52}$$

Fractional derivatives can also be introduced as interpolants in Laplace space, similarly to what we just did in Fourier space. Consider, for instance, the Laplace transform of the usual first time derivative of a function $f(t)$ (see Appendix 1):

$$L\left[\frac{df}{dt}\right] = s^1 \tilde{f}(s) - s^0 f(0), \tag{5.53}$$

where we have abused the identity $s^0 = 1$. in addition, the Laplace transform of $f(t)$ can also be written as,

$$L[f(t)] = s^0 \tilde{f}(s). \tag{5.54}$$

Then, one could define the **fractional derivative of order** $q \in (0,1)$ as the one whose Laplace transform verifies:

$$L\left[_0 D_t^q f\right] = s^q \tilde{f}(s). \tag{5.55}$$

The Laplace inverse of Eq. 5.55 can be computed and is given by[26]:

$$_0D_t^q f = \frac{1}{\Gamma(1-q)} \frac{d}{dt} \left[\int_0^t (t-t')^{-q} f(t') dt' \right].$$ (5.56)

This temporal fractional derivative will be useful to deal with the $s^{1-\beta}$ factor that appears in Eq. 5.42.

We proceed now to invert Eq. 5.42. We start by Fourier inverting with the help of Eq. 5.51 to get:

$$sn(x,s) - n_0(x) = s^{1-\beta} \left[D_{(\alpha,\beta)} \frac{\partial^\alpha n}{\partial |x|^\alpha}(x,s) \right] + S(x,s).$$ (5.57)

Next, we Laplace invert it by invoking Eq. 5.55 to obtain:

$$\frac{\partial n}{\partial t} = {}_0D_t^{1-\beta} \left[D_{(\alpha,\beta)} \frac{\partial^\alpha n}{\partial |x|^\alpha} \right] + S(x,t).$$ (5.58)

Equation 5.58 is known as a **fractional transport equation** (or **fTe**). It will play a role analogous to that of the classical diffusive equation (Eq. 5.3), but in cases where transport is self-similar and lacks characteristic scales. The fractional transport equation has seen a large number of applications in many fields, particularly in the last few decades [17–19]. Some of its main properties will be discussed in detail in Sect. 5.3.3.

5.3.1.3   Underlying Assumptions: Absence of Characteristic Scales

The lack of characteristic scales in an underlying "microscopic" transport process of additive nature (i.e., the CTRW) is made apparent by the two integro-differential fractional operators that appear in the fTe, one in space (the Riesz derivative) and one in time (the time fractional derivative). Indeed, the presence of these operators implies that, in order to calculate the rate of change of n at any point x at the current time t, one needs to take into account also the values of the density at *every other point in the domain*, $x' \neq x$, and at *every past time*, $t' < t$. Or, in other words, that all scales matter when it comes to setting the value of transport throughout the system. As a result, the effective "macroscopic" transport equation becomes **nonlocal** in space, and **non-Markovian** in time. Scale-invariance appears because of the power-law dependence of each of the kernels within the fractional operators. It is quantified in terms of the two **fractional transport exponents**, α and β, whose values classify systems into different classes.

---

[26]$_0D_t^q$ is in fact, a *left-sided Riemann-Liouville fractional derivative of order* $q \in (0,1)$, with starting point at $t = 0$ (see Appendix 1).

5.3.2 The Fractional Langevin Equation

The natural scale-invariant extension of the Langevin equation (Eq. 5.26) is provided by the fractional Brownian and Lévy motions already discussed in Sect. 3.3. Here, we will express them using a single, more concise expression, referred to as the **fractional Langevin equation** (fLe):

$$x^{fLe}(t) = x_0 + {}_0D_t^{-(H+1/\alpha-1)}\xi_\alpha, \quad H \in (0,1], \quad \alpha \in (0,2], \tag{5.59}$$

expressed in terms of the fractional derivatives we just introduced [22]. In the fLe, $\xi_\alpha(t)$ represents a sequence of noise uncorrelated in time and distributed[27] according to a symmetric Lévy distribution with tail-index α if $\alpha < 2$, or a Gaussian if $\alpha = 2$ for $\alpha = 2$. The equivalence of Eq. 5.59 with the definitions for fBm (Eq. 3.71) and fLm (Eq. 3.76) can be shown rather trivially by making the fractional derivatives (Eq. 5.56) explicit.

We already proved in Sect. 3.3 that the ensemble-averaged solutions of Eq. 5.59 were scale-invariant by showing that its propagator was. However, the propagator was not computed explicitly then. Here, we will not compute it either, but we will provide a few hints on how this can be done and point to appropriate references for those interested in doing the calculation themselves. For $\alpha = 2$, the propagator can be computed as we did for the Langevin equation: evaluating its moments and showing that they correspond to those of a Gaussian [23]. Another possibility is to compute the propagator through the application of path integral techniques to the trajectories of Eq. 5.59 [24]. For $\alpha < 2$, however, the moment method fails due to the lack of finite moments above the tail-index α. The path integral method is still applicable, though [25]. The final result is given by (compare to Eqs. 3.73 for fBm, and 3.77 for fLm):

$$G_{[\alpha,H]}^{fLang}(x,t|x_0,0) = L_{[\alpha,0,0,a(H,\alpha)\sigma_\xi t^H]}(x - x_0), \tag{5.60}$$

with the constant $a(H,\alpha)$ given by Eq. 3.78. Therefore, the propagator of the fractional Langevin equation is a Lévy law, with the same tail-index α that the noise driving it, whose scaling factor grows in time as $\sigma \propto t^H$. Its center remains unmoved and fixed at $x = x_0$ throughout the motion.

5.3.2.1 Fluid Limit of the Fractional Langevin Equation

In order to find a macroscopic transport equation for the fLe, we need to take the fluid limit (i.e., $k \to 0$, $s \to 0$) of its general solution, for arbitrary initial condition

---

[27]For the sake of conciseness, we will adopt the convention that $L_{[2,0,0,\sigma]}(x)$ becomes the Gaussian law with zero mean when $\alpha = 2$. This decision can be justified because the Fourier transform of the symmetric Lévy pdf, $\hat{L}_{[\alpha,0,0,\sigma]}(k) = \exp(-\sigma^\alpha|k|^\alpha)$ becomes the Fourier transform of the Gaussian, $\hat{N}_{[0,2\sigma^2]}(k) = \exp(-\sigma^2 k^2)$, for $\alpha = 2$.

and external source:

$$\bar{n}(k, s) = \bar{G}_{[\alpha,H]}^{\text{Lang}}(k, s)\bar{S}^{\text{aug}}(k, s). \tag{5.61}$$

This calculation requires the evaluation of the Fourier-Laplace transform of Eq. 5.60. The Fourier transform is very simple, since we are dealing with a Lévy law. It is given by the stretched exponential [20]:

$$F\left[G_{[\alpha,H]}^{\text{Lang}}(x, t|x_0, 0)\right] = \exp(-a^\alpha(H, \alpha)\sigma_\xi^\alpha t^{\alpha H}|k|^\alpha). \tag{5.62}$$

Regretfully, the Laplace transform of the stretched exponential lacks an analytical expression. For our purposes, it suffices with expressing it via the infinite series [26, 27]:

$$L\left[\exp(-(t/\tau)^b)\right] = \tau^b s^{b-1} \sum_{n=0}^{\infty} \frac{(-1)^n}{(1 + (\tau s)^b)^{n+1}} \tag{5.63}$$

$$\times \left\{\sum_{j=0}^{n} \binom{n}{j} (-1)^{n-j} \frac{\Gamma(bj + 1)}{\Gamma(j + 1)}\right\}, \quad 0 < b < 1.$$

Keeping only the leading order of this expression for $s \to 0$, one finds that the Fourier-Laplace transform we need scales as:

$$\bar{G}_{[\alpha,H]}^{\text{Lang}}(k, s) \sim \frac{s^{\alpha H - 1}}{a^\alpha(H, \alpha)\sigma_\xi^\alpha |k|^\alpha + s^{\alpha H}}, \quad s \to 0, \, k \to 0. \tag{5.64}$$

Inserting this expression into Eq. 5.61, and Fourier-Laplace inverting the result, as we did for the scale invariant CTRW (Eq. 5.42), we find that the fluid limit of the fractional Langevin equation is given by:

$$\frac{\partial n}{\partial t} = {}_0D_t^{1-\alpha H}\left[D'_{(\alpha,\alpha H)} \frac{\partial^\alpha n}{\partial |x|^\alpha}\right] + S(x, t), \tag{5.65}$$

with the coefficient $D'_{(\alpha,H)} = (a(H, \alpha)\sigma_\xi)^\alpha$. Therefore, the result is again a **fractional transport equation**, in which the lack of characteristic scales of the transport process is made apparent by the presence of fractional derivatives both in space and time. Interestingly, the comparison of Eq. 5.65 with the fluid limit of the scale-invariant CTRW (Eq. 5.58), suggests that the exponents β (from the scale-invariant CTRW) and the product αH (for the fLe) play essentially the same role with respect to the macroscopic transport dynamics.

5.3.3 The Fractional Transport Equation

We have just found that the fluid limits of both the scale invariant CTRW (Eq. 5.58) and of the fLe (Eq. 5.65), take the form of a fractional transport equation (fTe):

$$\frac{\partial n}{\partial t} = {_0}D_t^{1-\beta}\left[D\frac{\partial^\alpha n}{\partial |x|^\alpha}\right] + S(x,t), \tag{5.66}$$

where the actual value of the coefficient D depends on the model considered, and with $\beta = \alpha H$ in terms of the exponents that define the fLe. The general solution of the fTe is given by,

$$n(x,t) = \int_0^t dt' \int_{-\infty}^{\infty} dx' G_{[\alpha,\beta]}^{\text{fTe}}(x-x',t-t') S^{\text{aug}}(x',t'), \tag{5.67}$$

with the augmented source defined, as always, as $S^{\text{aug}} = S(x,t) + n(x,0)\delta(t)$.

The Fourier-Laplace transform of the fTe propagator is:

$$\bar{G}_{[\alpha,\beta]}^{\text{fTe}}(k,s) = \frac{s^{\beta-1}}{s^\beta + D|k|^\alpha}, \tag{5.68}$$

whose inverse does not have a closed analytical form, except in a few special cases. A useful approach is to express its Laplace inversion by introducing the **Mittag-Leffler** function, a generalization of the usual exponential defined by the infinite series,[28]

$$E_a(z) = \sum_{n=0}^{\infty} \frac{z^n}{\Gamma(an+1)}, \quad a > 0. \tag{5.69}$$

whose Laplace transform verifies,

$$L\left[E_a(cz^a)\right] = \frac{s^{a-1}}{s^a - c}. \tag{5.70}$$

Thanks to the similarity of this formula with Eq. 5.68, the Laplace inverse of the fTe propagator can be written in terms of the following Mittag-LeffLer function,[29]

$$\hat{G}_{[\alpha,\beta]}^{\text{fTe}}(k,t) = E_\beta\left(-D|k|^\alpha t^\beta\right). \tag{5.71}$$

---

[28]Clearly, $E_1(z) = \exp(z)$.

[29]This expression of the fTe propagator can be useful to avoid the always delicate evaluation of Laplace transforms at small values of s; instead, we just need to sum the Mittag-LeffLer function up to the value of the index n that ensures the desired level of convergence. However, the inverse of the Fourier transform of Eq. 5.71 must be performed numerically.

We discuss next some of the main features of Eq.5.66. They will clearly depend on the specific values of α and β [17–19, 28].

5.3.3.1 The Standard Diffusive Case: $\alpha = 2$, $\beta = 1$

It is worth noting first that, for $\alpha = 2$ and $\beta = 1$, the fTe reduces to the classical diffusive equation (Eq. 5.3). Naturally, its propagator is thus (Eq. 5.35):

$$G_{2,1}^{\text{f Te}}(x, t | x_0, 0) = \frac{1}{\sqrt{2\pi Dt}} \exp\left(-\frac{(x - x_0)^2}{2Dt}\right) \tag{5.72}$$

5.3.3.2 The Nonlocal, Superdiffusive Case: $\alpha < 2$, $\beta = 1$

In this case, the temporal fractional derivative in the fTe vanishes, and Eq. 5.66 becomes only **nonlocal in space**:

$$\frac{\partial n}{\partial t} = D\frac{\partial^\alpha n}{\partial |x|^\alpha} + S(x, t), \tag{5.73}$$

The propagator, in this case, can be easily calculated and found to be (see Prob. 5.4):

$$G_{\alpha,1}^{\text{f Te}}(x, t | x_0, 0) = L_{[\alpha, 0, 0, (Dt)^{1/\alpha}]}(x - x_0). \tag{5.74}$$

It corresponds to a symmetric Lévy with tail-index $0 < \alpha < 2$, whose scale parameter grows with time as $\sim t^{1/\alpha}$. We will always use the term **superdiffusive** to describe transport in cases like this, in which the propagator spreads with time faster than the diffusive scaling, $\sim t^{1/2}$.

It is also interesting to find an expression for the local flux in this case (see Prob. 5.5), that could be compared with Fick's law (Eq. 5.1). Using some basic properties of fractional derivatives (see Appendix 2), the flux is found to be:

$$\Gamma_{\alpha,1}(x, t) = \frac{1}{2\cos(\alpha\pi/2)}\left[-\infty D_x^{\alpha-1}n - {}^\infty D_x^{\alpha-1}n\right] = \frac{\Gamma^{-1}(k - \alpha - 1)}{2\cos(\alpha\pi/2)} \tag{5.75}$$

$$\cdot \frac{\partial^k}{\partial x^k}\left[\int_{-\infty}^x (x - x')^{k-\alpha-2}n(x', t) - \int_x^\infty (x' - x)^{k-\alpha-2}n(x', t)\right],$$

with the integer k defined as that satisfying $k \leq \alpha \leq k + 1$. **Spatially nonlocality** is apparent in Eq. 5.75. The flux at x, t is determined by the value of the density at x, t (or more precisely, its gradient) only for $\alpha = 2$, since Eq. 5.75 then reduces to $-\partial n/\partial x$. For $\alpha < 2$, on the other hand, *all points in the domain* contribute to setting the value of local flux at (x, t), as made explicit by the integrals. Indeed, the derivative starting at $-\infty$ collects contributions from $x' < x$, whilst the one ending at $x = \infty$ does the same for $x > x'$. This nonlocality has some non-intuitive consequences that are strange for those used to the classical diffusion equation. For instance, *the local flux at x could be non-zero even if the local gradient at x is zero at time t!*

5.3.3.3 The Non-Markovian, Subdiffusive Case: $\alpha = 2, \beta < 1$

We examine next the case defined by setting $\alpha = 2$ and $\beta = 1$. Equation 5.66 then becomes **non-Markovian** in time:

$$\frac{\partial n}{\partial t} = {}_0D_t^{1-\beta}\left[D\frac{\partial^2 n}{\partial x^2}\right] + S(x, t), \tag{5.76}$$

although it remains **local in space**. The propagator has, in this case, no closed analytical expression. It can be shown, however, that its asymptotic form takes the form of the following *stretched exponential* for large values of its argument [28]:

$$G_{2,\beta}^{\mathrm{fTe}}(x, t|x_0, 0) \sim \frac{1}{2D^{1/2}t^{\beta/2}}\left(\frac{|x - x_0|}{Dt^\beta}\right)^a \exp\left(-b\left[\frac{|x - x_0|}{Dt^\beta}\right]^c\right) \tag{5.77}$$

with the exponents given by,

$$a = \frac{(\beta - 1)}{2(2 - \beta)} < 0, \; b = (2 - \beta)2^{-\frac{2}{2-\beta}}\beta^{\frac{\beta}{2-\beta}} > 0 \; \text{and} \; c = \frac{2}{2 - \beta} > 0. \tag{5.78}$$

Its even moments can also be shown to grow with time as [28]:

$$\langle|x(t) - x_0|^n\rangle = \frac{\Gamma(n + 1)}{\Gamma(\frac{\beta}{2}n + 1)}t^{(\beta/2)n}. \tag{5.79}$$

Therefore, the propagator spreads slower than a diffusive Gaussian propagator (Eq. 5.72), whose moments grow as $\sim t^{n/2}$. Whenever this happens, we will say that transport becomes **subdiffusive**.

The local flux is given, in this case, by:

$$\Gamma_{2,\beta}(x, t) = -{}_0D_t^{1-\beta}\left[D\frac{\partial n}{\partial x}\right] \tag{5.80}$$

$$= -\frac{D}{\Gamma(\beta)}\frac{\partial}{\partial t}\left[\int_0^t (t - t')^{-(1-\beta)}\frac{\partial n}{\partial x}(x, t')\right]$$

that reduces to the local Fick's law (Eq. 5.1) only for $\beta = 1$. For $\beta < 1$, **non-Markovianity** become obvious from the fact that, in order to get the local flux at position x and time t, one needs to consider *all values of the gradient at x for $t' < t$*. Non-Markovianity also has important non-intuitive consequences for those used to the classical diffusion equation. For instance, that *the local flux at x can be non-zero at time t even if the value of the local gradient at x vanishes at that time!*

5.3.3.4 The General Case: $\alpha < 2$, $\beta < 1$

In the general case, when any choice satisfying $\alpha < 2$ and $\beta < 1$ is allowed, the dynamics become far richer. The propagator has, in general, no closed analytical expression. Probably, the best one can do is to express it as [28, 29]:

$$G_{\alpha,\beta}^{\text{fTe}}(x, t|x_0, 0) = \int_0^\infty dy\, y^{\beta/\alpha} L_{[\alpha,0,0,1]}\left(y^{\beta/\alpha} \frac{|x - x_0|}{D^{1/\alpha} t^{\beta/\alpha}} \right) L_{[\beta,-1,0,1]}(y), \tag{5.81}$$

that contains a symmetric Lévy of tail-index α and an extremal Lévy of tail-index β, which is very reminiscent of the choices we made for the scale-invariant CTRW in Sect. 5.3.1.

For our purposes, however, it is sufficient with knowing the behaviour of its even moments, that scale with time as[30]:

$$\langle |x(t) - x_0|^n \rangle \sim t^{(\beta/\alpha)n}. \tag{5.82}$$

This scaling implies that the transport described by the general fTe (see Fig. 5.6, left frame) can be **superdiffusive** when $2\beta > \alpha$ and **subdiffusive** transport if $2\beta < \alpha$. It can even scale **diffusively** if $2\beta = \alpha$, in spite of the fact that the dynamics are still non-Markovian in time and non-local in space. This classification is in fact simplified when made in terms of the exponent $H = \beta/\alpha$ (see Fig. 5.6, right frame) since **diffusion** needs $H = 1/2$, **subdiffusion** $H < 1/2$ and **superdiffusion** $H > 1/2$, as was the case for fractional Brownian/Lévy motions.[31]

The expression of the local flux is, in this general case,

$$\Gamma_{\alpha,\beta}(x, t) = -{}_0 D_t^{1-\beta}\left[\frac{D}{2\cos(\alpha\pi/2)} \left[{}_{-\infty}D_x^{\alpha-1} n - {}^\infty D_x^{\alpha-1} n \right] \right], \tag{5.83}$$

thus involving integrals over the previous history of the whole domain in order to find the flux at any point and any time.

---

[30]Due to the assumed symmetry of the underlying motion, all odd moments vanish.

[31]This exponent, H, corresponds in fact to the self-similarity exponent of both fBm and fLm (see Chap. 3). It is also the exponent used in the definition of the fractional Langevin equation.

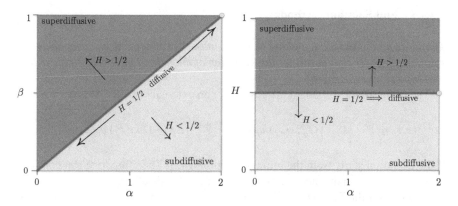

Fig. 5.6 Possible transport behaviours that the fTe can exhibit, as determined by the values of the exponents α and β (left), or the values of the exponents α and $H = \beta/\alpha$ (right). Diffusive behaviour (i.e., $H = 1/2$ or $2\beta = \alpha$) is represented by the red straight line. The classical diffusive behaviour ($\alpha = 2$, $\beta = 1$, $H = 1/2$)) is represented by a green circle. Superdiffusive areas shown in red, subdiffusive areas in yellow

5.4 Techniques for the Characterization of Fractional Transport

In order to characterize the transport across any system as fractional, at least in the asymptotic sense assumed in Sect. 5.3, one needs efficient methods to determine the fractional exponents that enter into the fTe (Eq. 5.66). That is, α and β (or $H = \beta/\alpha$). In this section, we will review several methods that may be useful in practical situations. We have classified the methods within two large families, **eulerian** and **lagrangian**, depending on the type of data that they require [3]. The discussion that follows will be restricted to one-dimensional systems, with the idea that the methods could be applied to any particular direction in a Cartesian system. In order to apply them to other types of coordinates, suitable modifications might be needed to take into account additional geometric factors.[32]

5.4.1 Eulerian Methods

A method to determine the fractional transport exponents is classified as **eulerian** if it only requires *information at the nodes of a fixed grid in both time and space*. These methods are often the best suited to deal with experimental situations, where measurements at fixed locations and fixed rates are usually more accessible.

---

[32]It is not known how to do this in most other cases. This is in fact an active area of current research that needs to be developed greatly in the next years.

5.4.1.1 Full Solution Methods

The more straightforward method to get fractional exponents is by *comparing the solution of the fTe that describes the evolution of the field of interest with the actual gridded data for the same field.*

The solution of Eq. 5.66, in the absence of an external source,[33] is given by,

$$n^{\mathrm{fTe}}(x,t) = \mathrm{F}^{-1}\left[\mathrm{L}^{-1}\left[G_{\alpha,\beta}^{\mathrm{fTe}}(k,s)\hat{n}_0(k)\right]\right] = \mathrm{F}^{-1}\left[E_\beta\left(-D|k|^\alpha t^\beta\right)n_0(k)\right], \quad (5.84)$$

if the evolution starts from the initial condition $n_0(x)$. Therefore, one can compare this expression with the measured evolution of the profile of interest, $n_{jk} = n(x_j, t_k)$, as recorded on the Eulerian mesh: $x_j = x_1 + (j-1)\Delta x$ and $t_k = (k-1)\Delta t$ (for some $\Delta x,\ \Delta t > 0$) and with the indices varying as $k = 1,\cdots N_t$ and $j = 1,\cdots,N_x$. The advance of Eq. 5.84 can be done by assuming $n_0(x_j) = n(x_j, t_1)$.

In order to obtain the fractional exponents, it is often convenient to use minimum chi-square parameter estimation techniques (see Sect. 2.5.3). That is, to minimize the target function,

$$\chi_{xt}^2(\alpha,\beta,D) = \sum_{j=1}^{N_x}\sum_{k=1}^{N_t} \frac{\left|n^{\mathrm{fTe}}(x_j,t_k) - n(x_j,t_k)\right|^2}{\left|n(x_j,t_k)\right|}, \quad (5.85)$$

with respect to α, β and D using the reader's preferred method.[34] Several variations of this method exist, depending on how the target function is built. For instance, one could also choose to build it in (k,s) space, or in any of the mixed spaces (k,t) or (x,s). The merit of each of these choices must however be weighed against the complications of performing accurate Laplace transforms on discrete data for $s \to 0$, as well as the possible aliasing or Gibbs contamination that the Fourier transform of discrete, often non-periodic and discontinuous data, may introduce.

5.4.1.2 Propagator Methods

Another popular family of methods tries first to **estimate the process propagator on the Eulerian grid**, and then compares it with the fTe propagator (Eq. 5.66). However, performing this comparison requires a larger degree of control over the system under examination, since one needs to consider *a very localized (in x) initial*

---

[33]The method is trivially extended to accommodate external sources, if they are present and information about them is available. However, the method is no good if the sources are unknown. One has to use then other approaches, such as the propagator method that is discussed next.

[34]For instance, one could use local methods such as the Levenberg-Marquardt algorithm, or global methods, such as genetic algorithms [30].

condition for n. On the plus side, *these methods can be made insensitive to the presence of any external sources.*[35]

The method goes like this. First, we will assume that a localized initial condition such as $n(x_j, t_1) \simeq \delta(x_j - x_L)$ can be made experimentally for some index $1 < L < N_x$, and that its evolution can be followed in time. The evolution of this localized initial seed, is, in essence, *a discrete approximation of the problem propagator.* That is, $n(x_j, t_k) \simeq G(x_j, t_k|x_L, t_1)$. The obtained discrete propagator can be then compared with the fTe propagator, and optimal parameters estimated, by using the minimum chi-square estimation technique (Sect. 2.5.3). It requires to minimize the quantity,

$$\chi^2_{\text{Prop,xt}}(\alpha, \beta, D) = \sum_{j=1}^{N_x} \sum_{k=1}^{N_t} \frac{\left|n(x_j, t_k) - G^{\text{fTe}}(x_j, t_k|x_L, t_1)\right|^2}{\left|G^{\text{fTe}}(x_j, t_k|x_L, t_1)\right|}, \tag{5.86}$$

where the fTe propagator is available from

$$G^{\text{fTe}}_{\alpha,\beta}(x, t|x_L, t_1) = \text{F}^{-1}\left[E_\beta\left(-D|k|^\alpha (t_k - t_1)^\beta\right)\right], \tag{5.87}$$

with k the wave vector associated to $x - x_L$. Or, if one wants to avoid the need to carry out Fourier transforms completely, one can also use the series representations of the fTe propagator that are available for $\alpha < 2$, $\beta < 1$ [28]. As happened in the case of the full solution, one could also build the target function directly in Fourier-Laplace space, if desired.

Figure 5.7 illustrates the propagator method just outlined. We have evolved $N = 10^6$ particles, all starting from $x = 0$, according to the scale invariant CTRW discussed in Sect. 5.3.1. We have chosen a step size pdf given by a symmetric Levy with tail-index $\alpha = 1.25$ and $\sigma = 2 \times 10^{-2}$, and a waiting-time pdf given by an extremal Lévy with tail-index $\beta = 0.75$ and $\tau = 5$. The target function used in the minimization is the same as in Eq. 5.86. The minimizing algorithm used was a standard Levenberg-Marquardt algorithm [31]. The resulting fractional exponents were $\alpha \simeq 1.28 \pm 0.05$ and $\beta = 0.76 \pm 0.03$, reasonably close to the nominal values.

5.4.1.3   Other (Simpler) Propagator Methods

It is also possible to take advantage of some of the scaling properties the fTe propagator (Eq. 5.68) and estimate at least some of the exponents with less hassle,

---

[35]Propagator methods require, in many cases, the use of some kind of **tracer field** or **tracer particles**, easily distinguishable from the system background but with similar dynamical behaviour. These tracers must be initialized in a very localized setup, and then followed in time without any further addition of tracers. In this way, one could get an estimate of the system transport exponents without having to consider external sources. We will say more about tracers soon, when discussing Lagrangian methods.

Fig. 5.7 Discrete propagators estimated by advancing $N = 10^6$ particles following a scale invariant CTRW with $p(\Delta x) = L_{[1.25,0,0,10^{-2}]}(\Delta x)$ and $\psi(\Delta t) = L_{[0.75,-1,0,5]}(\Delta t)$. In dashed red, the best fit provided by the minimization of the target function Eq. 5.86 using a Levenberg-Marquardt algorithm is also shown

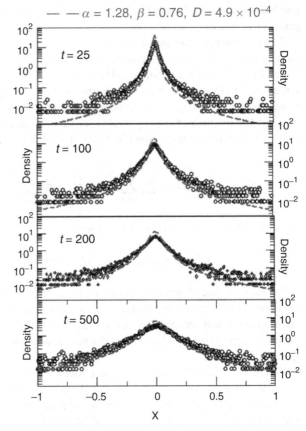

avoiding the need to carry out any minimization of any kind. As could be expected, the price to pay is that these methods are not so accurate and must thus be used with some care, since they do not use the global solution to estimate the exponents.

One possibility is to take advantage of Eq. 5.82, that tells us that the even (and finite) **moments of the propagator**, scale as $\sim t^{(\beta/\alpha)n} = t^{Hn}$. Thus, if we have at our disposal the temporal evolution of any initially-localized data such that $n(x_j, t_1) \simeq \delta(x_j - x_L)$ for some index $1 < L < N_x$, we can estimate those moments as:

$$\langle |x(t_k) - x_L|^n \rangle \simeq \frac{1}{N_x} \sum_{j=1}^{N_x} |n(x_j, t_k) - n(x_L, t_1)|^n \sim (t_k - t_1)^{nH} \tag{5.88}$$

This procedure thus allows to estimate the transport exponent H. Since $H = \beta/\alpha$, we need to combine it with some other method to get both α and β.

One simple method to estimate α and β is to take advantage of the **asymptotic behaviour of the fTe propagator** (Eq. 5.68). It can be shown that, for fixed t_c, it satisfies that [32, 33],

$$G^{\mathrm{fTe}}(x, t_c|x_0, 0) \sim |x - x_0|^{-(1+\alpha)}, \quad |x - x_0| \gg D^{1/\beta} t_c^{\beta/\alpha}, \tag{5.89}$$

whilst for fixed x_c, it satisfies,

$$G^{\mathrm{fTe}}(x_c, t|x_0, 0) \sim t^{-\beta}, \quad t \gg D^{1/\beta} x_c^{\alpha/\beta}, \tag{5.90}$$

and

$$G^{\mathrm{fTe}}(x_c, t|x_0, 0) \sim t^{\beta}, \quad t \ll D^{1/\beta} x_c^{\alpha/\beta}, \tag{5.91}$$

Thus, using again the same temporal evolution of the field of interest starting from $n(x_j, t_1) \simeq \delta(x_j - x_L)$, for some index $L \in (1, N)$, one could estimate α from the scaling of,

$$n(x_j, t_P) \sim |x_j - x_L|^{-(1+\alpha)}, \tag{5.92}$$

for $t_P \gg t_1$ and sufficiently large $|x_j - x_L|$. Similarly, β is available from the scalings,

$$n(x_P, t_k) \sim t_k^{-\beta}, \tag{5.93}$$

for $|x_P - x_L| \gg 0$ and *sufficiently large time*, or,

$$n(x_P, t_k) \sim t_k^{\beta}, \tag{5.94}$$

for $|x_P - x_L| \gg 0$ and *sufficiently short time*.

5.4.1.4 Kernel Methods

Another family of Eulerian methods are based on the computation of the integro-differential kernel introduced at the beginning of this chapter (Eq. 5.4), and that relates the local gradient and the local flux [3, 34].

In the case of the fTe, the kernel can be calculated analytically. Indeed, one starts by noticing first that the Fourier-Laplace transform of Eq. 5.4 is,

$$\bar{K}(k, s) = \frac{\bar{\Gamma}(k, s)}{-\imath k \bar{n}(k, s)}. \tag{5.95}$$

Then, we consider the Fourier-Laplace transform of the continuity equation,

$$\frac{\partial n}{\partial t} + \frac{\partial \Gamma}{\partial x} = S(x, t), \tag{5.96}$$

permits to express the flux, after introducing once again the augmented source $S^{\mathrm{aug}}(x, t) = S(x, t) + n_0(x)\delta(t)$, in the form,

$$\bar{\Gamma}(k, s) = \frac{\bar{S}^{\mathrm{aug}}(k, s) - s\bar{n}(k, s)}{-\mathrm{i}k}. \tag{5.97}$$

Finally, we combine Eqs. 5.95 and 5.97 to obtain:

$$\bar{K}(k, s) = \frac{s\bar{n}(k, s) - \bar{S}^{\mathrm{aug}}(k, s)}{k^2 \bar{n}(k, s)} = \frac{s\bar{G}^{\mathrm{fTe}}(k, s) - 1}{k^2 \bar{G}^{\mathrm{fTe}}(k, s)} = -Ds^{1-\beta}|k|^{\alpha-2}, \tag{5.98}$$

where we have that the solution of the fTe can be expressed as (see Eq. 5.67):

$$n(k, s) = G^{\mathrm{fTe}}(k, s)S^{\mathrm{aug}}(k, s), \tag{5.99}$$

with the fTe propagator given by Eq. 5.68.

The fractional exponents can then be obtained by comparing Eq. 5.98 and the estimation of the kernel on the Eulerian grid. This can be done, as we have already seen several times, by minimizing the target function,

$$\chi^2_{\mathrm{Kern, ks}}(\alpha, \beta, D) = \sum_{j=1}^{N_k} \sum_{k=1}^{N_s} \frac{\left| \bar{K}(k_j, s_k) - Ds_k^{1-\beta}|k_j|^{\alpha-2} \right|^2}{Ds_k^{1-\beta}|k_j|^{\alpha-2}}. \tag{5.100}$$

The experimental kernel is estimated on the Eulerian grid as,

$$K(k_j, s_k) = \frac{s_k \bar{n}(k_j, s_k) - \hat{n}_0(k) - \bar{S}(k, s)}{k^2 \bar{n}(k_j, s_k)} \tag{5.101}$$

Due to the need of carrying out Laplace and Fourier transforms of discrete data, the kernel method thus suffers from the same difficulties previously mentioned when discussing full solution methods. In particular, knowledge of the external sources is required, if these are present during the evolution.

5.4.2   Lagrangian Methods

Lagrangian methods are those that require transport information *along the trajectories of advection of the quantity of interest.* For instance, if our system is a fluid, and if one wants to study how the fluid mass, momentum or energy is transported, the information will be required along the trajectories dictated by the *characteristics of the flow motion*, also known as *Lagrangian trajectories*.[36]

In practice, Lagrangian trajectories are approximated by means of **tracer particles**.[37] That is, marked particles whose evolution, as they are advected by the flow, can be followed with ease.[38] Due to inertial effects, the actual trajectories of physical tracers do not coincide with the Lagrangian trajectories of the flow but, if chosen carefully, they can give us a good estimate of what is going on. Tracer particles are particularly useful in numerical simulations where many of these discrepancies can be made to go away, if desired, by adjusting their physical properties.[39]

Sometimes, however, one might be interested in the transport of say, a certain pollutant in the atmosphere, or of high-Z impurities in a plasma, or some other quantity that is different from the system carrying out the advection. In that case, retaining the full physical properties of the tracer is essential, since it is their trajectories that we will be interested on.

We will discuss next several Lagrangian methods that can be used to estimate the fractional transport exponents. To do so, we will assume that the successive positions along some Lagrangian trajectory are available. In particular, we will consider one Cartesian component of the position[40] of N particles, sampled at a constant rate Δt,

$$x^j(t_k), \quad j = 1, 2, \cdots, N, \quad t_k = (k-1)\Delta t, \quad k = 1, 2, \cdots, N_t. \tag{5.102}$$

All the methods discussed next will try to use these positions to estimate some of the elements that define one of the two "microscopic" models that we described as underlying the fTe: the scale-invariant continuous-time random walk (Sect. 5.3.1) or the fractional Langevin equation (Sect. 5.3.2).

5.4.2.1 CTRW Method

As will be remembered, the scale invariant CTRW was defined in Sect. 5.3.1 in terms of two pdfs, one for the jump sizes, $p(\Delta x) = L_{[\alpha,0,0,\sigma]}(\Delta x)$, with $0 < \alpha < 2$, the

---

[36]That is, if the flow velocity is defined by the field, $\mathbf{v}(\mathbf{r}, t)$, its Lagrangian trajectory that passes through $\mathbf{r}_0$ at time t_0 is the solution of the differential equation $\dot{\mathbf{r}} = \mathbf{v}(\mathbf{r}, t)$, with $\mathbf{r}(t_0) = \mathbf{r}_0$.

[37]Eulerian methods, on the other hand, often use **tracer fields**, such as oil or dye, in order to be able to follow the evolution in time of a perturbation advected by a system on a Eulerian grid. The use of the tracer field permits to ignore the presence of any external drive when needed to sustain the system and that might be unknown. Or to be able to taylor the initial perturbation at will, as it is required in order to estimate a propagator.

[38]Think, for instance, of radioactive isotopes or the polyethylene particles or oil droplets used in *particle image velocimetry* (PIV) studies of turbulence.

[39]In fluids, one might want to adjust their mass or buoyancy. In plasmas, on the other hand, one might also want to use chargeless tracer particles in order to avoid magnetic drifts.

[40]Since the fTe is one-dimensional, we will assume either that the displacements occur along one direction of interest, or that we are focusing on one particular component of the motion only.

second for the waiting-times, $\psi(\Delta t) = L_{\beta,1,0,\tau}(\Delta t)$, with $0 < \beta < 1$. However, in order to derive the fluid limit that connected the CTRW to the fTe, it was sufficient with having $p(\Delta x) \sim |\Delta x|^{-(1+\alpha)}$ for large $|\Delta x|$, and $\psi(\Delta t) \sim \Delta t^{-(1+\beta)}$ for large Δt.

One can easily use this information to build a Lagrangian method to estimate the transport exponents. All one needs to do is to *define suitable jump sizes and waiting-times for the process* under examination, that could be easily computed from the available $x^j(t_k)$ series (Eq. 5.102). Once these definitions are available, α and β could be determined from the *scaling behaviour of the tails of their respective pdfs*, computed according to any of the methods described in Chap. 2. In particular, since we are interested in the tails, it is to better to use either the CBC (Sect. 2.4.2) or SF/CDF (Sect. 2.4.3) methods.

The subtlety here is clearly to decide *how to define waiting-times and jumps from signals in which, in most cases, there is not a clear alternation of motion and rest*. We will illustrate this difficulty with one example. Let's consider the motion of tracer particles along a certain direction while advected by fluid turbulence. In this case particles are clearly always on the move. There is however is a difference between those periods of time during which the particle is trapped within a turbulent eddy, and those during which the tracer is travelling in between eddies. The former is reminiscent of a trapping period, in which the component of the velocity along any direction changes sign over a period of time of the order of the eddy turnover time. The latter, on the other hand, look much more like jumps, during which the sign of the tracer velocity is maintained. Thus, one could envision defining a jump size as *any displacement of the tracer along the direction of choice during which the velocity sign is maintained beyond a certain amount of time* (that must be prescribed before hand, and that should be a reasonable estimate of the average eddy turnover time); a waiting-time, on the other hand, would correspond to *the periods of time in between successive jumps*.

Diagnostics such as these have been proven useful many times in many different contexts, in particular in plasmas [35–37]. However, it is clear that they must rely on the good intuition of the researcher to find a suitable criterion to define jump sizes and waiting-times that makes sense for the system at hand. Because of this, in spite of its usefulness, the resulting exponents often lack the degree of objectivity that could be desired.

5.4.2.2   fLe Methods

The second Lagrangian method we will discuss is based in comparing the behaviour of the Lagrangian trajectories available, $x^j(t_k)$, with the properties of fractional Brownian/Lévy motion and their increments, fGn and fLn. The idea is to test whether the available trajectories resemble (on average) fBm/fLn for some values of the exponents α and H. This determination can be done by applying any of the methods already discussed in Chap. 3 and 4 to our trajectories. Finally, one can infer the temporal exponent of the fTe, β, using $\beta = \alpha H$ (Eq. 5.65).

For simplicity, we will restrict the discussion here to the investigation of whether the increments of the x^j series behave as either fGn or fLn. To do this, one simply needs to collect the statistics of the increments (for arbitrary spacing $h > 0$, as discussed in Sect. 3.4.2) of each Lagrangian trajectory,

$$\Delta_h x^j(t_j) := \frac{x^j(t_k + (h/2)\Delta t) - x^j(t_k - (h/2)\Delta t)}{h\Delta t}, \tag{5.103}$$

and characterize its distribution. In fact, one can exploit the fact that we are assuming translational invariance to improve statistics. Indeed, since all trajectories should be statistically equivalent, one can use the *increments of all trajectories* and build a unique pdf, thus reducing the uncertainties and facilitating a more precise determination of the tail exponent α. It is also advisable that, in order to better resolve the tail, methods such as the CBC (Sect. 2.4.2) or SF/CDF (Sect. 2.4.3) be used.

Regarding the estimation of the Hurst exponent H, one starts by determining H for each $x^j(t_k)$ series using the method of choice and averaging the result over all available trajectories. The Hurst exponent of each time series can be estimated in many different ways, as we have discussed extensively in Chaps. 3 and 4. A popular one is to apply the R/S analysis to each trajectory as described in Sect. 4.4.4, and then average the result over trajectories [38]. Care must be exercised, though, if the statistics of the increments are Lévy, since in that case $H^{R/S}$ does not coincide with the self-similarity exponent H. One must then use the modification of R/S analysis (Eq. 4.50) discussed in Sect. 4.4.4, or estimate the Hurst exponent using instead Eq. 4.57.

5.5   Case Study: The Running Sandpile

We will conclude the investigation of the sandpile dynamics that we have been doing at the end of each chapter by characterizing the nature of the transport of sand through the pile. We will do it by using some of the techniques presented in this chapter. However, it should be clear from the start that none of the *spatially symmetric, scale invariant models* that we have discussed (CTRW, fLe or fTe) will be adequate in this case, given the fact that transport in the running sandpile is **directed down the slope.** Thus, the sandpile will serve as illustration not only of the methods presented, but also of how to extend and adapt some of these ideas depending on the features of the problem of interest.

We will consider a sandpile of size $L = 1000$, critical slope $Z_c = 200$, toppling size $N_F = 20$, rain probability $p_0 = 10^{-4}$ and rain size $N_b = 10$, that has already been run to saturation and whose state is deep into the SOC state. The height of each sandpile cell is represented by h_i, with $1 < i < L$. It will be remembered that the sandpile is advanced, at each iteration, by first dropping N_b grains of sand on each location with probability p_0, and then relaxing the state by moving N_F grains from each unstable location i where $h_i - h_{i+1} > Z_c$ to the $i + 1$ cell. Therefore, motion in the sandpile SOC state affects *a narrow strip at its surface that has an average width of the order of* N_F. It is the propagator associated to transport taking place within this layer that we will try to estimate in what follows.

We will use N marked grains of sand that will be initially located close to the center of the pile. The temporal evolution of the marked population, as it is transported down the pile, will give us the propagator we are looking for. The m-th marked grain will be positioned, at some initial time, t_0^m, at an arbitrary cell i_m, chosen randomly from within a reduced number of cells near the top of the pile. The initial position of the m-th grain is then $x^m(0) = i_m$; its depth in the i_m column, as measured from its top, will be initially set to $d^m(0) = uN_F$, where u is a random number uniformly distributed in $[0, 1]$. As the sandpile is iterated, the position, x^m, and depth d^m of the marked grain of sand will change. Their values, at the k-th iteration, will be updated after finding out which of the following rules applies [39]:

1. the current cell is stable and no grains of sand have been dropped on it in the previous driving phase;
   then $d^m(k) = d^m(k-1); x^m(k) = x^m(k-1)$;
2. the current cell is stable, but N_b grains of sand have fallen on it in the previous driving phase;
   then $d^m(k) = d^m(k-1) + N_b; x^m(k) = x^m(k-1)$;
3. the current cell is stable, but the previous one is unstable and moves N_F grains over to the current cell;
   then $d^m(k) = d^m(k-1) + N_F; x^m(k) = x^m(k-1)$;
4. the current cell is stable, the previous one is unstable and, in the driving phase, N_b grains have fallen on the current cell;
   then $d^m(k) = d^m(k-1) + N_F + N_b; x^m(k) = x^m(k-1)$;
5. the current cell is unstable and N_F grains are thus moved to the following cell; no grains of sand have been dropped on the current cell in the driving phase;
   then, if $d^m(k-1) \leq N_F \longrightarrow d^m(k) = uN_F; x^m(k) = x^m(k-1) + 1$;
   if $d^m(k-1) > N_F \longrightarrow d^m(k) = d^m(k) - N_F; x^m(k) = x^m(k-1)$;
6. the current cell is unstable and N_F grains are thus moved to the following cell; at the previous driving phase, N_b grains of rain have fallen on the current cell:
   then, $d^m(k-1) \leq N_F - N_b \longrightarrow d^m(k) = uN_F; x^m(k) = x^m(k-1) + 1$;
   if $d^m(k-1) > N_F - N_b \longrightarrow d^m(k) = z^m(k) - N_F; x^m(k) = x^m(k-1)$;

The majority of these rules are rather self-explanatory. Basically, they state that, when it is time to move N_F particles to the next cell, our marked particle will be transported within that bunch only if its depth in the cell is at most N_F. In that case, the particle will reset its depth at the new cell to a new value, randomly chosen between 0 and N_F.[41] If the particle is however deeper than N_F, it remains at the current cell. In the (relatively rare) case that sand has been dropped during the

---

[41]It should be remembered that u is a random number uniformly distributed in $[0, 1]$.

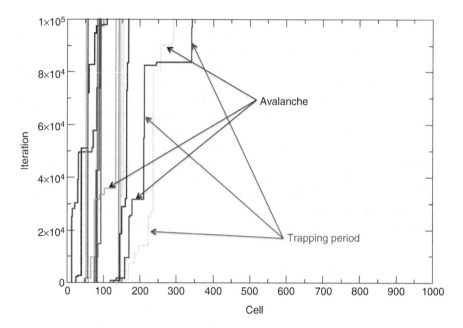

Fig. 5.8 Motion across the sandpile of size $L = 1000$ of fifteen selected particles, with initial locations randomly chosen within the central half of the pile. The vertical parts of the trajectories correspond to periods in which the particle is at rest on some cell; the (almost) horizontal parts, to periods of time in which the particle is transported radially, carried away by passing avalanches

previous driving phase on the same cell where the marked particle sits, the depth of the marked particle is increased by N_b.

Figure 5.8 shows the motion across the sandpile of fifteen marked particles for the first 10^5 iterations. As can be appreciated, the particles alternate radial displacements when carried by a passing avalanche—that appear as nearly horizontal segments,[42]—with periods of rest at the same cell, while the particle remains trapped there –that appear as vertical segments.

The recorded positions of the marked particles can be used to build a discrete version of the sandpile propagator. All that is needed is to calculate, at each iteration, the pdf of the particle displacements with respect to their respective initial locations, $p(\Delta x; i)$. Clearly, each marked particle will contribute with the displacement value $\Delta x^m(i) = x^m(i) - x^m(0)$ $(m = 1, \cdots, N)$ at the iteration i. Since $p(\Delta x; i)$ thus gives the probability of a particle having been displaced a distance Δx in a time i, averaged over its initial location, we can write that,

$$p(\Delta x; i) \simeq \langle G(x_0 + \Delta x, i | x_0, 0) \rangle_{x_0}, \tag{5.104}$$

---

[42]In fact, they are not horizontal since particles only advance one position per iteration. However, the scale of the temporal axis used in the figure makes them look so.

which is the estimate of the propagator we are looking for.[43]

However, there are some limitations to the goodness of our estimate for the propagator. They are due to the fact that markers will eventually reach the end of the sandpile. This implies that Eq 5.104 must no longer be used beyond the average number of iterations required for the marked particles to reach the pile edge. In addition, we also need to consider that marked particles are initialized at different locations in order to improve statistics. Therefore, each particle travels a different distance to reach the edge of the sandpile. To avoid distortions, Eq. 5.104 will also be disregarded for any Δx larger than the minimum of these distances.

Figure 5.9 shows several snapshots of the propagator obtained for our sandpile using $N = 60,000$ marker particles, initialized in cells with $i_m \leq 150$. The resulting propagator exhibits a power law tail very close to $p(\Delta x) \sim (\Delta x)^{-2}$, that becomes distorted at times of the order of 30,000 iterations and above. This number corresponds to the number of iterations for which a sizeable amount of marked

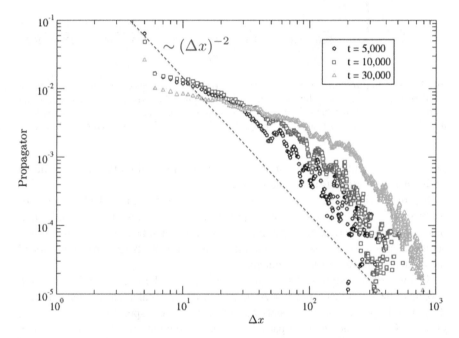

Fig. 5.9 Estimate of the sandpile propagator for times corresponding to 5000, 10,000 and 30,000 iterations using 60,000 tracer particles for the $L = 1000$ sandpile with $Z_c = 200$, $N_F = 30$, $N_b = 10$ and $p_0 = 10^{-4}$. The tail seems to exhibit a power-law decay with an exponent that is best fit by $P(\Delta x) \sim (\Delta x)^{-2.09 \pm 0.23}$. The scaling line $(\Delta x)^{-2}$ is shown to guide the eye

---

[43]We have implicitly assumed that the range of cells were the marked particles were initially dropped have similar dynamics, which is the case for the running sandpile.

particles has already reached the sandpile edge. From it, we can infer a value $\alpha \sim 1$ for the spatial fractional exponent (see Eq. 5.92).

We can also characterize the temporal exponent of the propagator by characterizing the initial growth (Eq. 5.94) and later decay (Eq. 5.93) of the propagator at a fixed location. In Fig. 5.10, this evolution of the propagator is shown at two different displacement values, $\Delta x = 160$ and $\Delta x = 240$. For the smallest displacement value, the propagator appears to grow locally as $p(\Delta t) \sim (\Delta t)^{0.3}$ for $\Delta t < 10^4$, and then decrease as $p(\Delta t) \sim (\Delta t)^{-0.4}$ for $\Delta t \gg 10^4$. The decay phase is however not seen for the largest displacement value within the 10^6 iterations examined. This is due to the fact that it takes more iterations for the propagator to build up at so large a distance from the seeding region.

From Figs. 5.9 and 5.10, one might naively conclude that transport down the sandpile slope is indeed self-similar, with fractional exponents close to $\alpha \sim 1$ and $\beta \sim 0.35$. However, this transport cannot be described by the symmetric fTe (Eq. 5.66), since the propagator is only defined for $\Delta x > 0$. As we mentioned at the beginning, this is due to the fact that sand always travels downwards, towards the edge of the pile. One needs to take this fact into account when deriving a proper fractional transport equation for this case.

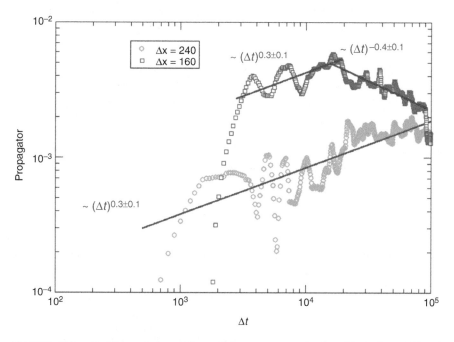

Fig. 5.10 Estimate of the evolution of the sandpile propagator at fixed positions $\Delta x = 160$ and $\Delta x = 240$ cells as a function of time using $60,000$ tracer particles for the $L = 1000$ sandpile with $Z_c = 200$, $N_F = 30$, $N_b = 10$ and $p_0 = 10^{-4}$. As expected, the local value first grows and then decays with power-law with an exponent that is best fit by $P(\Delta t) \sim (\Delta t)^{\pm 0.35}$

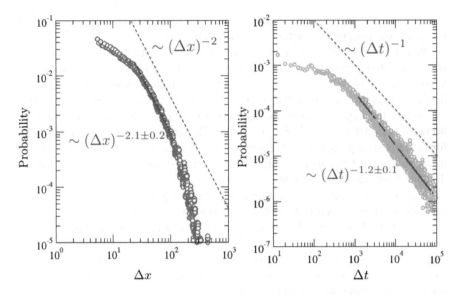

Fig. 5.11 Pdfs for the jumps (left) experienced by $60,000$ tracer particles, and the waiting times (right) they remain at rest for the $L = 1000$ sandpile with $Z_c = 200$, $N_F = 30$, $N_b = 10$ and $p_0 = 10^{-4}$. The jump-size pdf exhibits a tail $P(\Delta x) \sim (\Delta x)^{-2.1\pm0.2}$ for sizes still far from the system size. The waiting-time pdf also exhibits a power law tail $\psi(\Delta t) \sim (\Delta t)^{-1.2\pm0.1}$

We will show how to find this equation in what follows by invoking the CTRW formalism and considering the (almost) horizontal marker displacements as instantaneous jumps, and the vertical segments as waiting times. We have calculated the statistics of the jump sizes and waiting-times experienced by the markers in the sandpile are shown in Fig. 5.11. As can be seen, $p(\Delta x) \sim (\Delta x)^{-2.1}$ only for $\Delta x > 0$ and $\psi(\Delta t) \sim (\Delta t)^{-1.2}$, which yields values for $\alpha \sim 1.1$ and $\beta \sim 0.2$, not too far from those obtained with the propagator estimate.

5.5.1 *fTe for the Directed Running Sandpile*

Since both pdfs are only defined for positive values of the argument, we will consider a CTRW *defined in terms on extremal distributions for both jumps and waiting times,*

$$p(\Delta x) = L_{[\alpha,1,0,\sigma]}, \quad \psi(\Delta t) = L_{[\beta,1,0,\tau]}. \tag{5.105}$$

As will be remembered, the Laplace transform of these extremal Lévy pdfs is given by (Eq. 5.40),

$$\tilde{L}_{[\beta,1,0,\tau]}(s) \sim 1 - \frac{\tau^\beta s^\beta}{\cos(\pi\beta/2)}, \quad s \to 0 \tag{5.106}$$

Its Fourier transform also verifies that (Eq. 2.32),

$$\hat{L}_{[\alpha,1,0,\sigma]}(k) \sim 1 - \sigma^\alpha |k|^\alpha (1 - \imath\mathrm{sgn}(k)\tan(\pi\alpha/2)), \quad |k| \to 0. \tag{5.107}$$

Inserting these asymptotic behaviours in the general solution of the CTRW (Eq. 5.19) one obtains,

$$\bar{n}(k,s) \simeq \frac{s^{\beta-1}\bar{S}^{\mathrm{aug}}(k,s)}{\left(s^\beta + \cos(\pi\beta/2)\frac{\sigma^\alpha}{\tau^\beta}|k|^\alpha \left[1 - \imath\mathrm{sgn}(k)\tan\left(\frac{\pi\alpha}{2}\right)\right]\right)}. \tag{5.108}$$

that can be easily reordered as,

$$s\bar{n}(k,s) - \hat{n}(k,0) \simeq - s^{\beta-1}\left(\tilde{D}_{\alpha,\beta}|k|^\alpha\left[1 - \imath\mathrm{sgn}(k)\tan\left(\frac{\pi\alpha}{2}\right)\right]\right)\bar{n}(k,s) +$$
$$+ \bar{S}(k,s), \tag{5.109}$$

where we have defined the coefficient $\tilde{D}_{\alpha,\beta} := \cos(\pi\beta/2)\sigma^\alpha/\tau^\beta$.

This equation can be now Fourier-Laplace inverted to yield,

$$\frac{\partial n}{\partial t} = {}_0D_t^{1-\beta}\left[\tilde{D}(\alpha,\beta)\frac{\partial^{\alpha,1}n}{\partial|x|^{\alpha,1}}\right] + S(x,t) \tag{5.110}$$

where we have used yet another fractional derivative: one of the two *completely asymmetrical Riesz-Feller fractional derivatives* that are discussed in Appendix 3. In spite of its apparent complexity, the obtained result makes complete physical sense. Indeed, it should be noted that the new fractional derivative, when written explicitly, becomes,

$$\frac{\partial^{\alpha,1}n}{\partial|x|^{\alpha,1}} \propto {}_{-\infty}D_x^\alpha n \propto \frac{d^k}{dx^k}\left[\int_{-\infty}^x \frac{n(x')\,dx'}{(x-x')^{\alpha-k+1}}\right], \tag{5.111}$$

with the integer k defined as the one satisfying $k - 1 < \alpha < k$. As can be seen, only points $x' < x$ contribute to the integral, in complete agreement with the fact that net transport in the sandpile comes (almost) always from the left. Therefore, Eq. 5.110 should provide a good effective description for the down-the-slope sandpile transport in the long times, large scale limit. Given our previous findings, we should then choose $\alpha \simeq 1$ and $\beta \simeq 0.35$ for the particular sandpile realization analyzed here.

5.6 Final Considerations

In this chapter we have reviewed the basics of the modern theory of fractional transport. We have also argued that the fractional transport equation (that is, Eq. 5.66) and its variations are the appropriate framework to describe transport in many complex systems, at least those in which the underlying processes are of an additive nature. For the sake of simplicity, we have limited the majority of our discussion to unbounded, homogeneous, one-dimensional systems, in which the underlying motion is symmetric (with the exception of the sandpile study, where full asymmetry was considered). This is, of course, an oversimplification. It is thus worth to discuss some of the issues that one might have to deal with when applying these concepts to real systems.

Symmetry

The assumption of symmetric underlying motion is sometimes too narrow. We have seen one such example when characterizing the transport dynamics of the running sandpile. However, the sandpile example has served as an illustration of the fact that asymmetric motion can be easily accommodated, what has lead us to the formulation of the asymmetric fTe (Eq. 5.110). Similar modifications might be needed before one can apply these ideas to other systems.

Boundedness

Real systems are usually bounded in space. This issue has often been addressed directly at the fTe level,[44] by setting the starting and ending points of the fractional derivatives to finite values that represent the system boundaries. That is, if the system lives within the $[a, b]$ interval, one makes the changes $-\infty \to a$ and $\infty \to b$ in the fractional derivatives of the fTe.

The restriction of the fTe to a bounded system introduces yet another problem related to the fact that RL fractional derivatives are singular at the starting/ending points. This is particularly problematic when deriving the kind of discrete approximations of the fractional derivatives that might be used for the numerical integration of these equations. To avoid this problem, it is customary to regularize the RL derivative at the starting/ending points by replacing it by the closely related **Caputo fractional derivative** [21]. Caputo fractional derivatives share many of the more important properties of RL derivatives, but not all. In particular, they are regular at the starting/ending points. A detailed account of some these technical aspects, that are essential in order to use these operators in numerical calculations, is given in Appendix 4.

Dimensionality

Fractional derivatives and integrals have been extensively studied in one dimension [21, 44, 45]. Regretfully, higher dimensional versions are scarce, although higher-dimensional versions of the symmetric Reisz fractional derivative exist.

---

[44]The consequences of boundedness have also been addressed within the CTRW and fTe frameworks by considering truncated Lévy distributions [19, 40–43]. It is however much more complicated to establish a connection with fractional transport equations that in the unbounded case, since one can no longer rely on the advantages of the Fourier representation to take the long-time, large-distance limit.

Clearly, there is still much work to do in this regard. This is an area where a lot of activity should be expected in the future.

Physical Basis for Fractional Transport

A never-ending discussion among transport modellers is whether the fTe is actually a good *physical* model for transport in any real system. In particular, it is often mentioned that the fact that, for $\alpha < 2$, any location in the system can influence the level of transport at any other location, necessarily implies that *information is transported through the system with infinite velocity*. Another criticism that is often made is that *all microscopic physical laws are local in space and time*, so how could non-Markovianity and non-locality appear in the first place.[45]

The best answer that one can provide against this criticism is that the fTe (and its variations) is not an exact representation of transport in any system, but only an effective transport model that is a good approximation of the real behaviour, under certain conditions (i.e., the lack of characteristic scales), in the asymptotic limit of long times and large distances. Very much like the classical diffusion equation is in many cases! In that limit, any interaction that traverses a finite region in a certain amount of time, would appear instantaneous when looked at from scales that exceed the largest/longest allowed by the system dynamics. Take, for instance, the running sandpile. Transport events take place through avalanches, whose linear size and duration is limited only by the system size. This was apparent, for instance, in the R/S analysis of the sandpile activity examined in Sect. 4.5, where the region $R/S \sim \tau$ that appeared at the shortest timescales was a reflection of the self-awareness of avalanches. For the sandpile examined there, with $L \sim 200$, the region extended up to timescales of the order of $\tau \simeq 200$ (see Fig. 4.18). Memory, however, survives in the sandpile for timescales much longer than the largest duration of a single avalanche. In the R/S plot in Fig. 4.18, these timescales corresponded to the intermediate region scaling as $R/s \sim \tau^{0.8}$, that extends up to $\tau \sim 5 \times 10^5$. When looked at from scales that large, single avalanches do indeed appear instantaneous.[46] As a result, the local flux effectively becomes determined by contributions from all the values of the gradient over the whole spatial domain and the past history. The same argument can be used to explain how the local interactions that determine the sand toppling can become effectively non-local and non-Markovian when looked at from the longest and largest scales.

---

[45]In fact, no rigorous derivation of any fTe exists, to the best of our knowledge, that obtains the values of the fractional exponents from the physical equations of motion based on Newton's law or Hamiltonian dynamics.

[46]This is, in fact, why the marked particle jumps appeared as (almost) horizontal lines in Fig. 5.8!

Appendix 1: The Laplace Transform

The **Laplace transform**[47] of a function $f(t)$, $t > 0$, is defined as [46]:

$$L[f(t)] := \tilde{f}(s) \equiv \int_0^\infty f(t)e^{-st}dt. \tag{5.112}$$

The main requirement for $f(t)$ to have a Laplace transform is that it cannot grow as $t \to \infty$ as fast as an exponential. Otherwise, the Laplace integral would not converge.[48] A list of common Laplace transforms is given in Table 5.1.

The Laplace transform is a linear operation that has interesting properties. A first one has to do with the *translation of the independent variable in Laplace space*. That is,

$$\tilde{f}(s - a) = L\left[\exp(at)f(t)\right]. \tag{5.113}$$

Another one, with the *scaling of the independent variable*,

$$\tilde{f}(s/a) = aL\left[f(at)\right]. \tag{5.114}$$

The *Laplace transform of the n-th order derivative of a function* satisfies:

$$L\left[\frac{d^n f}{dt^n}\right] = s^n \tilde{f}(s) - s^{n-1}f(0) - s^{n-2}f'(0) - \cdots - f^{(n-1)}(0), \tag{5.115}$$

Table 5.1 Some useful Laplace transforms

$f(t)$	$\tilde{f}(s)$	Restrictions	$f(t)$	$\tilde{f}(s)$	Restrictions
1	$\dfrac{1}{s}$	$s > 0$	$\exp(at)$	$\dfrac{1}{s-a}$	$s > a$
t	$\dfrac{1}{s^2}$	$s > 0$	$\exp(\iota\omega t)$	$\dfrac{1}{s-\iota\omega}$	$s > 0$
$H(t - t_0)$	$\dfrac{\exp(-t_0 s)}{s}$	$t_0 \geq 0, s > 0$	$\cos(\omega t)$	$\dfrac{s}{s^2 + \omega^2}$	$s > 0$
t^a	$\dfrac{\Gamma(a+1)}{s^{a+1}}$	$a > -1, s > 0$	$\sin(\omega t)$	$\dfrac{\omega}{s^2 + \omega^2}$	$s > 0$

$H(x)$ represents the Heaviside step function

---

[47] We will always use the tilde (i.e., $\tilde{f}$) to represent the Laplace transform of a function (i.e., $f(t)$) throughout this book. Also, the Laplace variable will always be represented by the letter s.

[48] Note that it is not even required that $f(t)$ be continuous for it to have a Laplace transform.

whilst the *Laplace transform of the integral of a function*,

$$L\left[\int_0^t f(t')dt'\right] = \frac{\tilde{f}(s)}{s}. \tag{5.116}$$

These properties are very useful to solve linear systems of ordinary differential equations [13]. Another interesting result is known as the **convolution theorem**. It refers to the *temporal convolution* of two functions [46],

$$h(t) = \int_0^t f(t')g(t-t')dt', \tag{5.117}$$

whose Laplace transform is equal to the product of the Laplace transforms of the two functions,

$$\tilde{h}(s) = \tilde{f}(s) \cdot \tilde{g}(s). \tag{5.118}$$

We will also discuss a property particularly useful in the context of self-similar functions. The Laplace transform of a power law, $f(t) = t^a$, $t \geq 0$ is given by $\tilde{f}(s) = \Gamma(a+1)/s^{1+a}$, for $a > -1$, where $\Gamma(x)$ is Euler's gamma function [46]. This result, can also be extended to any function that asymptotically scales as $(a > -1)$,

$$f(t) \sim t^a, \quad t \to \infty, \quad \Longleftrightarrow \quad \tilde{f}(s) \sim s^{-(1+a)}, \quad s \to 0. \tag{5.119}$$

Finally, we will mention that the **inverse Laplace transform** is given by [46]:

$$f(t) = \frac{1}{2\pi i} \int_{c-i\infty}^{c+i\infty} \tilde{f}(s)e^{st}ds, \tag{5.120}$$

where s varies along an imaginary line, since c is real and must be chosen so that the integral converges.

Appendix 2: Riemann-Liouville Fractional Derivatives and Integrals

Fractional integrals and derivatives were introduced as interpolants between integrals and derivatives of integer order [21, 44, 45]. In this book, we will always use the Riemann-Lioville (RL) definition of fractional integrals and derivatives.

Riemann-Liouville Fractional Integrals

The **left-sided RL fractional integral of order** $p > 0$ of a function $f(t)$ is defined as [21]:

$$_aD_t^{-p}f(t) \equiv \frac{1}{\Gamma(p)} \int_a^t (t-\tau)^{p-1}f(\tau)d\tau, \tag{5.121}$$

A negative superscript, $-p$, is used to reveal that we are dealing with a RL fractional integral. a is known as the **starting point** of the integral.[49] As advertised, it can be shown that Eq. 5.121 reduces, for $p = n > 0$, to the usual integral of order n.[50] But they have some other interesting properties. For instance, they satisfy a **commutative composition rule** [21]. Indeed, one has that for $p, q > 0$:

$$_aD_t^{-p} \cdot \left[_aD_t^{-q} \cdot f(t)\right] =_a D_t^{-q} \cdot \left[_aD_t^{-p} \cdot f(t)\right] =_a D_t^{-(p+q)}f(t) \tag{5.124}$$

The **Laplace transform of the RL fractional integral**, if the starting point is $a = 0$, is particularly simple [21]. It is given by:

$$L\left[_0D_t^{-p} \cdot f(t)\right] = s^{-p}\tilde{f}(s). \tag{5.125}$$

This result naturally follows from the fact that, for $a = 0$, the RL fractional integral becomes a temporal convolution of $f(t)$ with a power law, $t^p/\Gamma(p)$. Thus, Eq. 5.125 follows from applying the convolution theorem (Eq. 5.118).

A simple relation also provides the **Fourier transform** (see Appendix 1 of Chap. 2) of the RL fractional integral if the starting point is $a = -\infty$. It is given by[51]:

$$F\left[_{-\infty}D_t^{-p} \cdot f(t)\right] = (-i\omega)^{-p}\hat{f}(\omega). \tag{5.127}$$

---

[49]Right-sided RL fractional integrals can also be defined:

$$^bD_t^{-p}f(t) \equiv \frac{1}{\Gamma(p)} \int_t^b (\tau-t)^{p-1}f(\tau)d\tau, \tag{5.122}$$

b is known as the **ending point**.

[50]An integral of order n is equivalent to carrying out n consecutive integrals on f:

$$I^n(f) = \int_0^t dt_1 \int_0^{t_1} dt_2 \cdots \int_0^{t_n} f(t_n)dt_n. \tag{5.123}$$

[51]In the case of the right-sided RL fractional integral, the Fourier transform, for ending point $b = +\infty$, is given by:

$$F\left[^\infty D_t^{-p} \cdot f(t)\right] = (-i\omega)^{-p}\hat{f}(\omega). \tag{5.126}$$

Riemann-Liouville Fractional Derivatives

The **left-sided RL fractional derivative of order** $p > 0$ of a function $f(t)$ is defined as [21]:

$$_aD_t^p f(t) \equiv \frac{1}{\Gamma(k-p)} \frac{d^k}{dt^k} \int_a^t (t-\tau)^{k-p-1} f(\tau) d\tau, \tag{5.128}$$

where the integer k satisfies that $k - 1 \leq p < k$. Note that they are simply a combination of normal derivatives and RL fractional integrals: $_aD_t^p = (d/dt)^k \cdot_a D_t^{-(k-p)}$. Again, it turns out that, for $p = n$, the RL fractional derivative reduces to the standard derivative of order n.[52]

RL fractional derivatives have interesting, but somewhat not intuitive properties. The most striking property is probably that *the fractional derivative of a constant function is not zero*. Indeed, using the fact that the derivative of a power law can be calculated to be [21]:

$$_aD_t^p \cdot (t-a)^\nu = \frac{\Gamma(1+\nu)}{\Gamma(1+\nu-p)} (t-a)^{\nu-p}, \quad p > 0, \ \nu > -1, \ t > 0, \tag{5.130}$$

it is clear that choosing $\nu = 0$ does not yield a constant, but $(t-a)^{-p}/\Gamma(1-p)$.

RL fractional derivatives can be combined with other derivatives (fractional or integer) and derivatives. But the combinations are not always simple. One of the simplest cases is when a *fractional derivative of order $p > 0$ acts on a fractional integral of order $q > 0$*:

$$_aD_t^p \left(_aD_t^{-q} f(t)\right) =_a D_t^{p-q} f(t). \tag{5.131}$$

However, the action of a RL fractional integral of order $q > 0$ on a RL fractional derivative of order $p > 0$ is given by a much more complicated expression [21].

If one sets $p = q$ in Eq. 5.131, one finds that the inverse (from the left) of a RL fractional derivative of order $p > 0$ is the fractional integral of order $p > 0$:

$$_aD_t^p \left(_aD_t^{-p} f(t)\right) = f(t). \tag{5.132}$$

---

[52] Again, right-sided RL fractional derivatives can also be defined:

$$^bD_t^p f(t) \equiv \frac{1}{\Gamma(k-p)} \frac{d^k}{dt^k} \int_t^b (\tau-t)^{k-p-1} f(\tau) d\tau, \tag{5.129}$$

Their properties are analogous to the left-sided counterpart.

However, note that the RL fractional derivative of order $p > 0$ is not the inverse from the left of the RL fractional integral of order p. Instead, one has that [21]

$$_aD_t^{-p}\left(_aD_t^pf(t)\right) = f(t) - \sum_{j=1}^{k}\left[-_aD_t^{p-j}f(t)\right]_{t=a}\frac{(t-a)^{p-j}}{\Gamma(p-j+1)}. \qquad (5.133)$$

The action of normal derivatives on RL fractional derivatives is also simple to express[53]:

$$\frac{d^m}{dt^m}\cdot {}_aD_t^pf(t) =_a D_t^{p+m}f(t). \qquad (5.135)$$

But again, the action of the RL fractional derivative on a normal derivative is much more complicated, and given by the expression:

$$_aD_t^p\cdot\frac{d^m}{dt^m}f(t) =_a D_t^{p+m}f(t) - \sum_{j=0}^{m-1}\frac{f^{(j)}(a)(t-a)^{j-p-m}}{\Gamma(1+j-p-m)}. \qquad (5.136)$$

Finally, the composition of RL fractional derivatives of different orders is given by a rather complex expression, that is never equal to a fractional derivative of a higher order, except in very special cases [21],

Relatively simple expressions also exist for the **Laplace transform of the left-sided RL fractional derivative of order** p if the starting point is $a = 0$:

$$L\left[_0D_t^p\cdot f(t)\right] = s^p\tilde{f}(s) - \sum_{j=0}^{k-1}s^j\left[_0D_t^{p-j-1}\cdot f(t)\right]_{t=0}. \qquad (5.137)$$

This expression is very reminiscent of the one obtained for normal derivatives (Eq. 5.115). Similarly, the **Fourier transform of the left-sided RL fractional derivative** satisfies a very simple relation,[54] but only for starting point $a = -\infty$:

$$F\left[_{-\infty}D_t^p\cdot f(t)\right] = (i\omega)^p\hat{f}(\omega). \qquad (5.139)$$

---

[53]For the right side RL derivatives, this property becomes:

$$(-1)^m\frac{d^m}{dt^m}\cdot {}^bD_t^pf(t) =_a D_t^{p+m}f(t). \qquad (5.134)$$

[54]For the right-sided fractional integral with ending point $b = \infty$, the Fourier transform is given by:

$$F\left[{}^\infty D_t^p\cdot f(t)\right] = (-i\omega)^p\hat{f}(\omega). \qquad (5.138)$$

Appendix 3: The Riesz-Feller Fractional Derivative

The **Riesz fractional derivative of order** α is defined by the integral[55]:

$$\frac{\partial^\alpha f}{\partial |x|^\alpha} := -\frac{1}{2\Gamma(\alpha)\cos(\alpha\pi/2)} \int_{-\infty}^{\infty} dx' \frac{f(x')}{|x - x'|^{\alpha+1}} dx'. \tag{5.140}$$

The most remarkable property of this derivative, and one of the reasons why it appears so often in the context of transport (see Sect. 5.3.1), has to do with its Fourier transform, which is given by [47]:

$$F\left[\frac{\partial^\alpha f}{\partial |x|^\alpha}\right] = -|k|^\alpha \hat{f}(k). \tag{5.141}$$

Using now the complex identity ($\imath = \sqrt{-1}$),

$$(\imath k)^\alpha + (-\imath k)^\alpha = 2\cos\left(\frac{\pi\alpha}{2}\right)|k|^\alpha, \tag{5.142}$$

it is very easy to prove that the Riesz derivative can also be expressed as a symmetrized sum of two (one left-sided, one right-sided) RL fractional derivatives of order α [21],

$$\frac{\partial^\alpha f}{\partial |x|^\alpha} = -\frac{1}{2\Gamma(\alpha)\cos(\alpha\pi/2)} \left(_{-\infty}D_x^\alpha + {}^\infty D_x^\alpha\right) \tag{5.143}$$

It is also possible to define an asymmetric version of the Riesz-Feller derivative, often known as the **Riesz-Feller fractional derivative of order** α **and asymmetry parameter** λ [48]. This can be done through its Fourier transform, that is given by:

$$F\left[\frac{\partial^{\alpha,\lambda} f}{\partial |x|^{\alpha,\lambda}}\right] = -|k|^\alpha \left[1 - \imath\lambda \operatorname{sgn}(k)\tan(\pi\alpha/2)\right]\hat{f}(k). \tag{5.144}$$

The indices are restricted, in this case, to $\alpha \in (0, 2)$ and $\lambda \in [-1, 1]$. For $\lambda = 0$, the standard symmetric Riesz derivative is recovered. It can also be shown that the Riesz-Feller derivative can also be expressed as an asymmetric sum of the same two

---

[55]The attentive reader will note that we have changed the name of the independent variable (now x) to represent that the range is no $(-\infty, \infty)$ instead of the range $(0, \infty)$ used when defining the RL fractional derivatives. Similarly, we will referred to the Fourier variable as k, instead of ω.

RL fractional derivatives of order α [49]:

$$\frac{\partial^{\alpha,\lambda} f}{\partial |x|^{\alpha,\lambda}} = -\frac{1}{2\Gamma(\alpha)\cos(\alpha\pi/2)}\left(c_-(\alpha,\lambda)_{-\infty}D_x^\alpha + c_+(\alpha,\lambda)^\infty D_x^\alpha\right) \tag{5.145}$$

with the coefficients defined as:

$$c_\pm(\alpha,\lambda) := \frac{1 \mp \lambda}{1 + \lambda\tan(\pi\alpha/2)} \tag{5.146}$$

Thus, in the limit of $\lambda = 1$, only the left-sided α-RL derivative $_{-\infty}D_x^\alpha$ remains, whilst for $\lambda = -1$, only the right-side one $^\infty D_x^\alpha$ does.

Appendix 4: Discrete Approximations for Fractional Derivatives

In order to be useful for actual applications, discrete representations of fractional derivatives are needed that could be easily implemented in a computer [44, 50–52]. We discuss here one possible way to do this [51]. The main difficulty has to do with the singularities of the Riemann-Lioville definition, since it turns out that it becomes singular at the starting point, $t = a$, *whenever a is finite*. This will certainly be the case in practical cases, since a computer must always deal with finite intervals.

Case $1 < p < 2$
One can made these singularities explicit by rewriting the fractional derivative $_aD_t^p f$, $1 < p < 2$, as the infinite series [52]:

$$\frac{1}{\Gamma(1-p)}\frac{f(a)}{(t-a)^p} + \sum_{k=1}^\infty \frac{f^{(k)}(a)(t-a)^{k-p}}{\Gamma(k+1-p)} \tag{5.147}$$

simply by Taylor expanding $f(t)$ around $t = a$ and using Eq. 5.130 to derive the different powers of x. Clearly, the first two terms of Eq. 5.147 are singular. At least, unless one sets $f(a) = 0$, and $f'(a) = 0$ as well for $1 < p < 2$, which puts too much of a restriction in some cases.

One way to circumvent this problem is to replace in practice the Riemann-Liouville derivative by the so-called **Caputo fractional derivative** [21]:

$$_a^C D_t^p f := {_aD_t^p}\left[f - \sum_{i=0}^{\text{int}(p)} f^{(i)}(a)(t-a)^i \right], \tag{5.148}$$

where int(p) is the integer part of p.[56] The Caputo definition removes the singularities so that the expression is now well defined and can be easily discretized. To do it, we give its analytical expression [21],

$$
{}^C_a D^p_t f(t) \equiv \frac{1}{\Gamma(k-p)} \int_a^t (t-\tau)^{k-p-1} f^{(k)}(\tau) d\tau, \tag{5.150}
$$

with $k-1 \le p < k$, which is very similar to the RL definition (Eq. 5.128) but *with the integer derivatives acting inside of the integral, instead of outside.* It can be shown that RL and Caputo derivatives are identical, for most functions for $a \to -\infty$ and $t \to \infty$. However, they are different for finite starting points. For starters, it should be noted that the Caputo derivative of a constant is now zero!

We proceed now to find discrete expressions for the Caputo derivative for $1 < p < 2$ on the discrete regular mesh, $t_i = (i-1)\Delta t$, with $i = 0, 2, \cdots, N_t$. The integral from $t = t_1$ up to $t = t_i$ is then discretized[57] following the scheme [44, 51],

$$
{}^C_a D^p_t f = \frac{1}{\Gamma(2-p)} \int_a^t (t-\tau)^{1-p} f^{(2)}(\tau) d\tau \tag{5.151}
$$

$$
\simeq \frac{1}{\Gamma(2-p)} \sum_{j=0}^{i-1} \int_{t_j}^{t_{j+1}} \frac{f''(t-s)}{s^{p-1}} ds
$$

$$
\simeq \frac{1}{\Gamma(2-p)} \sum_{j=0}^{i-1} \left[\frac{f(t_i - t_{j+1}) + f(t_i - t_{j-1}) - 2f(t_i - t_j)}{(\Delta t)^2} \int_{t_j}^{t_{j+1}} \frac{ds}{s^{p-1}} \right]
$$

$$
\simeq \sum_{j=0}^{i-1} \left[\frac{(f(t_i - t_{j+1}) + f(t_i - t_{j-1}) - 2f(t_i - t_j)) \left[(j+1)^{2-p} - j^{2-p} \right]}{\Gamma(3-p)(\Delta t)^p} \right].
$$

This expression is more conveniently expressed as,

$$
{}^C_a D^p_t f = \sum_{j=-1}^i \frac{W_j^{p>1}}{\Gamma(3-p)(\Delta t)^p} f(t_i - t_j), \tag{5.152}
$$

---

[56]For the right-sided derivatives, the divergence of the RL derivative happens at the ending point, $t = b$. The Caputo fractional derivative is defined very similarly:

$$
{}^b_C D^p_t f := {}^b D^p_x \left[f - \sum_{i=0}^{\text{int}(p)} f^{(i)}(b)(b-t)^i \right]. \tag{5.149}
$$

[57]Note that we have defined t_{-1} as the first point in the series. This is done on purpose, since we will need it to discretize the integral that starts at t_0.

in terms of the weights, $W_j^{p>1}$,

$$
W_j^{p>1} = \begin{cases}
1, & j = -1 \\
2^{2-p} - 3, & j = 0 \\
(j+2)^{2-p} - 3(j+1)^{2-p} + 3j^{2-p} - (j-1)^{2-p}, & 0 < j < i-i \\
-2i^{2-p} + 3(i-1)^{2-p} - (i-2)^{2-p}, & j = i-1 \\
i^{2-p} - (i-1)^{2-p}, & j = i
\end{cases}
\tag{5.153}
$$

Case $0 < \alpha < 1$

When $p < 1$ there is only one singularity at the starting point a of the fractional derivative. We can make it explicit by expressing again the fractional derivative $_aD_t^p f$, as the infinite series:

$$
\frac{1}{\Gamma(1-p)} \frac{f(a)}{(t-a)^p} + \sum_{k=1}^{\infty} \frac{f^{(k)}(a)(t-a)^{k-p}}{\Gamma(k+1-p)},
\tag{5.154}
$$

done by Taylor expanding $f(t)$ around $t = a$ and using Eq. 5.130 to derive the different powers of x. Clearly, the first term of Eq. 5.154 is singular. At least, unless one sets $f(a) = 0$.

The solution is again to use the Caputo derivative (Eq. 5.150) instead. Now, $\text{int}(p) = 0$ and the integral to discretize is:

$$
{}_a^C D_t^p f(t) \equiv \frac{1}{\Gamma(1-p)} \int_a^t (t-\tau)^{-p} f^{(k)}(\tau) d\tau,
\tag{5.155}
$$

For $0 < p < 1$, on the other hand, we discretize the integral from $t = t_0$ up to $t = t_i$ as,

$$
{}_a^C D_t^p f = \frac{1}{\Gamma(1-p)} \int_a^t (t-\tau)^{-p} f^{(1)}(\tau) d\tau
\tag{5.156}
$$

$$
\simeq \frac{1}{\Gamma(1-p)} \sum_{j=0}^{i-1} \int_{t_j}^{t_{j+1}} \frac{f'(t-s)}{s^p} ds
$$

$$
\simeq \frac{1}{\Gamma(1-p)} \sum_{j=0}^{i-1} \left[\frac{f(t_i - t_{j+1}) - f(t_i - t_{j-1})}{2\Delta t} \int_{t_j}^{t_{j+1}} \frac{ds}{s^p} \right]
$$

$$
\simeq \sum_{j=0}^{i-1} \left[\frac{(f(t_i - t_{j+1}) - f(t_i - t_{j-1})) \left[(j+1)^{1-p} - j^{1-p} \right]}{2\Gamma(2-p)(\Delta t)^p} \right].
$$

Again, we can express this formula more conveniently by introducing weights, $W_j^{p<1}$,

$$
{}_a^C D_{t}^p f = \sum_{j=-1}^{i} \frac{W_j^{p<1}}{2\Gamma(2-p)(\Delta t)^p} f(t_i - t_j),
\tag{5.157}
$$

which are now given by the expressions,

$$
W_j^{p<1} =
\begin{cases}
-1, & j = -1 \\
-(2^{1-p} - 1), & j = 0 \\
j^{1-p} - (j-1)^{1-p} - (j+2)^{1-p} + (j+1)^{1-p}, & 0 < j < i-i \\
(i-1)^{1-p} - (i-2)^{1-p}, & j = i-1 \\
i^{1-p} - (i-1)^{1-p}, & j = i
\end{cases}
\tag{5.158}
$$

Equations 5.152 and 5.157 are not however, the only possible discrete representation of the Caputo fractional derivative of order p. These formulas are based on central differencing. But other discretizations also exist, that may sometimes offer either better accuracy over some ranges of p or higher orders of discretization [44, 50–52]. Equations 5.152 and 5.157 are however sufficient for the purposes of this book.

Problems

5.1 CTRW: Fluid Limit

Write a code that evolves in time a one-dimensional CTRW that has a Gaussian jump size distribution, $p(\Delta x) = N_{[0,\sigma^2]}(\Delta x)$, and an exponential waiting time pdf, $\psi(\Delta t) = E_{\tau_0}(\Delta t)$. In order to generate values for the jumps and waiting times at run-time, use the algorithms described in Appendix 1 of Chap. 2. Use the code to calculate numerically the CTRW propagator for $\sigma^2 = 2$ and $\tau_0 = 1$, using $N = 100,000$ particles. Compare the results with its fluid limit: $G(x - x_0, t) = N_{[0,2t]}(x - x_0)$.

5.2 Langevin Equation: Propagator

Write a code that evolves in time a collection of N particles according to the Langevin equation (Eq. 5.26). Use a uniform noise with autocorrelation given by Eq. 5.27 with $D = 2$. Use $N = 100,000$ particles to calculate the propagator of the Langevin equation and compare it with its analytical solution: $G(x - x_0, t) = N_{[0,2t]}(x - x_0)$.

5.3 Langevin Equation: Moments of the Propagator

Compute all moments of the propagator of the Langevin equation and show that all odd moments vanish and all even moments $n > 2$ are given by $m_n = (Dt)^n (n-1)!!$.

5.4 Fractional Transport Equation: Propagator for $\alpha < 2$, $\beta = 1$

Calculate the propagator of the fractional transport equation (Eq. 5.66) in the case in which $\beta = 1$.

5.5 Scale-Invariant Generalization of Fick's Law

Find the expression that gives the local particle flux for the fractional transport equation. To do it, recast Eq. 5.66 into the form $\partial n / \partial t + \nabla \cdot \Gamma = S$, using the properties of fractional derivatives discussed in Appendices 2 and 3.

5.6 Advanced Problem: Propagator of the Running Sandpile

Write a code that uses the rules discussed in Sect. 5.5 to advances an arbitrary number of tracers on the height profile evolution calculated by the sandpile code previously developed (see Prob 1.5). Use the code to estimate the numerical propagator of the running sandpile for a sandpile with $L = 2000$, $Z_c = 200$, $N_f = 20$, $N_b = 10$ and $p_0 = 10^{-4}$.

5.7 Advanced Problem: Numerical Integration of the fTe

Write a code that integrates the fractional transport equation (Eq. 5.66) for $\beta = 1$ and arbitrary α, arbitrary initial condition, $n_0(x)$, and external source, $S(x, t)$. Use, for that purpose, the discrete expressions of the Caputo fractional derivative (Eqs. 5.152 and 5.157) given in Appendix 4.

References

1. Fick, A.: Über Diffusion. Annalen der Physik 94, 59 (1855)
2. Balescu, R.: Aspects of Anomalous Transport in Plasmas. Institute of Physics, Bristol (2005)
3. Sanchez, R., Newman, D.E.: Topical Review: Self-organized-criticality and the Dynamics of Near-marginal Turbulent Transport in Magnetically Confined Fusion Plasmas. Plasma Phys. Controlled Fusion 57, 123002 (2015)
4. Field, S., Witt, J., Nori, F., Ling, X.: Superconducting Vortex Avalanches. Phys. Rev. Lett. 74, 1206 (1995)
5. Isichenko, M.: Percolation, Statistical Topography and Transport in Random Media. Rev. Mod. Phys. 64, 961 (1992)
6. Bazant, Z.P.: Scaling of Quasibrittle Fracture: Hypotheses of Invasive and Lacunar Fractality, Their Critique and Weibull Connection. Int. J. Fract. 83, 41 (1997)
7. Montroll, E.W., Weiss, G.H.: Random Walks on Laticess II. J. Math. Phys. 6, 167 (1965)
8. Einstein, A.: On the Movement of Small Particles Suspended in Stationary Liquids Required by the Molecular-Kinetic Theory of Heat. Annalen der Physik 17, 549 (1905)
9. Shugard, W., Reiss, H.: Derivation of the Continuous-Time Random Walk Equation in Non-homogeneous Lattices. J. Chem. Phys. 65, 2827 (1976)
10. Barkai, E., Metzler, R., Klafter, J.: From Continuous Time Random Walks to the Fractional Fokker-Plack Equation. Phys. Rev. E 61, 132 (2000)
11. van Milligen, B.P., Sanchez, R., Carreras, B.A.: Probabilistic Finite-Size Transport Models for Fusion: Anomalous Transport and Scaling Laws. Phys. Plasmas 11, 2272 (2004)

12. Sanchez, R., Carreras, B.A., van Milligen, B.P.: Fluid Limit of Nonintegrable Continuous-Time Random Walks in Terms of Fractional Differential Equations. Phys. Rev. E 71, 011111 (2005)
13. Morse, P.M., Feshbach, H.: Methods of Mathematical Physics. McGraw Hill, New York (1953)
14. Langevin, P.: Sur la theorie du mouvement brownien. C.R. Acad. Sci. (Paris) 146, 530 (1908)
15. Gardiner, C.W.: Handbook of Stochastic Methods. Springer, New York (1997)
16. Compte, A.: Statistical Foundations of Fractional Dynamics. Phys. Rev. E 53, 4191 (1996)
17. Metzler, R., Klafter, J.: The Random Walk's Guide to Anomalous Diffusion: A Fractional Dynamics Approach. Phys. Rep. 339, 1 (2000)
18. Zaslavsky, G.M.: Chaos, Fractional Kinetics, and Anomalous Transport. Phys. Rep. 371, 461 (2002)
19. Metzler, R., Klafter, J.: The Restaurant at the End of the Random Walk: Recent Developments in the Description of Anomalous Transport by Fractional Dynamics. J. Phys. A 37, R161 (2004)
20. Samorodnitsky, G., Taqqu, M.S.: Stable Non-Gaussian Processes. Chapman & Hall, New York (1994)
21. Podlubny, I.: Fractional Differential Equations. Academic Press, New York (1998)
22. Laskin, N., Lambadaris, I., Harmantzis, F.C., Devetsikiotis, M.: Fractional Lévy Motion and Its Application to Network Traffic Modelling. Comput. Netw. 40, 363 (2002)
23. Mandelbrot, B.B., van Ness, J.W.: Fractional Brownian Motions, Fractional Noises and Applications. SIAM Rev. 10, 422 (1968)
24. Calvo, I., Sanchez, R.: The Path Integral Formulation of Fractional Brownian Motion for the General Hurst Exponent. J. Phys. A 41, 282002 (2008)
25. Calvo, I., Sanchez, R.: Fractional Lévy Motion Through Path Integrals. J. Phys. A 42, 055003 (2009)
26. Weiss, G.H., Dishon, M., Long, A.M., Bendler, J.T., Jones, A.A., Inglefield, P.T., Bandis, A.: Improved Computational Methods for the Calculation of Kohlrausch-Williams/Watts Decay Functions. Polymers 35, 1880 (1994)
27. Medina, J.S., Prosmiti, R., Villarreal, P., Delgado-Barrio, G., Aleman, J.V.: Frequency Domain Description of Kohlrausch Response Through a Pair of Havriliak-Negami-Type Functions: An Analysis of Functional Proximity. Phys. Rev. E 84, 066703 (2011)
28. Mainardi, F., Luchko, Y., Pagnini, G.: The Fundamental Solutions for the Fractional Diffusion-Wave Equation. Appl. Math. Lett. 9, 23 (1996)
29. Uchaikin, V.V.: Montroll-Weiss Problem, Fractional Equations and Stable Distributions. Int. J. Theor. Phys. 39, 2087 (2000)
30. Nocedal, J., Wright, S.J.: Numerical Optimization. Springer, Heidelberg (2006)
31. Press, W.H., Teukolsky, S.A., Vetterling, W.T., Flannery, B.P.: Numerical Recipes. Cambridge University Press, Cambridge (2007)
32. Saichev, A., Zaslavsky, G.M.: Fractional Kinetic Equations: Solutions and Applications. Chaos 7, 753 (1997)
33. del-Castillo-Negrete, D., Carreras, B.A., Lynch, V.E.: Fractional Diffusion in Plasma Turbulence. Phys. Plasmas 11, 3854 (2004)
34. Dif-Pradalier, G., Diamond, P.H., Grandgirard, V., Sarazin, Y., Abiteboul, J., Garbet, X., Ghendrih, Ph., Strugarek, A., Ku, S., Chang, C.S.: On the Validity of the Local Diffusive Paradigm in Turbulent Plasma Transport. Phys. Rev. E 82, 025401 (2010)
35. Balescu, R.: Anomalous Transport in Turbulent Plasmas and Continuous-Time Random Walks. Phys. Rev. E 51, 4807 (1995)
36. Carreras, B.A., Lynch, V.E., Zaslavsky, G.M.: Anomalous Diffusion and Exit Time Distribution of Particle Tracers in Plasma Turbulence Model. Phys. Plasmas 8, 5096 (2001)
37. Mier, J.A., Sanchez, R., Newman, D.E., Garcia, L., Carreras, B.A.: On the Nature of Transport in Near-critical Dissipative-Trapped-Electron-Mode Turbulence: Effect of a Subdominant Diffusive Channel. Phys. Plasmas 15, 112301 (2008)
38. Mier, J.A., Sanchez, R., Newman, D.E., Garcia, L., Carreras, B.A.: Characterization of Nondiffusive Transport in Plasma Turbulence via a Novel Lagrangian Method. Phys. Rev. Lett. 101, 165001 (2008)

39. Carreras, B.A., Lynch, V.E., Newman, D.E., Zaslavsky, G.M.: Anomalous Diffusion in a Running Sandpile Model. Phys. Rev. E 60, 4770 (1999)
40. Mantegna, R., Stanley, H.E.: Stochastic Process with Ultraslow Convergence to a Gaussian: The Truncated Lévy Flight. Phys. Rev. Lett. 73, 2946 (1994)
41. Chechkin, A., Gonchar, V., Klafter, J., Metzler, R.: Fundamentals of Lévy Flight Processes. Adv. Chem. Phys. 133B, 439 (2006)
42. Cartea, A., del-Castillo-Negrete, D.: Fluid Limit of the Continuous-Time Random Walk with General Lévy Jump Distribution Functions. Phys. Rev. E 76, 041105 (2007)
43. van Milligen, B.P., Calvo, I., Sanchez, R.: Continuous Time Random Walks in Finite Domains and General Boundary Conditions: Some Formal Considerations. J. Phys. A 41, 215004 (2008)
44. Oldham, K.B., Spanier, J.: The Fractional Calculus. Academic Press, New York (1974)
45. Miller, K.S., Ross, B.: An Introduction to the Fractional Calculus and Fractional Differential Equations. Wiley, New York (1993)
46. Schiff, J.L.: The Laplace Transform: Theory and Applications. Springer, Heidelberg (1999)
47. Samko, S.G., Kilbas, A.A., Marichev, O.I.: Fractional Integrals and Derivatives: Theory and Applications. Gordon and Breach, New York (1993)
48. Feller, W.: An Introduction to Probability Theory and Its Applications. Wiley, New York (1968)
49. Gorenflo, R., Mainardi, F., Moretti, D., Pagnini, G., Paradisi, P.: Discrete Random Walk Models for Space-Time Fractional Diffusion. Chem. Phys. 284 521 (2002)
50. Carpinteri, A., Mainardi, F.: Fractals and Fractional Calculus in Continuum Mechanics. Springer, New York (1997)
51. Lynch, V., Carreras, B.A., del-Castillo-Negrete, D., Ferreira-Mejias, K.M., Hicks, H.R.: Numerical Methods for the Solution of Partial Differential Equations of Fractional Order. J. Comput. Phys. 192, 406 (2003)
52. del-Castillo-Negrete, D.: Fractional Diffusion Models of Nonlocal Transport. Phys. Plasmas 13, 082308 (2006)

Part II
Complex Dynamics in Magnetized Plasmas

Several instances of possible complex phenomena in magnetized plasmas will be discussed in the second part of this book. These examples have been selected from within various fields, such as fusion laboratory plasmas, solar plasmas and Earth magnetospheric plasmas. After introducing each problem and clarifying their physics at an introductory level, some of the analysis tools that were introduced in the first part of this book will be used to characterize the dynamics in these systems. The approach is not intended to be comprehensive by any means, but illustrative. That is, it will be used to point out the strengths and weaknesses of the methods, as well as to suggest the capabilities that these tools offer for the analysis of complex dynamics in whatever systems might be of interest to our readers, either related to similar plasmas or to something completely different.

Part II
Complex Dynamics in Magnetized Plasma

Chapter 6
Laboratory Fusion Plasmas: Dynamics of Near-Marginal Turbulent Radial Transport

6.1 Introduction

We start our brief journey across the world of complex behaviours in plasmas by focusing on those confined by magnetic fields in order to produce fusion energy on Earth. The field is often referred to as *magnetic confinement fusion* (MCF). For those readers unfamiliar with it, we will start by providing a broad overview of MCF. Those already acquainted with the field are more than welcomed to move directly to Sect. 6.4, where the basic physics that appear to be responsible for one of the more important complex behaviours observed in MCF plasmas will be discussed.

The goal of MCF is to reproduce the fusion processes that power the stars in an Earthly-based reactor and to generate power in an efficient, clean and economically viable way. It is the expectation of many that the near future of our world will be a fusion-powered society, perhaps as soon as by the end of the twenty first century. However, the enterprise of building such a reactor is extremely complicated. MCF tries to confine a hot, moderately dense plasma inside a toroidal volume, for a sufficiently long time, by means of external magnetic fields. These toroidal plasmas are prone to instabilities and are often dominated by strong turbulence. It is precisely because of turbulence, being the highly nonlinear process that we all know and love, that complex dynamics can emerge in them in various fashions.

In this chapter, we will examine one of these situations. In fact, it is one that might probably be of relevance for the operation of any future MCF reactor. The discussion will focus on the processes that govern the transport of energy and particles across and out of the confined plasma. These processes are the ones that ultimately determine for how long energy can be confined, and how large and expensive these reactors will need to be in order to become feasible power plants. There are reasons to believe that complex dynamics might be at work behind these processes, particularly at the near-marginal regimes in which toroidal MCF reactors will probably be operated. The meaning of this last sentence will become much clearer as the chapter advances.

© Springer Science+Business Media B.V. 2018
R. Sánchez, D. Newman, *A Primer on Complex Systems*,
Lecture Notes in Physics 943, https://doi.org/10.1007/978-94-024-1229-1_6

We will conclude the chapter by applying some of the analysis tools described in the first part of the book to a turbulent dataset obtained with a local (Langmuir) probe at the edge of a MCF plasma. We will search in this dataset for any evidence that could support the idea that complex dynamics is indeed governing the radial transport dynamics in these plasmas. Along the way, we will also point the interested readers to the abundant literature that exists in this area and that describes many independent efforts to test these ideas in various fashions and multiple devices. For now, however, let's start by discussing what fusion is and how energy might be obtained from it on Earth in a controlled manner.

6.2   Nuclear Fusion Processes

Fusion processes have been powering the stars for thousands of million of years. The idea was probably first put forward by Arthur Eddington, as early as 1920 [1]. In nuclear fusion, light nuclei fuse together to produce heavier nuclei while releasing energy in the process. Naturally, the net amount of energy released is the difference in mass between the fusion reactants and the fusion products, given by Einstein's famous formula [2]:

$$E_{\text{released}} = (m_r - m_p) \cdot c^2. \tag{6.1}$$

The fusion reaction chains that power the stars were first proposed in the late 1930s [3, 4]. They involve four protons (i.e., hydrogen nuclei, ${}_1^1\text{H}$) that fuse, either directly[1] or via catalysts,[2] to yield a ${}_2^4\text{He}$ nucleus (i.e., an α-particle), other things

---

[1]The most significant direct proton fusion cycle occurring in our Sun is in fact:

$$2 \times \left[{}_1^1\text{H} + {}_1^1\text{H} \rightarrow {}_1^2\text{D} + \beta^+ \right] \tag{6.2}$$

$$2 \times \left[{}_1^2\text{D} + {}_1^1\text{H} \rightarrow {}_2^3\text{He} + \gamma \right] \tag{6.3}$$

$${}_2^3\text{He} + {}_2^3\text{He} \rightarrow {}_2^4\text{He} + 2{}_1^1\text{H} \tag{6.4}$$

The cycle consumes four protons (i.e., ${}_1^1\text{H}$), although six are necessary in order to produce the two ${}_2^3\text{He}$ that will fuse to yield the final ${}_2^4\text{He}$ and release 26.7 MeV of energy. It is believed that this cycle (known as the pp-I branch) is dominant for temperatures in the range $(10–14) \times 10^6$ K and accounts for more than 85% of the fusion energy produced in our Sun. It is also worth noting that the third reaction of this cycle (Eq. 6.4) is not the only one possible, and more complicate reactions involving lithium or beryllium also happen, although at much lower probabilities [5].

[2]In this case, the reaction requires the contribution of carbon, nitrogen and oxygen, that act as catalysts, forming the so-called CNO cycle [5]. The fusion reaction that ultimately dominates the process depends on the temperature of the star, with the catalystic route being important only for stars that are much more massive and hotter than our Sun. Higher temperatures than those of the Sun are also required for the direct proton fusion reactions involving lithium or beryllium to dominate. And even higher temperatures are needed if heavier elements are to be fused.

that include photons and neutrinos, and an abundant amount of energy. Our Sun[3] carries out approximately 10^{38} fusion reactions per second, generating a net power of the order of 10^{18} GW. This enormous power output[4] is what keeps the central part of the Sun at a balmy temperature of about 10^7 K.

It is easy to understand why such large temperatures are needed if fusion processes are to happen significantly. Nuclei, being positively charged, repel each other at sufficiently large distances ($d \gg 10^{-15}$ m), but they attract each other instead when within the range of the nuclear (strong) forces ($d \sim 10^{-15}$ m). Therefore, nuclei must attain sufficient relative kinetic energy as to overcome the electrostatic barrier (or more exactly, tunnel through it) and allow the strong force to engage, for fusion to take place. The probability of such process ending in a successful fusion reaction is usually quantified in terms of the so-called *cross-section* of the process, $\sigma(v)$, that sort of measures the probability of a successful reaction as a function of the relative velocity of the reactants, v. Using these cross-sections, the energy production of any reaction that fuses two reactants, A and B, is given by (per unit of volume):

$$P_{AB} = n_a n_b \cdot \langle \sigma_{AB} \cdot v \rangle \cdot E_{AB}. \tag{6.5}$$

Here, v is the relative velocity between the reactants, E_{AB} is the energy released in one reaction, and n represents their number density (i.e., number of particles per unit volume). $< \sigma \cdot v >$, the *reactivity* of the reaction, is the average of the product $\sigma \cdot v$ over the probability density function (i.e., its pdf) of such relative velocities. $< \sigma v >$ is a strong function of the plasma temperature.

Figure 6.1 shows the temperature dependence of the reactivity of several fusion reactions. The reactivity is very low for any reaction it the temperature is smaller than a million degrees. The reaction whose reactivity peaks at the lowest temperature corresponds to the fusion of the two isotopes of hydrogen, *deuterium* (*D*) and *tritium* (*T*) and requires temperatures of the order of fifty to a hundred million degrees. At such high temperatures matter is in a **plasma** state. That is, it has become *an ionized gas formed by free positively charged nuclei and negatively charged electrons in such proportions so that there is no overall net electric charge, and that interact with each other via electric and magnetic fields.* Interestingly, none of the main fusion reactions that take place in the Sun are included in Fig. 6.1. The fusion of four protons (either directly or via catalysts) requires even larger temperatures to reach its reactivity peak. In fact, at the Sun's temperature of roughly 10^7 K, the rate of the reaction is so low that it takes about 10^9 years for a particular proton to undergo a fusion reaction. However, the Sun is so massive and large that it

---

[3]An extensive introduction to the Sun, its features and its dynamics will be given in Chap. 7.

[4]To help putting this number in perspective, one should consider that the largest modern power plant on Earth is the Three Gorges Dam in China, that produces about 22.5 GW of power; the largest nuclear power plant, located in Japan, produces about 8.5 GW.

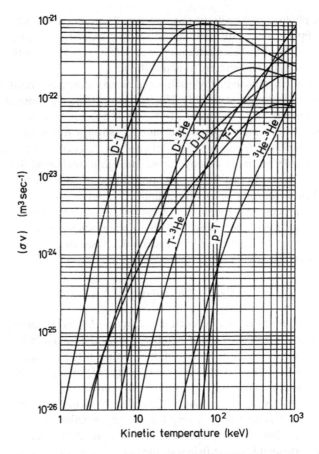

Fig. 6.1 Plots showing the dependence of the reactivity of several fusion reactions as a function of temperature. Temperature is given in kinetic units. That is, as $K_B T$. Here, K_B is Boltzman constant, that is equal to $K_B = 1.38 \times 10^{-23} \, \text{m}^2 \, \text{kg/s}^2 \, \text{K}$, or $K_B = 8.6 \times 10^{-5} \, \text{eV/K}$. The temperature T is measured in the absolute (Kelvin) scale. Therefore, roughly speaking, 1 KeV $\simeq 11.6 \times 10^6 \, \text{K}$ [Credits: Kaye & Laby online site at the National Physical Laboratory: http://www.kayelaby.npl.co.uk/]

may operate very far from its most efficient temperature and still be able to maintain the thermal pressure required to overcome its gravitational self-implosion.

Achieving an efficient production of energy via fusion in an Earth-based laboratory is however a complete different story since both mass and volume are strongly limited. For those reasons, an alternate approach is required. Naturally, the large majority of them rely on the deuterium-tritium (in short, $D - T$) reaction:

$$^2_1 D + ^3_1 T \rightarrow ^4_2 He \, (3.5 \, \text{MeV}) + n \, (14.1 \, \text{MeV}). \tag{6.6}$$

One can easily work out the conditions that will be required within any Earth-based reactor that intends to sustain itself using the energy released from $D - T$ fusion. Such a state is often referred to as one of **ignition**. In it, the fraction of the energy produced by the $D - T$ fusion reactions that can be used to maintain the internal temperature,[5] E_{heat}, must balance all energy losses. The quantitative criterion, first formulated by John Lawson in the late 1950s [6], can be written as[6]:

$$n_D n_T \langle \sigma_{DT} \cdot v \rangle E_{heat} \geq \frac{3(n_D + n_T)K_B T}{\tau_E}. \tag{6.7}$$

Here, n_D and n_T represent the deuterium and tritium number densities and the constant $K_B = 8.6 \times 10^{-5}$ eV/K is the usual Boltzmann constant. τ_E is known as the **energy confinement time**. It characterizes the quality of the energy confinement achieved by a particular device, being itself a function of plasma parameters and the magnetic configuration.

Assuming equal proportions of deuterium and tritium, this inequality can be recast in terms of the so-called *triple product*, that is often used as the figure-of-merit that characterizes the performance of any fusion reactor candidate. The triple product for $D - T$ reaches its minimum at a temperature of $T \sim 150 \times 10^6$ K, where the Lawson criterion becomes,

$$n \cdot T \cdot \tau_E(n, T, \cdots) \geq 3 \times 10^{21} \text{ KeV s/m}^{-3}. \tag{6.8}$$

Since the plasma temperature is fixed, two knobs remain with which one can play to reach ignition: the plasma (number) density n and the energy confinement time, τ_E. The two most promising approaches play with these knobs differently. **Magnetic confinement fusion (MCF)** [7], in which the plasma is confined by means of magnetic fields and heated until suitable conditions are reached, aims at achieving a moderate[7] density $n \sim (10^{20}-10^{21})$ m^{-3} and confinement times of the order of $\tau_E \sim (1-10)$ s. **Inertial confinement fusion (ICF)** [8], in which the plasma is compressed by means of powerful lasers, aims for much larger densities, $n \sim (10^{30}-10^{31})$ m^{-3}, and much shorter confinement times, $\tau_E \sim (10^{-9}-10^{-10})$ s.

---

[5]In the case of $D - T$ fusion, this energy is the one carried by the α-particles produced, that amounts to $E_{heat} = 3.5$ MeV per reaction. The high energy neutrons (14.1 MeV) produced, lacking any electrical charge, cannot be confined and leave the plasma barely interacting with it.

[6]On the right of the inequality, energy losses are expressed as the total plasma energy, given by $3(n_D + n_T)K_B T$, divided by the energy confinement time. The plasma total energy is the sum over all species of $(3/2)nK_B T$, being n the species density and T its temperature. For simplicity, we assume a quasi-neutral plasma in which both species have the same temperature. The part of the energy carried by the electrons (one per each ion!) has been absorbed into that of the ions by dropping the 2 factor in the denominator.

[7]To put these number densities in perspective, it is worth mentioning that the number density of air is 2×10^{25} m^{-3}, that of water is approximately 3×10^{28} m^{-3}, and that of diamond is 1.7×10^{29} m^{-3}.

6.3   Primer on Magnetic Confinement Fusion

Magnetic confinement fusion (MCF) tries to generate fusion power by relying on magnetic fields to confine a $D - T$ plasma. The basic principle of MCF is that any plasma constituent has a net electric charge, either positive or negative, and will thus describe approximate helical trajectories around the magnetic field lines of the configuration due to the action of Lorenz forces. Thus, MCF aims at providing a volume where these magnetic field lines are confined, so that plasma particles, that can freely move along field lines but not perpendicularly to them, can be accumulated and heated to the conditions required for fusion.

One of the easiest ways to create one such configuration is to dispose coils toroidally in a way similar to how a toroidal solenoid is constructed (see Fig. 6.2, left frame). The resulting magnetic field lines are purely toroidal, closing on themselves after a complete toroidal turn. The problem is that this setup, however, fails to confine the plasma. It turns out that, due to the $1/R$-decrease in field strength forced by Maxwell equations in toroidal geometry, a particle drift appears in the $\nabla B \times \mathbf{B}$ direction that separates vertically the ions and electrons in the confined plasma. This charge separation creates a vertical electric field that originates a second particle drift, in the direction of $\mathbf{E} \times \mathbf{B}$, that ultimately pushes all charged particles radially out of the confining volume. The easiest way to avoid this undesired ending is to force the magnetic field lines to rotate poloidally as well, as they advance in the toroidal direction. Or, in other words, to add a poloidal component to the magnetic field. If this is done preserving the rotational symmetry of the solenoid in the

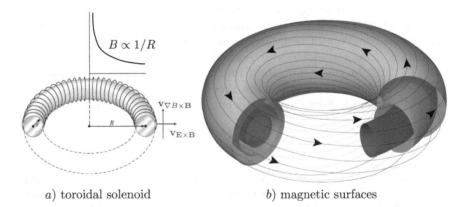

$a)$ toroidal solenoid $b)$ magnetic surfaces

Fig. 6.2 (a) **Toroidal coil setup** leading to a purely toroidal magnetic field; the direction of the magnetic drift (reversed, for negative charges), that appears because of the radial decay of the magnetic field amplitude required by Maxwell equations, is shown in red; the electrostatic drift that ensues, that is independent of charge sign, pushes the plasma out of the device (shown in blue). (b) by adding a poloidal component to the magnetic field, field lines (in black) wrap around nested, toroidal **magnetic surfaces** (in different blue tones). The most central, singular magnetic surface is a closed curved known as the **magnetic axis** [Credits: courtesy of Estefanía Cuevas]

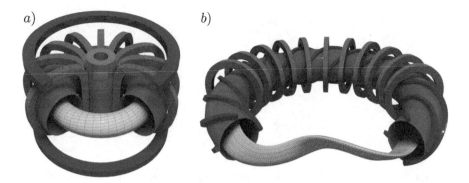

Fig. 6.3 (**a**) **Tokamak configuration**: the poloidal magnetic component is added by making the plasma the secondary of a large transformer represented by the inner cylindrical piece at the centre of the device (in orange); the resulting confined plasma is toroidally symmetric to a large approximation (in blue, another set of coils is shown used to keep the plasma from vertical instabilities). (**b**) **Stellarator configuration**: both the toroidal and poloidal components of the magnetic fields are generated by external coils. The shape of the confined plasma becomes however much more complicated, and no longer toroidally symmetric [Credits: courtesy of Estefanía Cuevas]

toroidal direction, each individual magnetic field lines will be contained in one of many possible nested toroidal surfaces, that are known as **magnetic surfaces** (see Fig. 6.2, right frame). Depending on the local relation between poloidal and toroidal components, a field line may close on itself after a few toroidal transits (in this case, the magnetic surface that contains it is called a **rational surface**) or it may fill the magnetic surface ergodically. If toroidal symmetry is not preserved, more complicated topologies are possible. In particular, magnetic topologies known as **magnetic islands** and **stochastic volumes** may appear [9], both of which deteriorate confinement.[8] Be it as it may, the fact is that magnetic field lines do now connect the top and bottom of the torus, effectively short-circuiting any charge separation, and making the confinement of the plasma possible. The two main MCF toroidal configurations, tokamaks and stellarators, follow the principles just discussed, mainly differentiating themselves through the way in which the poloidal twist of the magnetic field lines is achieved.

6.3.1 Tokamaks

Tokamaks (see Fig. 6.3, left frame) generate the poloidal component of the magnetic field by driving a large toroidal current, of the order of several millions

---

[8]However, the degree to which the nested magnetic surface topology described is broken depends on how far from exact toroidal symmetry each configuration is. Most current devices are purposely designed to avoid, or at least restrict as much as possible, the appearance of magnetic islands and stochastic regions within the confining volume.

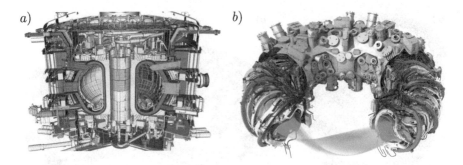

Fig. 6.4 (**a**) the **ITER tokamak** is currently under construction in France. Its operation is planned to start in the early 2020s. The toroidal plasma confined in this device will have a major radius of about 6 m and a minor radius of about 1.5 m. All coils are superconducting and create a magnetic field of about 13T inside the plasma volume. The toroidal plasma current flowing inside is 15 MA; (**b**) the **W7-X stellarator** has started operating in Greifswald (Germany) in 2016. It is the largest stellarator ever built, with a major radius of 5.5 m and a minor radius of 0.6 m. The configuration is formed by superconducting coils that generate an interior magnetic field of 3 T. The configuration has been heavily optimized so that the toroidal plasma current is negligible [Credits: copyright of ITER image belongs to the ITER organization (http://www.iter.org); copyright of the W7-X image belongs to IPP (http://www.ipp.mpg.de)]

of Amperes, inside of the confined plasma [10]. The poloidal magnetic field created by the toroidal plasma current is sufficient to achieve the required twist, but is still significantly smaller than the toroidal component. The current is driven by making the plasma act as the secondary of a gigantic transformer. As a result, tokamaks are intrinsically pulsed devices.[9]

The existence of this large plasma current is, at the same time, good and bad news. On the plus side, the resulting confined plasma is toroidally symmetric to a large approximation. Thus, magnetic islands and stochastic are a relatively small problem. In addition, as it will be remembered from basic Mechanics, the existence of symmetries is always associated to the conservation of a related quantity.[10] These conservation laws strongly restrict the possible motions of the confined particles. As a result, tokamaks exhibit good confinement properties. In fact, they confine much better than any other existing configuration, which is why tokamaks spearhead current efforts to prove that efficient fusion power production is feasible. The largest fusion experiment ever planned, the ITER tokamak [13] (see Fig. 6.4, left frame), is

---

[9]A large body of research has been carried out over the last three decades, and still continues, that explores the possibility of operating tokamaks in a continuous mode by applying different techniques of current drive [11]. These type of scenarios are often known as *advanced tokamak scenarios*, and will play an important role in the operation of ITER, the next-step tokamak experiment currently under construction in Southern France that must prove the feasibility of the tokamak way of generating energy via fusion. Current drive schemes are also pursued in order to help control tokamak instabilities, among other things.

[10]For instance, invariance under temporal translations leads to the conservation of energy. Invariance under spatial translations, to the conservation of linear momentum. Invariance under rotations, to the conservation of angular momentum [12].

currently under construction in Southern France, funded by a large international consortium that includes the majority of the developed world. The operation of ITER is planned to start in the early 2020s and, if successful, will be the first device ever to confine an ignited fusion plasma. On the negative side, such a large current provides a large source of free energy for the development of instabilities. In particular, disruptive events (known as **disruptions** [10]) may be excited, which requires all tokamak operations to be carefully monitored and controlled. In unchecked, disruptions may lead to a large fraction of the stored plasma energy being dumped against the reactor walls in a very short period of time, which might have rather catastrophic consequences for devices of the size of ITER [14]. They might also lead to the formation of large beams of so-called **runaway electrons**, that reach energies of the order of tens of MeV, and that could be dangerous for the physical integrity of the plasma-facing components of the reactor [15–18].

6.3.2 Stellarators

In **stellarators** (see Fig. 6.3, right frame), on the other hand, both components of the magnetic field are created by means of external coils. As an illustration, in the stellarator configuration shown in Fig. 6.3, the blue coils shown are the ones responsible for providing the poloidal magnetic field. Many different stellarator configurations are possible depending on the how coils are laid out: heliotrons, torsatrons, heliacs and so forth [19]. Since the current flowing toroidally in the plasma is usually very small, stellarators can be operated continuously and avoid large current-driven MHD instabilities such as disruptions. These features would apparently point at stellarators as a better choice (than tokamaks) for a first fusion reactor. However, a large drawback of any stellarator configuration is that the confined plasma is no longer toroidally symmetric. Topologies such as magnetic islands and stochastic regions must now be actively avoided. Otherwise, confinement properties can degrade significantly. In addition, the lack of toroidal symmetry makes particle motion less constrained and, as a result, confinement properties are rather worse than those of a tokamak for similar parameters.[11] The situation has been somewhat improved in the last decades by relying on heavy numerical optimization and with the advent of new design concepts, such as the idea of *quasi-symmetries* [20]. The largest member of the coming generation of optimized devices, that will have to prove the validity of these new design approaches, is the superconducting W7-X stellarator [21], that started operation in 2016 in Greifswald, Germany (see Fig. 6.4, right frame).

---

[11]That is why the tokamak has remained the king since the late 1970s, although this might change in future generations of reactors.

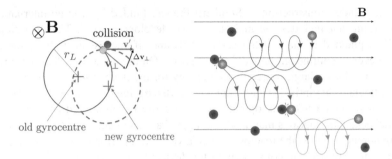

Fig. 6.5 Left: when a charged particle (shown in orange) suffers a collision with another particle (in blue) its velocity vector changes its orientation (or pitch). In the presence of a constant magnetic field, the particle trajectory is a helix along the field line, whose projection onto the plane perpendicular to the magnetic field is a circle or radius $r_L = mv_\perp/qB$ centered at the gyrocenter. After the collision, the gyrocenter is displaced as a result of the change in the velocity vector. **Right:** collisions thus may drive transport across the magnetic field. However, since the change in the velocity pitch is essentially random, there is the same probability of moving the gyrocenter in any direction. Thus, net particle transport only ensues from regions with more particles to those with less. That is, in the presence of a density gradient [Credits: courtesy of Estefanía Cuevas]

6.3.3 Main Transport Processes in Toroidal MCF Plasmas

The term *transport* is used, in the context of fusion toroidal plasmas, to refer to any process responsible for the exchange of *mass*, *energy* and *momentum* among regions of the confined plasma. The most relevant direction, though, is the one perpendicular to the magnetic surfaces (i.e., the *radial* direction), since radial transport is the one that leads to losses through the system boundaries. The understanding and control of radial transport processes is essential when trying to push the triple product of any configuration towards the ignition threshold (Eq. 6.8). Two are the processes that are the main dynamical players in toroidal MCF plasmas: *collisions* and *turbulence*.

6.3.3.1 Collisions

Collisional transport is due to the change in orientation of the velocity vector of a particle (ion or electron) caused by the collision with another one [22]. Due to the long-range nature of Coulomb interactions, these changes are very small, but accumulate quickly over time.[12] When collisions take place in the presence of a magnetic field, transport in the direction perpendicular to the magnetic field (and, therefore, across magnetic surfaces) quickly ensues if a gradient exists in that direction. The fundamentals of the transport process are illustrated in Fig. 6.5, for

---

[12]Collisional processes are a good example of *additive transport processes*, that were discussed first in Sect. 2.3.1 while introducing the central limit theorem, and then again in Chap. 5.

a uniform magnetic field B. Charged particles describe helical trajectories along the magnetic field that, when projected onto the plane perpendicular to the field, become circles with a (Larmor) radius given by $r_L = mv_\perp/qB$. Here, m and q are the mass and charge of the particle, and $v_\perp$ is the magnitude of its velocity perpendicular to the field. The location of the centre of the circle, or **gyrocenter**, is set by the initial orientation of the velocity. After a collision takes place, the sudden change in velocity orientation translates into the gyrocenter being displaced and, subsequently, to the particle moving across the magnetic field (see Fig. 6.5, right frame). These gyrocenter displacements are however essentially random in nature, as are the changes in velocity orientation and the collisions that originate them.[13] Therefore, gyrocenters are displaced on average with equal probability in any direction. For that reason, a net flux of particles only takes place from the regions with more particles to those with less particles (i.e., in the presence of a local gradient). The same principle would also apply for momentum or energy transport, that are also locally conserved quantities. The difference here is that it is no longer the number of particles that must vary from one location to another, but the fraction of them with a given amount of energy or momentum [22].

6.3.3.2 Turbulence

Most plasmas in the universe are turbulent. By this, it is meant that they are strongly non-linear, high-dimensional systems. The root of this behaviour must be sought in the fact that all plasma constituents have electrical charge, and thus interact among themselves via the electric and magnetic fields they create, and that are in turn quickly affected by the particle motion induced by these interactions. As a result, small fluctuations in the fields may become unstable and grow, or be quickly damped away. The resulting state of this complicated zoo of interactions is usually known as *plasma turbulence*.

When a plasma is confined by a strong external magnetic field, as it is the case in a tokamak or stellarator, the turbulent nature of the plasma remains active. The presence of strong gradients in density, pressure and temperature across magnetic surfaces and the large plasma currents that might exist provide with abundant free energy to feed a myriad of possible instabilities [23]. If unchecked, this state of things may permit the growth of magnetic perturbations that might even lead to the destruction of the confining magnetic field. In most cases, however, a careful design prevents the appearance of large instabilities, and perturbations remain at a relatively small amplitude and size. Their main impact is then the enhancement of

---

[13]The process is very reminiscent of the continuous time random walk (CTRW) that we discussed at length in Sect. 5.2.1, and that provided a microscopic model for diffusive transport.

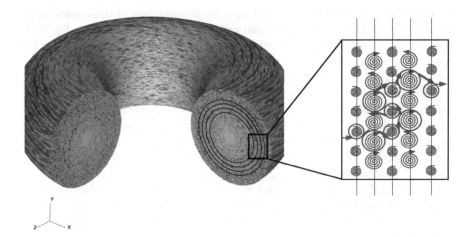

Fig. 6.6 Left: snapshot of a numerical simulation of ITG turbulence in a tokamak with concentric, circular magnetic surfaces (shown in black, for reference purposes). A colour map of electrostatic potential fluctuations is shown (blue corresponds to negative fluctuations; red, to positive). **Right:** sketch showing the formation of local vortical structures, perpendicular to the magnetic field, associated to the largest (in absolute value) potential fluctuations via $\mathbf{E} \times \mathbf{B}$ drifts with $\tilde{\mathbf{E}} = -\nabla\tilde{\phi}$. In red, the enhanced transport across magnetic surfaces driven by these vortical structures is illustrated, stressing its random nature. Vertical black lines represent magnetic surfaces. The alignment of the vortical structures with the magnetic surfaces is however exaggerated

transport across the system, including the one taking place along the radial direction that leads directly out of the device.[14]

Figure 6.6 illustrates the kind of enhanced transport across magnetic surfaces that plasma turbulence leads to. On the left, a colour map of electrostatic potential fluctuations is shown for a snapshot of a tokamak turbulent simulation [24] in which the dominant unstable mode is the so-called *ion-temperature-gradient* (ITG) mode.[15] These electrostatic potential fluctuations lead to tube-like vortical structures that are aligned with the local magnetic field[16] that help move particles and energy

---

[14]This behaviour is very similar to what takes place in fluid turbulence. In some fluid systems, enhanced turbulent transport may often be used to our advantage, though. For instance, to make the combustion of fuel more efficient in car engines. Or to facilitate mixing processes in industrial environments. For MCF plasmas, however, turbulence is often bad news. Enhanced transport across magnetic surfaces leads to larger energy losses, which degrades plasma confinement and brings the triple product value down, thus making power generation much more difficult and expensive.

[15]The physics of the ITG mode are irrelevant for the point we are trying to make here. It suffices to say that it is a toroidal mode that feeds from the free energy stored in the large ion temperature radial gradient that exists between the **magnetic axis** (the innermost, degenerate magnetic surface) and the plasma edge. It is also widely accepted that the ITG is one of the dominant modes that govern ion heat transport in current tokamaks [23].

[16]Electrostatic potential fluctuations lead to local vortical structures via $\mathbf{E} \times \mathbf{B}$ drifts, where $\tilde{\mathbf{E}} = -\nabla\tilde{\phi}$, being $\tilde{\phi}$ the local potential. These fluctuations form vortical structures centred at

across magnetic surfaces. If these local fluctuations are random in nature, as it is traditionally assumed, the cross-magnetic-surface motion induced would be similar to a CTRW process (see discussion in Chap. 5). In that case, these turbulent fluctuations would not drive a net radial flux unless a radial gradient exists, since there would be the same probability of being displaced in any (radial) direction. These gradients are certainly present in MCF toroidal plasmas.

6.4 Are MCF Plasmas Complex Systems?

Radial turbulent transport in MCF plasmas (particularly in tokamaks, but also in stellarators) has been traditionally described (and by many, also understood) in the terms outlined in the previous section. Accordingly, turbulence-enhanced transport of energy, heat or mass has been often modelled in terms of effective transport coefficients (i.e., diffusivities, conductivities or viscosities), implicitly assuming that an underlying additive, CTRW-like process is a good approximation to what actually happens microscopically in these plasmas, at least over sufficiently large and long scales.[17]

As discussed at length in Sect. 5.2, any ("microscopic") CTRW process can be well represented by its fluid limit at the "macroscopic" level. This limit corresponds to the classical diffusion equation (Eq. 5.3) with an effective transport coefficient that can be estimated once the characteristic length and time scales of the "microscopic" transport process are known. In the case of collisional transport, the typical displacement of the gyrocenter after a collision is of the order of the orbit radius, r_L. The typical time in between collisions can be estimated by the inverse collision frequency, v_c^{-1}. Thus, one can infer that the effective collisional diffusivities and conductivities will be of the order of[18]:

$$D^{\mathrm{coll}}, \chi^{\mathrm{coll}} \propto r_L^2 v_c, \tag{6.9}$$

being v_c the proper collision frequency.

---

the locations of the extrema of the fluctuating potential, with the orientation of the local rotation (mostly contained within the plane perpendicular to the local magnetic field) being determined by whether the extremum is a maximum or a minimum.

[17]These ideas are also the basis of the pinch-diffusion phenomenological models that are so often used in MCF [25]. These models consider local fluxes that, in the case of particle transport, take the general form $\Gamma_n = Vn - D\nabla n$, where V is an effective pinch velocity and D an effective diffusivity. Similar expressions are also used to describe momentum and energy transport.

[18]It turns out that things are of course much more complicated than this. The toroidal geometry of MCF plasmas plays a huge role in collisional transport, and a full-fledged theory must be developed (known as **neoclassical theory** [22, 26, 27]) to estimate these transport coefficients properly. Miraculously enough, it turns out that the simple random walk estimates discussed here are in the right ballpark.

In the case of turbulence transport, it seems reasonable to assume that the characteristic length of the process will be of the order the typical size of the vortical structures in the radial direction, Δ_r (see Fig. 6.6). The characteristic time, on the other hand, should be of the order the time that a local fluid parcel spends in the neighbourhood of each vortical structure. It can be estimated in different ways such as the average lifetime of the structures, τ_L, the inverse of the linear growth rate of the instability that creates them, γ_L^{-1} and others. Therefore, one could estimate,

$$D^{\text{turb}}, \chi^{\text{turb}} \propto \Delta_r^2 \gamma_L, \ \Delta_r^2 / \tau_L, \cdots \qquad (6.10)$$

It is also possible to determine effective coefficients phenomenologically after one has accepted that Fick's law (Eq. 5.1) provides a proper description. Then, it is sufficient to simultaneously measure (in experiments or simulations) the local radial flux and local radial gradient of the quantity of interest, and to obtain the transport coefficient from its ratio. In this way, one could infer effective (radial) transport diffusivities and conductivities as (usually by averaging over many realizations),

$$D \sim -\left\langle \frac{\Gamma_n}{\nabla_r n} \right\rangle, \quad \chi \sim -\left\langle \frac{q}{n \nabla_r T} \right\rangle, \qquad (6.11)$$

being Γ_n and q the radial particle and heat fluxes, respectively.

However, there are some scenarios in MCF plasmas in which turbulence probably does not lead to the type of CTRW-like transport just described. Instead, a more complex type of radial transport dynamics seems to be at work. This should not come as a surprise since these plasmas, being the strongly coupled systems they are, contain many of the typical ingredients of complex systems that were listed in Sect. 1.2.2: they are open and driven systems, are strongly non-linear, have a very large dimensionality, and contain a plethora of local thresholds that govern the onset of a myriad of instabilities. In order to understand better when and how complex dynamics may come to dominate radial transport in MCF plasmas, it is worth reviewing how the confinement properties of tokamaks and stellarator have been experimentally found to change as the external power, that is used to heat up the confined plasma, is increased.

6.4.1 Tokamak Transport Phenomenology

We provide here a brief (and simplified) description of the phenomenology of radial transport as has been experimentally observed in many tokamaks. By radial, once again, we mean transport across magnetic surfaces, that is roughly equivalent to transport along the direction of the minor radius of the torus (see Fig. 6.7). It has been found over the years that tokamaks go through a variety of confinement regimes as the external power that heats the plasma is increased.

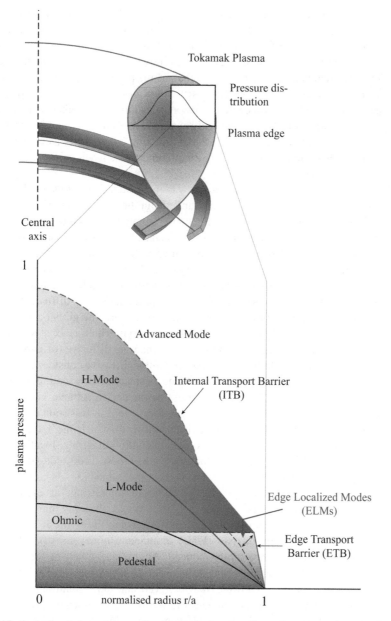

Fig. 6.7 Typical radial pressure profiles for the main tokamak confinement regimes reached as external power is increased: ohmic (black), L-mode (blue), H-mode (red, solid) and advanced modes (red, dashed). The transition from *L*-mode to *H*-mode takes place when the edge transport barrier develops, above a certain power threshold. The transition to advanced confinement modes happens when an internal transport barrier is formed at some intermediate radial location [Credits: courtesy of Estefanía Cuevas]

Naturally, the first heating method used in tokamaks is **Ohmic heating**. The principle is simple. Since a large toroidal current flows in the plasma, the heat dissipated via the Joule effect effectively heats it, building up the (radial) temperature and pressure profiles. Ohmic heating must however be soon complemented by additional external heating due to the fact that, as the plasma temperature increases, its electrical resistivity diminishes (its resistivity scales as $\eta \sim T^{-3/2}$ [10]). **Wave heating** (at the ion or electron cyclotron resonances mostly, but other methods also exist) and **neutral beams** (in which highly energetic neutral deuterium is injected in the plasma centre, mostly along the toroidal direction, where it ionizes, increasing the plasma density and transferring its kinetic energy to it via collisions) are the most usual choices.

As the power injected into the plasma was increased, it was observed that the confinement behaviour of the tokamak changed significantly from that exhibited in most Ohmic discharges. This new confinement regime was called the *L*-**mode**, and it exhibited some remarkable properties. For instance, the radial plasma profiles—mostly, temperature and pressure—were remarkably stiff in the sense that their shape was rather insensitive to the strength and location of the external power (except at the very edge of the plasma) [28]. In addition, locally-excited perturbations (both hot and cold pulses) were observed to propagate super-diffusively across magnetic surfaces, instead of diffusively [29]. Furthermore, the global confinement time, τ_E, exhibited a scaling (known as Bohm scaling) in which $\tau_E \propto a^\epsilon$ with $\epsilon \sim 1$ (*a* is the minor radius of the torus), much worse than the diffusive (or gyro-Bohm) scaling, $\tau_E \propto a^2$, that would be expected from simple diffusive considerations [30]. A Bohm scaling would have been really bad news since, if such scaling truly held, much larger (and expensive) tokamaks would be needed to achieve viable fusion reactors. In order to explain these unexpected observations, it was soon conjectured that some critical threshold probably existed, close to which the radial plasma profiles were forced to stay by the intermittent excitation of strong, locally-excited turbulent transport. Direct evidence of the existence of these thresholds, particularly for electron and ion temperature gradients, was soon obtained [31–34].

Things kept changing, and getting much more interesting, at even higher external powers. Above a certain power threshold, it was observed that the plasma transitioned to a new confinement regime that is known as the *H*-**mode** [35]. In the *H*-mode, an **edge transport barrier** (ETB) spontaneously develops (see Fig. 6.7) where radial transport is importantly reduced. As a result, much larger gradients must be established across this barrier in order to drive the level of transport needed to balance the external drive. It is believed that the formation of the edge barrier is associated with the spontaneous development of a strong poloidal and toroidal plasma flow, with a large radial shear, very close to the plasma edge.[19]

---

[19]The appearance of a transport barrier at the edge of tokamaks has been likened to the presence of the tachocline in the Sun, both being thin radial regions sustaining strong sheared-flows, probably driven by turbulent dynamics in a self-organized way. We will discuss the tachocline and the role it plays in solar dynamics later in this book, in Sect. 7.3.1. We will also discuss transport across shear flows in tokamaks in Chap. 9.

This zonal flow might be driven, among other possibilities, by turbulence itself via the Reynolds stresses or by radial electric fields associated with several effects such as ion-orbit losses and others [36]. Curiously, though, the plasma confined radially inside of the barrier region appears to behave very similarly to its L-mode counterpart, still exhibiting strong profile stiffness and superdiffusive propagation of perturbations. The central temperature is however much higher than in L-mode. The discovery of the H-mode was very good news for the MCF international program. The presence of the edge barrier not only increased the confinement time τ_E, but it also brought its scaling much closer to a gyro-Bohm scaling, thus making the predicted size of future reactors smaller, and much more attractive from an economical standpoint. The change in global scaling is probably due to the fact that radial transport across the zonal flow is probably much more diffusive in nature (although we will have a lot more to say about this in Chap. 9).

Some bad news also come with the formation of the ETB in H-mode tokamaks, though. The edge barrier confines the plasma so well that impurities and helium ash from fusion reactions might accumulate in the inner plasma, reducing the overall efficiency of the fusion process. At the same time, the edge pressure gradients gets so steep that they end up hitting an instability threshold[20] that leads to a large percentage of the energy contained in the barrier being released towards the reactor walls. These quasi-periodic bursts are known as **type-I Edge Localized Modes** or, in short, **type-I ELMs**. ELMs, if left unchecked, might become quite dangerous for the reactor first wall and its divertor, particularly at sizes as large or beyond that of ITER. Therefore, a careful control of the discharge is needed to avoid them, or to keep their size sufficiently small (at the expense of a higher frequency) as to be manageable [39].

Finally, it is also worth noting that additional advanced confinement regimes can be reached starting from either the L-mode or the H-mode. Access to them requires the excitation of **internal transport barriers** (ITBs). That is, radial regions of reduced transport that appear in the inner part of the plasma. The formation and placement of ITB can be accomplished in a variety of ways, as profusely described in the literature [40, 41].

6.4.2 Stellarator Confinement Phenomenology

Due to the lack of toroidal symmetry, stellarators have much larger collisional (neo-classical) losses than tokamaks. In fact, neoclassical diffusion probably dominates transport for a large portion of the central part of currently existing stellarators, with turbulence dominating transport over the rest of the outer radius [42]. As a result of

---

[20]These instabilities are probably related to the excitation of a family of modes known as **ideal peeling-ballooning modes** [37]. ELMs are explosive instabilities that have been likened by some authors to the ones responsible for the formation of solar flares in our Sun [38].

the much larger diffusive losses, stellarators exhibit a much weaker profile stiffness than comparable tokamaks, and only at the largest temperatures achieved [34]. This situation might change, however, as larger, hotter and better confining stellarators are brought online. Both internal and edge barriers have also been observed in stellarators, the latter giving access to sort of an H-mode, that even showed edge relaxations reminiscent of tokamak ELMs [42–44].

6.4.3 Self-organized Criticality and Toroidal MCF Plasmas

It was in the mid 1990s, while the tokamak community was still trying to make sense of the transport phenomenology of the L-mode, that the concept of self-organized criticality (SOC) first appeared in the context of MCF plasmas [45–48]. The observation of stiffness in the profiles and the super-diffusive radial propagation of perturbations found in L-mode, that we discussed in the previous subsection, were soon connected to the possibility that some plasma profiles might stay rather close to some local instability threshold in this confinement regime. SOC dynamics may become important in situations of closeness to a threshold, particularly if it happens in an open, driven system in steady state in which there is a sufficiently large time separation between local drive and local relaxation processes, and in the presence of some local inertia (see Sect. 1.3). It turns out that all of these elements are present in tokamak plasmas.

In order to identify which processes or elements present in MCF toroidal plasmas could play these roles, an analogy was drawn between the confined tokamak plasma and the running sandpile that we introduced in Sect. 1.3.1. Table 6.1 shows this correspondence, as made by the original proposers [46]. The assignments are based on the fact that most modes that can be excited in MCF plasmas are often radially localized around specific magnetic surfaces (the rational surfaces we mentioned in Sect. 6.3), that would act as the sandpile cells. These local modes

Table 6.1 Analogies between the running sandpile and a MCF plasma drawn in [46]

Turbulent MCF plasma	Running sandpile (Sect. 1.3.1)
Localized instability (eddy)	Cell
Critical threshold for local instability	Local sand slope (Z_c)
Heating noise/background fluctuations	Intermittent drive (p_0)
Local eddy-induced transport	Sand toppling (N_f)
Inertia of relaxation	$N_f > 1$
Turbulence able to flatten profiles against drive	$N_f > 2p_0L$
Mean plasma temperature/density profiles	Sand height profile
Transport event	Sand avalanche
Total energy/particle content	Total sand in the sandpile
Energy/particle flux	Sand flux

would become linearly unstable first, after some threshold condition was locally overcome, typically a gradient or a characteristic length of the surface-averaged plasma temperature, density or pressure. The unstable mode (or modes) would then grow and saturate, locally flattening the profile in the process by transporting the relevant quantity to adjacent radial locations until the local critical condition is no longer exceeded. Inertia would continue the flattening process beyond the local threshold value before the instability disappeared, leaving behind a footprint that would impact future relaxations. In addition, the modifications of the profile at adjacent locations could be sufficiently large as to make the profile unstable there as well, so that the whole process would repeat itself, starting a transport event or radial avalanche. Avalanches would propagate radially until they reached some end location, where the induced modification of the profile would not be sufficiently large as to turn it unstable. Finding an intermittent component on top of the average drive in a MCF plasma that could trigger these processes is rather easy, due to the large number of fluctuating sources and sinks present in the system (for instance, beam, wave and Ohmic heating coupling efficiencies all fluctuate as the local density or temperature values quickly change due to the heating itself, the effect of background fluctuations or many other processes).

The details of which instability is responsible for the unstable mode—and what its local dynamics are like—would be irrelevant for self-organized criticality to develop over longer and larger scales, just as one does not care about what happens within a single sandpile cell. Thus, it could be driven by any of the well-known tokamak micro-instabilities such as the ion-temperature-gradient (ITG), the dissipative-trapped electron (DTEM) or the resistive ballooning mode. The critical element here is that the competition between drive and relaxation must be such that it keeps the plasma profiles wandering around marginal values, which means that *classic SOC behaviour will only appear in certain regimes*. Indeed, if the drive is too strong relative to the excited turbulent fluxes, the system could stay supercritical most or all of the time, making almost impossible for SOC dynamics to be established. In the opposite limit, that of a very weak drive, the running sandpile would still exhibit SOC dynamics. This is because avalanches are the only transport process available, although it would take longer for the profile to become locally critical again since avalanches would become more sparsely distributed in time. This would not be the case in a MCF toroidal plasma, though. Other transport channels do exist in these plasmas, such as collisional classical/neoclassical diffusion or turbulent diffusion associated with one or more additional supermarginal instabilities.[21] If any of these processes can drive the transport needed to balance the external drive while keeping the profiles below the local threshold associated to the instability of interest, SOC dynamics could be importantly modified, and in some cases, even completely absent.

---

[21]Additional losses that might be important are ionization/recombination processes near the edge, radiation, or even plasma fuel spent in fusion reactions at the core.

Given the variety of regimes we touched upon earlier (Ohmic, L-mode, H-mode, etc), where should SOC dynamics be expected? Can SOC provide an explanation of the observed phenomenology in tokamaks? We will just provide some hints about these questions in the next couple of paragraphs, but all interested readers can find many more details, both theoretical and experimental, in a recent review that we wrote to describe the current state of research in this area [49].

For starters, it is clear that L-mode discharges exhibit the most favourable conditions to exhibit SOC dynamics, due to the near-marginality that might be inferred from the observed global profile stiffness, that is usually absent in most Ohmic regimes. The Bohm global energy scaling and the superdiffusive pulse propagation observed in L-mode discharges are also qualitatively consistent with SOC-like dynamics. SOC might be operative over the stiff part of the profile, excluding probably just the most central part. However, the lack of a global stiff profile does not necessarily exclude SOC dynamics from being dominant at some locations within lower-powered Ohmic discharges. It is only natural to expect that profiles would start to approach marginality from the plasma edge, where gradients are usually the largest,[22] with the near-critical region reaching further inwards as power increases. In fact, the type of driving, be it Ohmic or external, is irrelevant, since only the amount of power deposited matters. Therefore, regions dominated by SOC dynamics smaller than in the L-mode cases, that extend inwards from the edge, could also be expected in sufficiently hot Ohmic discharges.

In the case of H-mode tokamak discharges, it could be argued that the higher temperature inside (relative to the L-mode) should make profiles inside the edge transport barrier even stiffer, since turbulent fluxes increase with temperature, thus becoming more efficient in flattening the profiles and bringing them back below marginal. SOC dynamics could thus dominate radial transport across the plasma core. However, the edge transport barrier appears to effectively decorrelate any SOC-related transport event coming from the core. Therefore, a much more diffusive-like transport could probably be taking place across the barrier.[23] It is the combination of the transport coming through the barrier, together with the prompt losses associated to the quasi-periodic type-I ELMs, that must balance the external drive. Since the global confinement scaling depends on the transport properties of the region closest to the last closed flux surface (or LCFS), a gyro-Bohm-like confinement should be expected, even when the inner plasma could still be dominated by SOC-like dynamics all the way to the edge barrier. However, the effect of the most interior SOC dynamics could still be felt outside: for instance, since ELMs are ultimately driven by the amount of transport coming from the inner plasma into the barrier, the intermittency and/or memory (or lack of it) of the inner transport might condition the peak size, the size distribution and the frequency of

---

[22]These edge gradients might even drive local instabilities different than those that dominate transport processes further inside.

[23]It must be said, however, that this diffusive view of transport through the shear flow at the ETB might provide a too simplistic picture, as we will discuss in Chap. 9.

ELMs, that impose important requirements on first wall components. In tokamak enhanced confinement regimes, that may include one or more ITBs, each barrier region would behave similarly to the H-mode edge transport barrier just discussed. That is, it would effectively decorrelate any transport event crossing through it from any other. In the radial areas in between barriers, however, transport may still exhibit SOC-like features if profiles are close to marginal.

There has been a large effort, in the last decades, to try to search for evidence of SOC-like behaviour in magnetically confined fusion plasmas. The available evidence to this day certainly points in this direction, at least in certain regimes. Most of the experimental data available comes from the plasma edge, though, where the temperature is sufficiently low as to permit the introduction of local probes [51–53]. This is unfortunate, though, since SOC is a global phenomenon, but one naturally expects that the scale-invariance and long-term memory associated to it should be reflected somehow in local measurements. There is also some evidence, particularly in tokamaks, that has not been obtained with local probes. For instance, a glimpse of radial avalanching was reported in some L-mode discharges from the DIII-D tokamak in the USA. It was detected by the ECE radial array diagnostic that is used to measure the plasma temperature along different lines of view, that showed clear evidence of avalanche radial propagation of both hot and cold pulses [54]. We conclude this section by strong recommending all interested readers to browse through the review that we mentioned previously [49]. A large amount of additional information about the many studies carried out in this area during the last three decades can be found in it, including an extensive list of relevant references.

6.5 Case Study: Analysis of Turbulent Fluctuations from the Edge of the W7-AS Stellarator

In this section we will examine, using some of the techniques discussed in the first part of this book, a dataset obtained with a Langmuir probe just inside the plasma edge of the W7-AS stellarator, that was in operation in Germany until the mid-2000s [55]. Langmuir probes are the ones most often used in MCF plasmas, typically in a single or triple-tip configuration [56]. A Langmuir probe is a conducting rod whose voltage (relative to the plasma) can be varied externally. When its potential is set at a smaller value than that of the plasma, plasma ions will flow to the probe; if larger, electrons will flow to the probe and, if equal, there is no current. By operating the tip in each of these regimes one can easily get estimates of the ion and electron local densities from the saturation currents measured in the probe, or of the local electrostatic potential.[24] When used in single-tip configuration, the probe is usually

---

[24]Some strong assumptions must be accepted though, particularly in what refers to the effect of the local temperature, and to the degree of intrusiveness of the probe itself, that undoubtedly perturbs its surroundings considerably.

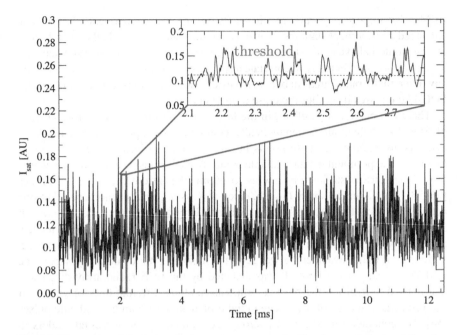

Fig. 6.8 Time series of the ion saturation current measured with a Langmuir probe at the edge of the plasma edge in W7-AS discharge No. 35425 [50]. The ion saturation signal is a surrogate, under certain assumptions, of the local density fluctuations. The length of the dataset is of about 12 ms, with the decorrelation time of the signal being of the order of 20–40 μs

set to measure local ion density. The triple-tip configuration, on the other hand, usually has its central tip prepared to measure the local ion density, and the other two tips measuring the local potential at neighbouring poloidal locations of the same magnetic surfaces. In this way, the local radial turbulent flux can be estimated.[25]

6.5.1   Statistics

The data that we will examine is shown in Fig. 6.8, corresponding to the time series of the ion saturation current measured by the probe, that acts as a surrogate of the local ion density. There are 25,000 values within the record, that was sampled at a frequency of 2 MHz. The first thing we will discuss is the pdf of the data, that is shown in Fig. 6.9. It has been computed using both the constant-bin-size (CBS;

---

[25]Here, it is assumed that the dominant turbulent velocity fluctuations come from local $\mathbf{E} \times \mathbf{B}$ drifts. Since the dominant magnetic field in MCF plasmas is toroidal, the main contribution of the electric field to the radial flux is via its poloidal component, that can be estimated as the difference of the two potential values measured by the probe.

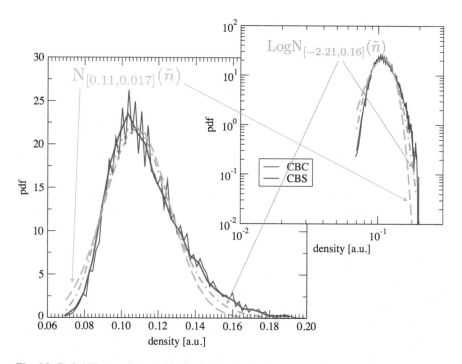

Fig. 6.9 Probability density function for the ion density time series shown in Fig. 6.8, both in lin-lin (left) and log-log (inset, right) scales. 100 bins were used for the CBS method; 15 events were used to define each bin in the CBC one. In green, best Gaussian (light green) and Log-normal (dark green) fits are also shown

see Sect. 2.4.1) and constant-bin-content (CBC; see Sect. 2.4.2) methods that we discussed in Chap. 2. Since the plasma density is a positive quantity, the pdf has naturally a non-zero mean. The obtained pdf does not seem to exhibit any power-law tail, though. We have fitted it against both a normal (Eq. 2.30) and log-normal (Eq. 2.45) law using a minimum chi-square method (see Sect. 2.5.3). Although both shapes provide relatively good fits, the log-normal one seems to be a bit better.[26] The difference is however not sufficiently large as needed to distinguish between a possible additive or multiplicative underlying process, though (see discussion in Sect. 2.3.2).

In order to reveal self-similar features we need to do better than the pdf of the dataset, though. If we are expecting avalanches to dominate radial transport, their passing by the probe should appear as extended periods of time over which the magnitude of the measured saturation current is larger than average. It is the statistics of these 'burst events' that should exhibit extended power-laws. One

---

[26]The goodness-of-the-fit parameter χ^2 is about twice as large for the Gaussian fit, compared with the log-normal fit.

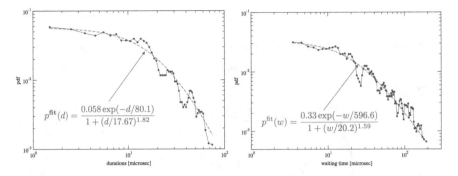

Fig. 6.10 PDFs for the 'burst' durations (left) and the waiting-time in between successive events (right) calculated using the SF method (see Sect. 2.4.3). The best fit to Eq. 1.8 is also included (in dashed red)

can try to separate these bursts by introducing an adequate threshold value,[27] and defining an 'event' as any period of time when the signal remains above the threshold, and a 'waiting time' as the lapse of time when the signal is below threshold. We will do just that on our signal. By choosing a threshold value[28] of 0.115 in the (arbitrary= units of the dataset (see Fig. 6.8), 471 distinct burst events can be identified. The pdfs of their durations and waiting times are shown in Fig. 6.10. The results are now much more interesting. For the event durations, we find an extended decaying power-law scaling roughly as $p(d) \sim d^{-1.82}$ that extends for about a decade, from 10 to a 100 ms. The value of the exponent is sufficiently small (i.e., between 1 and 2) to suggest that a characteristic scale for event duration is absent, consistent with critical behaviour (see Sect. 1.3.2). The fall-off at the end of the duration power-law suggests that burst durations are probably limited by finite-size effects (see discussion in Chap. 1). The pdf of the waiting-time between bursts also exhibits a decaying power-law scaling $p(w) \sim w^{-1.59}$ that extends for a bit more than a decade, from about 20 to 200 ms. The fact that the waiting-times do not follow an exponential power law implies that their triggering is not random, and the small value of the exponent points gain towards some kind of critical behaviour (see Sect. 4.4.5). We have gone even further and defined an "effective burst size" as the area subtended by the signal over the duration of each event. The pdf of these sizes is shown in Fig. 6.11 using a log-log scale. A very clear power-law scaling $p(s) \sim s^{-1.75}$ becomes apparent, extending for a bit more than a decade, thus pointing once more to a divergent size scale in the sense discussed in Sect. 1.3.2.

---

[27]It is also possible to do it by using wavelet techniques [50].

[28]The value for the threshold that we have chosen is, naturally, relatively arbitrary. We have chosen 0.115 because, after a quick visual inspection of the dataset, this number worked very well and introduced very few artefacts. Various practical criteria have been proposed in the literature to choose the threshold in this context. Among them, it is rather popular to calculate first the mean μ and standard deviation of the signal, σ, and then choose the threshold to be μ plus a few times σ.

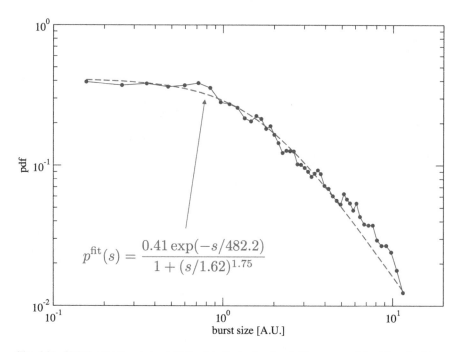

$$p^{\text{fit}}(s) = \frac{0.41 \exp(-s/482.2)}{1 + (s/1.62)^{1.75}}$$

Fig. 6.11 PDF for the 'burst sizes' (defined in the text) estimated by means of the SF method (see Sect. 2.4.3) shown in log-log scale

6.5.2   Power Spectrum

Next, we show the power spectrum of the data, calculated according to Eq. 4.20. The power-spectra of probe data has been often used to search for possible evidences of the type of scale-invariance that SOC demands. In particular, the observation of power-law regions in the power spectra [52, 53, 57], as well as the empirical self-similarity of different spectra measured in different devices have been often cited as pieces of evidence that support the SOC interpretation of the confined plasma dynamics. It has also been the subject of recent controversies, with some authors [58] claiming that the observed power spectra were better fit by exponential laws (see Table 2.1); other authors, including ourselves, have shown that this is not what is found in data from tokamaks and stellarators at sufficiently large power, where power-laws are indeed apparent [50, 59].

The power spectrum of the W7-AS dataset is shown in Fig. 6.12. The best fit to an exponential law obtained using a minimum chi-square method (see Sect. 2.5.3) is also shown (in dashed-red). Also, to help guide the eye, straight lines shown the ω^{-1} and ω^{-2} slopes are included. It seems clear to us that a power-law region, with an exponent close to (but slightly larger than) -1 is indeed present in the data, across the frequency range [1–10] KHz. The power-law scaling is however not as robust as one would desire, in part due to the fact that is not extended beyond a

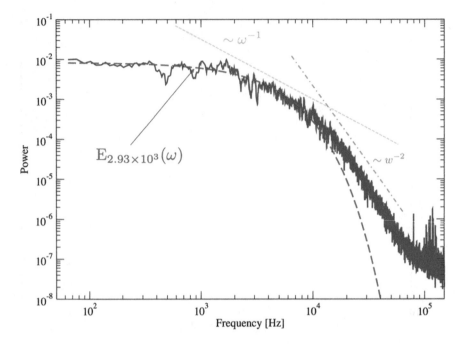

Fig. 6.12 Power spectrum of the ion density time series shown in Fig. 6.8 in log-log scale. The best exponential fit to the spectrum is also included (in red), corresponding to a mean frequency of approximately $\omega_0 \sim 2.9 \, \text{KHz}$

single decade. Therefore, the evidence this particular power spectrum offers is more suggestive than probatory. It would be certainly stronger if scale-invariance could be confirmed by other diagnostic tools.

6.5.3   R/S Analysis

Therefore, we proceed now to look for scale-invariant, long-term correlations in the dataset using the RS analysis (see Sect. 4.4.4 for details). The rescaled range (Eq. 4.50) is shown, as a function of the time lag, in Fig. 6.13. As previously seen with the running sandpile (see analysis in Sect. 4.5), several distinct power-law regions are apparent in it. There are three different regions. The first one spans from the shortest time lag up to $\sim (50\text{--}70)\,\mu\text{s}$, where the rescaled range grows almost linearly with lag. This timescale coincides with the decorrelation time of the signal, $\tau_d \sim 60\,\mu\text{s}$, that is often used to estimate the typical duration of the bursts it contains. Then, a second region exists (i.e., a mesorange, using the terminology introduced in Chap. 1) that spans the range [0.07–2] ms, over which $R/S \sim \tau^{0.61}$. Finally, a last scaling behaviour appears at larger lags than 2 ms, although the

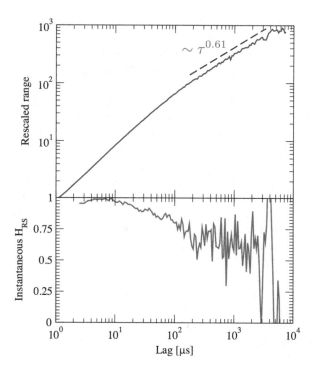

Fig. 6.13 Above: rescaled range as a function of temporal lag for the data set shown in Fig. 6.8 in log-log scale. Below: instantaneous Hurst exponent for the same data

unavailability of data makes it difficult to determine the exponent there.[29] The extension and the value of the Hurst exponent over each region is more clearly seen in the lower frame of Fig. 6.13, where the **instantaneous Hurst exponent** is shown. This quantity is defined as:

$$H(\tau) := \frac{d(\log(R/S))}{d\log(\tau)}. \tag{6.12}$$

The instantaneous Hurst exponent is a very useful tool because, if the rescaled range scales as $R/S \sim \tau^H$ over a certain range of scales, $H(\tau) \sim H$ over those same scales. Thus, any regions of scale-invariance present in the data would appear as flattish regions when plotting $H(\tau)$ as a function of the time lag. In the case of the W7-AS, this is certainly the case. A flattish region is seen over range of timescales $(0.07–2)$ ms with a value of $H \sim 0.6–0.65$. As it will be remembered, any Hurst exponent $H > 0.5$ means, for a close-to-Gaussian distributed dataset, that

---

[29]In fact, the change of scaling itself might be related to the lack of statistics, and not to a change in the dominant physics. Whether this is the case or not, it is impossible to say with the amount of data available in the dataset.

persistent long-term, scale-invariant correlations are present in the data. Particularly, since these scales are much longer than the typical local turbulent scales, previously estimated by the turbulent decorrelation time, $\tau_d \sim 0.06$ ms. Interestingly, the range where persistence is found here overlaps with the [1–10] KHz frequency range (that corresponds to the temporal range [0.1–1] ms) where a (possibly) decaying power-law scaling was previously identified in the power spectrum. This fact reinforces the interpretation given to this scaling region in previous paragraphs.

The R/S analysis has been frequently applied to MCF fluctuation data, mostly gathered at the plasma edge but sometimes also from the plasma core [49]. Results similar to what we have found here have been previously reported for many different devices, both tokamaks and stellarators, with persistent, scale-invariant regions found for timescales starting at around (50–100) μs, and extending for one or more decades [60–62]. The values of the Hurst exponent found are usually in the range $H \sim 0.6$–0.8 in most of these devices, dropping to $H \sim 0.5$ in cases in which a strong, radially-sheared poloidal flow is present, or in regions or devices with open field lines. In the first of these cases,[30] it is believed that the presence of radially-sheared flows can decorrelate radial avalanches as they pass through, effectively breaking any long-term correlations [46]. In the second, it has been theorized that the possibility of losses along the open-field lines provides a dominant memory-erasing mechanism that erases away any inhomogeneities in the radial profile that would act as a memory reservoir, in the sense discussed in Sect. 4.5.

6.5.4 Multifractal Analysis

Finally, we will proceed to characterize the multifractal properties of the W7AS dataset using the techniques discussed in Sect. 3.3.7. The degree of multifractality of probe data in MCF plasmas has also been the subject of long discussions and controversies among MCF researchers. Some authors have reported a significant degree of multifractality [63, 64], with singularity spectra much wider than what one would expect from a monofractal time-series (see Sect. 3.3.7). This has been suggested as proof of the irrelevance of SOC in this context. Other authors, however, report instead that the degree of monofractality is much larger by restricting the multifractal analysis to the same range of scales (i.e., the mesorange) over which the R/S analysis or the power-spectra exhibits a clear power-law [65]. Both approaches are, of course, legitimate. However, we think that it is the latter interpretation that makes sense from a physical point of view. In our opinion, it simply reflects the fact that any instance of scale-invariance must be limited by finite-size effects in real systems. In the case of our Langmuir probe signals, scale-invariance may only appear beyond the local timescale of the turbulence, which is usually of the order of

---

[30]We will revisit transport across sheared flows again in Chap. 9, since there are some interesting dynamics there that are worth looking at with the tools of fractional transport discussed in Chap. 5.

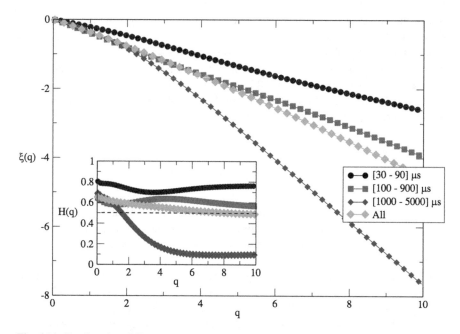

Fig. 6.14 Family of multifractal exponents for the W7-AS data. They have been computed considering different range of box sizes, as shown in the labels. In the inset, the associated generalized Hurst exponent is shown for each case

a few tens of μs. If tests for scale-invariance are however extended over all possible timescales, scale-invariance would simply not appear.

In the case of the W7-AS data, we have calculated the family of multifractal exponents $\xi(q)$ (defined in Eq. 3.97), that are shown in Fig. 6.14. In the same figure, the generalized Hurst exponent $H(q)$ (defined in Eq. 3.98) is shown in the inset for $q > 0$. As it will be remembered, any monofractal time series should have $H(q) = H_0, \forall q$. It turns out that, if $H(q)$ is computed using the techniques described in Sect. 3.3.7 for all possible box sizes, the result for the W7-AS dataset (shown in green) is clearly not mono-fractal, with values of H varying between 0.7 and 0.47. However, the R/S analysis previously discussed already warned us about the fact that the dynamics are rather different for timescales shorter than a few tens of μs, where local fluctuations dominate, compared with scales larger than 70 μs, where some scale-invariant behaviour seems indeed present. Thus, we have calculated the family of multifractal exponents and their associated generalized Hurst exponent limiting the analysis to boxes with sizes within three different ranges. The results are also shown in Fig. 6.14. For the shortest range, with boxes of length up to 90 μs included, $H(q) \sim 0.8$ over a wide range of q. This behaviour is clearly related to the auto-correlation dynamics of turbulent fluctuations with themselves, that we know are strongly persistent (see discussion in Chap. 4). In the intermediate range of box sizes between (0.1–0.9) ms, $H(q) \sim 0.6$ is however found for all q's, that is consistent with the value of the Hurst exponent previously found in the R/S analysis. Finally, for boxes with sizes longer than 1 ms, a much more multifractal behaviour is apparent, that varies from $H \sim 0.7$ at the lowest q's to an antipersistent $H \sim 0.1$ for $q > 5$. It is interesting to note that a similar trend towards a very anti-persistent Hurst exponent was also apparent in the R/S analysis for time lags larger than (3–4) ms (see Fig. 6.13). Whether this anti-persistent behaviour has a real physical meaning or not,[31] it remains to be seen.

6.6 Conclusions

In this chapter we have discussed situations in toroidal MCF plasmas in which complex dynamics are probably at play. In particular, we have focused the discussion on the dynamics of radial turbulent transport in conditions in which the turbulence is near-marginal. By that, it is meant that those plasma profiles whose gradients feed the turbulence sit close to the local thresholds for the onset of instabilities. As repeatedly said throughout this chapter, these conditions might be of relevance for the operation of future MCF reactors, since the larger temperatures expected (compared with that of the tokamaks and stellarators currently in operation) will be able to drive stronger turbulent fluxes, that should in turn be more capable of quickly relaxing plasma profiles below threshold. Therefore, the understanding of this type of complex dynamics is important not only from a fundamental perspective, but also urgent from an applied point of view.

We have also illustrated the usefulness and the dangers of some of the tools available to us to detect and characterize complex behaviour. The analysis of W7-AS edge turbulent data carried out in Sect. 6.5 is complementary to the analysis done in Chap. 4, using similar Langmuir data from the edge of the TJ-II stellarator. The most important message to take home is that a good knowledge of the data and of the underlying physics is essential to correctly interpret the results that these tools may provide. There is an innate tendency in all of us to expect that features be present to an extent that only mathematical models can satisfy. At best, many of these properties are often apparent in real systems over a vert limited range of scales, that is what we referred to repeatedly as the mesorange. Due to this limitation, approaches that are equivalent in mathematical models (such as the one-to-one relationships between exponents in the power spectra and in the rescaled range, that are exact only for monofractal systems) simply do not hold in real systems. If this fact is not kept in mind, the analysis of the data and the interpretation of the results may be obscured and valuable information about the underlying dynamics, ignored.

---

[31] It might be perhaps related to the anti-correlation dynamics that we found in the running sandpile at the longest timescales, due to its finite size (see Sect. 5.5). It could also be due to lack of meaningful statistics at the longest scales.

References

1. Eddington, A.S.: The Internal Constitution of the Stars. The Observatory 43, 341 (1920)
2. Einstein, A.: Ist die Tragheit eines Korpers von seinem Energieinhalt abhängig? Ann. Phys. 18, 639 (1905)
3. von Weizsacker, C.F.: Über Elementumwandlungen in Innern der Sterne II. Phys. Z. 39, 633 (1938)
4. Bethe, H.A.: Energy Production in Stars. Phys. Rev. 55, 434 (1939)
5. Clayton, D.D.: Principles of Stellar Evolution and Nucleosynthesis. University of Chicago Press, Chicago (1984)
6. Lawson, J.D.: Some Criteria for a Power Producing Thermonuclear Reactor. Report for the U.K. Atomic Energy Research Establishment (1955)
7. Freidberg, J.P.: Plasma Physics and Fusion Energy. Cambridge University Press, Cambridge (2008)
8. Pfalzner, S.: An Introduction to Inertial Confinement Fusion. CRC, New York (2006)
9. Grad, H.: Toroidal Containment of a Plasma. Phys. Fluids 10, 137 (1967)
10. Wesson, J.: Tokamaks. Oxford University Press, Oxford (2004)
11. Gormenzano, C., et al.: Progress in the ITER Physics Basis - Chapter 6: Steady State Operation. Nuclear Fusion 47, S285 (2007)
12. Goldstein, H., Poole, G., Safko, J.: Classical Mechanics. Addison-Wesley, New York (2000)
13. Shimada, M., et al.: Progress in the ITER Physics Basis - Chapter 1: Overview and Summary. Nuclear Fusion 47, S1 (2007)
14. Hender, T.C., et al.: Progress in the ITER Physics Basis - Chapter 3: MHD Stability, Operational Limits and Disruptions. Nuclear Fusion 47, S128 (2007)
15. Dreicer, H.: Electron and Ion Runaway in a Fully Ionized Gas. Phys. Rev. 115, 238 (1959); ibid. Phys. Rev. 117, 329 (1959)
16. Rosenbluth, M.N., Putvinski, S.C.: Theory for Avalanche of Runaway Electrons in Tokamaks. Nuclear Fusion 37, 1335 (1997)
17. Martin-Solis, J.R., Alvarez, J.D., Sanchez, R., Esposito, E.: Momentum-Space Structure of Relativistic Runaway Electrons. Phys. Plasmas 5, 2370 (1998)
18. Loarte, A., Riccardo, V., Martin-Solis, J.R., Paley, J., Huber, A., Lehnen, M.: Magnetic Energy Flows During the Current Quench and Termination of Disruptions with Runaway Current Plateau Formation in JET and Implications for ITER. Nuclear Fusion 51, 073004 (2011)
19. Wakatani, M.: Stellarator and Heliotron Devices. Oxford University Press, Oxford (1998)
20. Nührenberg, J., Lotz, W., Gori, S.: Varenna International Workshop on the Theory of Fusion Plasmas. Editrice Compositori, Bologna (1994)
21. Beidler, C., Grieger, G., Herrnegger, F., Harmeyer, E., Kisslinger, J., Lotz, W., Maassberg, H., Merkel, P., Nührenberg, J., Rau, F., Sapper, J., Sardei, F., Scardovelli, F., Schlüter, A., Wobig, A.: Physics and Engineering Design for Wendelstein 7-X. Fusion Technol. 17, 148 (1990)
22. Helander, P., Sigmar, D.J.: Collisonal Transport in Magnetized Plasmas. Cambridge University Press, Cambridge (2002)
23. Horton, W.: Turbulent Transport in Magnetized Plasmas. World Scientific, Singapore (2012)
24. Sanchez, R., Newman, D.E., Leboeuf, J.N., Decyk, V.K., Carreras, B.A.: On the Nature of Radial Transport Across Sheared Zonal Flows in Electrostatic Ion-Temperature-Gradient Gyrokinetic Tokamak Plasma Turbulence. Phys. Plasmas 16, 055905 (2008)
25. Yankov, V.V.: The Pinch Effect Explains Turbulent Transport in Tokamaks. Pisma Zh. Eksp. Theor. Fiz. 60, 169 (1994)
26. Hinton, F.L., Hazeltine, R.D.: Theory of Plasma Transport in Toroidal Confinement Systems. Rev. Mod. Phys. 48, 239 (1976)
27. Hirshman, S.P., Sigmar, D.F.: Neoclassical Transport of Impurities in Tokamak Plasmas. Nuclear Fusion 21, 1079 (1981)
28. Petty, C.C., Luce, T.C.: Inward Transport of Energy During Off-Axis Heating on the DIII-D Tokamak. Nuclear Fusion 34, 121 (1994)

29. Gentle, K.W., Bravenec, R.V., Cima, G., Gasquet, H., Hallock, G.A., Phillips, P.E., Ross, D.W., Rowan, W.L., Wootton, A.J., Crowley, T.P., Heard, J., Ouroua, A., Schoch, P.M., Watts, C.: An Experimental Counterexample to the Local Transport Paradigm. Phys. Plasmas 2, 2292 (1995)
30. Petty, C.C., Luce, T.C., Burrell, K.H., Chiu, S.C., Degrassie, J.S., Forest, C.B., Gohil, P., Greenfield, C.M., Groebner, R.J., Harvey, R.W., Pinsker, R.I., Prater, R., Waltz, R.E., James, R.A., Wroblewski, D.: Non-dimensional Transport Scaling in DIII-D: Bohm Versus Gyro-Bohm Resolved. Phys. Plasmas 2, 2342 (1995)
31. Becker, G.: Electron Temperature Pro Le Invariance in OH, L- and H-mode Plasmas and Consequences for the Anomalous Transport. Nucl. Fusion 32, 81 (1992)
32. Greenwald, M., Schachter, J., Dorland, W., Granetz, R., Hubbard, A., Rice, J., Snipes, J.A., Stek, P., Wolfe, S.: Transport Phenomena in Alcator C-Mod H-Modes. Plasma Phys. Controlled Fusion 40, 789 (1998)
33. Baker, D., Greenfiel, C.M., Burrell, K.H., DeBoo, J.C., Doyle, E.J., Groebner, R.J., Luce, T.C., Petty, C.C., Stallard, B.W., Thomas, D.M.: Thermal Diffusivities in DIII-D Show Evidence of Critical Gradients. Phys. Plasmas 8, 4128 (2001)
34. Ryter, F., Angioni, C., Beurskens, M., Cirant, S., Hoang, G.T., Hogeweij, G.M.D., Imbeaux, F., Jacchia, A., Mantica, P., Suttrop, W.: Experimental Studies of Electron Transport. Plasma Phys. Controlled Fusion 43, A323 (2001)
35. Wagner, F.: A Quarter-Century of H-Mode Studies. Plasma Phys. Controlled Fusion 49, B1 (2007)
36. Terry, P.W.: Suppression of Turbulence and Transport by Sheared Flow. Rev. Mod. Phys. 72, 109 (2000)
37. Connor, J.W., Hastie, R.J., Wilson, H.R., Miller, R.L.: Magnetohydrodynamic stability of tokamak edge plasmas. Phys. Plasmas 5, 2687 (1998)
38. Cowley, S.C., Wilson, H., Hurricane, O., Fong, G.: Explosive Instabilities: From Solar Flares to Edge Localized Modes in Tokamaks. Plasma Phys. Controlled Fusion 45, A31 (2003)
39. Loarte, A., Saibene, G., Sartori, R., Campbell, D., Becoulet, M., Horton, L., Eich, T., Herrmann, A., Matthews, G., Asakura, N.: Characteristics of Type I ELM Energy and Particle Losses in Existing Devices and Their Extrapolation to ITER Plasma. Plasma Phys. Controlled Fusion 45, 1549 (2003)
40. Levinton, F.M., Zarnstorff, M.C., Batha, S.H., Bell, M., Bell, R.E., Budny, R.V., Bush, C., Chang, Z., Fredrickson, E., Janos, A.: Improved Confinement with Reversed Magnetic Shear in TFTR. Phys. Rev. Lett. 75, 4417 (1995)
41. Strait, E.J., Lao, L.L., Mauel, M.E., Rice, B.W., Taylor, T.S., Burrell, K.H., Chu, M.S., Lazarus, E.A., Osborne, T.H., Thompson, S.J.: Enhanced Confinement and Stability in DIII-D Discharges with Reversed Magnetic Shear. Phys. Rev. Lett. 75, 4421 (1995)
42. Wagner, F., Stroth, U.: Transport in Toroidal Devices-The Experimentalist's View. Plasma Phys. Controlled Fusion 35, 1321 (1993)
43. Hirsch, M., Amadeo, P., Anton, M., Baldzuhn, J., Brakel, R., Bleuel, D.L., Fiedler, S., Geist, T., Grigull, P., Hartfuss, H., Holzhauer, E., Jaenicke, R., Kick, M., Kisslinger, J., Koponen, J., Wagner, F., Weller, A., Wobig, H., Zoletnik, S.: Operational Range and Transport Barrier of the H-Mode in the Stellarator W7-AS. Plasma Phys. Controlled Fusion 40, 631 (1998)
44. Garcia-Cortes, I., de la Luna, E., Castejon, F., Jimenez, J.A., Ascasibar, E., Brañas, B., Estrada, T., Herranz, J., Lopez-Fraguas, A., Pastor, I., Qin, J., Sanchez, J., Tabares, F.L., Tafalla, D., Tribaldos, V., Zurro, B.: Edge-Localized-Mode-Like Events in the TJ-II Stellarator. Nuclear Fusion 40, 1867 (2000)
45. Diamond, P.H., Hahm, T.S.: On the Dynamics of Turbulent Transport Near Marginal Stability. Phys. Plasmas 2, 3640 (1995)
46. Newman, D.E., Carreras, B.A., Diamond, P.H., Hahm, T.S.: The Dynamics of Marginality and Self-Organized Criticality as a Paradigm for Turbulent Transport. Phys. Plasmas 3, 1858 (1996)
47. Carreras, B.A., Lynch, V.E., Newman, D.E., Diamond, P.H.: A Model Realization of Self-Organized Criticality for Plasma Confinement. Phys. Plasmas 3, 2903 (1996)
48. Dendy, R.O., Helander, P.: Sandpiles, Silos and Tokamak Phenomenology: A Brief Review. Plasma Phys. Controlled Fusion 39, 1947 (1997)

49. Sanchez, R., Newman, D.E.: Self-Organized Criticality and the Dynamics of Near-Marginal Turbulent Transport in Magnetically Confined Fusion Plasmas. Plasma Phys. Controlled Fusion 57, 123002 (2015)

50. van Milligen, B.P., Sanchez, R., Hidalgo, C.: Relevance of Uncorrelated Lorentzian Pulses for the Interpretation of Turbulence in the Edge of Magnetically Confined Toroidal Plasmas. Phys. Rev. Lett. 109, 105001 (2012)

51. Carreras, B.A., van Milligen, B.P., Pedrosa, M.A., Balbin, R., Hidalgo, C., Newman, D.E., Sanchez, E., Frances, M., Garcia-Cortes, I., Bleuel, J., Endler, M., Riccardi, C., Davies, S., Matthews, G.F., Martines, E., Antoni, V., Latten, A., Klinger, T.: Self-similarity of the Plasma Edge Fluctuations. Phys. Plasmas 5, 3632 (1998)

52. Pedrosa, M.A., Hidalgo, C., Carreras, B.A., Balbin, R., Garcia-Cortes, I., Newman, D., van Milligen, B.P., Sanchez, E., Bleuel, J., Endler, M., Davies, S., Matthews, G.F.: Empirical Similarity of Frequency Spectra of the Edge-Plasma Fluctuations in Toroidal Magnetic-Confinement Systems. Phys. Rev. Lett. 82, 3621 (1999)

53. Rhodes, T.L., Moyer, R.A., Groebner, R, Doyle, E.J., Lehmer, R., Peebles, W.A., Rettig, C.L.: Experimental Evidence for Self-organized Criticality in Tokamak Plasma Turbulence. Phys. Lett. A 253, 181 (1999)

54. Politzer, P.: Observation of Avalanche-Like Phenomena in a Magnetically Confined Plasma. Phys. Rev. Lett. 84, 1192 (2000)

55. Hirsch, M., Baldzuhn, J., Beidler, C., Brakel, R., Burhenn, R., Dinklage, A., Ehmler, H., Endler, M., Erckmann, V., Feng, Y., Geiger, J., Giannone, L., Grieger, G., Grigull, P., Hartfuß, H.-J., Hartmann, D., Jaenicke, R., König, R., Laqua, H.P., Maaßberg, H., McCormick, K., Sardei, F., Speth, E., Stroth, U., Wagner, F., Weller, A., Werner, A., Wobig, H., Zoletnik, S., and for the W7-AS Team. Major Results from the Stellarator Wendelstein 7-AS. Plasma Phys. Controlled Fusion 50, 053001 (2008)

56. Hidalgo, C., Balbin, R., Pedrosa, M.A., Garcia-Cortes, I., Anabitarte, E., Sentíes, J.M., San Jose, M., Bustamante, E.G., Giannone, L., Niedermeyer, H.: Edge Plasma Turbulence Diagnosis by Langmuir Probes. Contrib. Plasma Phys. 36, 139 (1996)

57. Wang, W.H., Yu, C.X., Wen, Y.Z., Xu, Y.H., Ling, B.L., Gong, X.Z., Liu, B.H., Wan, B.N.: A Signature of Self-organized Criticality in the HT-6M Edge Plasma Turbulence. Chin. Phys. Lett. 18, 396 (2001)

58. Maggs, J.E., Morales, G.J.: Generality of Deterministic Chaos, Exponential Spectra, and Lorenzian Pulses in Magnetically Confined Plasmas. Phys. Rev. Lett. 107, 185003 (2011)

59. dos Santos Lima, G.Z., Iarosz, K.C., Batista, A.M., Caldas, I.L., Guimarães-Filho, Z.O., Viana, R.L., Lopes, S.R., Nascimento, I.C., Kuznetsov, Y.K.: Self-Organized Criticality in MHD Driven Plasma Edge Turbulence. Phys. Lett. A 376, 753 (2012)

60. Carreras, B.A., van Milligen, B.P., Pedrosa, M.A., Balbin, R.: Long-Range Time Correlations in Plasma Edge Turbulence. Phys. Rev. Lett. 80, 4438 (1998)

61. Gilmore, M., Yu, C.X., Rhodes, T.L., Peebles, W.A.: Investigation of Rescaled Range Analysis, the Hurst Exponent, and Long-Time Correlations in Plasma Turbulence. Phys. Plasmas 9, 1312 (2002)

62. Xu, Y.H., Jachmich, S., Weynants, R.R.: On the Properties of Turbulence Intermittency in the Boundary of the TEXTOR Tokamak. Plasma Phys. Controlled Fusion 47, 1841 (2005)

63. Budaev, V.P., Takamura, S., Ohno, N., Masuzaki, S.: Superdiffusion and Multifractal Statistics of Edge Plasma Turbulence in Fusion Devices. Nuclear Fusion 46, S181 (2006)

64. Neto, C.R., Guimarães-Filho, Z.O., Caldas, I.L., Nascimento, I.C., Kuznetsov, Y.K.: Multifractality in Plasma Edge Electrostatic Turbulence. Phys. Plasmas 15, 082311 (2008)

65. Carreras, B.A., Lynch, V.E., Newman, D.E., Balbin. R.: Intermittency of Plasma Edge Fluctuation Data: Multifractal Analysis. Phys. Plasmas 7, 3278 (2000)

Chapter 7
Space Plasmas: Complex Dynamics of the Active Sun

7.1 Introduction

The second example of complex plasma dynamics we will discuss takes us to our own Sun. The suspicion that complex dynamics might govern some of the inner workings of the Sun has been rooted, over the years, in many observations gathered with both telescopes and satellites. For example, in the fact that the Sun sustains a self-generated, strong magnetic field that plays a central role in most of the solar dynamics. Solar flares and coronal mass ejections also exhibit scale-invariant features over many decades. In addition, the active areas on the surface of the Sun also appear to be distributed following some kind of fractal-like patterns.

We will start by providing a brief introduction to the physics of the Sun. Those readers already familiar with the subject can choose to jump directly to Sect. 7.3, where some of the experimental evidence that points to complex behaviour is briefly presented, mostly in the context of solar flaring. We will conclude the chapter, as always, by applying some of the analysis tools described in the first part of this book to a dataset that contains information about the size, duration and speed of the coronal mass ejection events (or CMEs) that have been detected by the SOHO satellite between 1996 and 2016. We will try to look for evidence of complex behaviour in this data while, simultaneously, discussing the state of research in this field and pointing the interested reader to several relevant references.

7.2 Our Own Star: The Sun

The study of our star, the Sun, is an activity that humans have pursued since the dawn of history. Although the Sun is a fairly ordinary star by the universe standards,

© Springer Science+Business Media B.V. 2018
R. Sánchez, D. Newman, *A Primer on Complex Systems*,
Lecture Notes in Physics 943, https://doi.org/10.1007/978-94-024-1229-1_7

it has the unique charm of being our own.[1] Solar eclipses fascinated the ancient Greeks, Chinese and Egyptians. The Sun played an important role in Aztec and Mayan cultures. It was the center of Copernicus' famous plight against the Catholic Church in the sixteenth century. In the 1800s it was first suggested that the Sun might be a gaseous sphere, its famous 11-year cycle was proposed by Samuel Schwabe, and solar prominences were properly identified as such. In the 1900s, the important role that magnetic fields play in the Sun dynamics became more clearly understood, the fusion of hydrogen was identified as the engine powering the Sun, and the first laboratory experiments attempting to generate fusion energy in a controlled environment were started.

The Sun is a star of around 5000 million years of age that is at the top tier of the *yellow dwarf* family [1, 2]. Its mass is about 2×10^{30} kg and its radius is approximately 700,000 km.[2] It is, like all stars, a massive ball of hot plasma confined by its own gravitational attraction.[3] It is essentially made of hydrogen (about 90%) and helium (the remaining 10%), mostly in a fully ionized state.[4] The central temperature of the Sun is about ten million degrees, which is sufficient to drive the fusion processes that power it. In fact, about five million tons of hydrogen are fused in the solar core each second. On the surface, the temperature is much lower, about 5000–6000°. In what follows, we provide a brief introduction to its structure and some of its more salient dynamical features.[5]

7.2.1 Structure of the Sun

The structure of the Sun is sketched in Fig. 7.1. There are three main regions, each of them named after the physical process that dominates the local transfer of the fusion energy produced at the core [1, 2]. These regions are the **core**, that extends for approximately the first fourth of the radius of the Sun; the **radiative zone** extending

---

[1]Paraphrasing Prof. Philip Scherrer of Stanford University, *"The Sun is also the only star known to grow vegetables!"*

[2]In comparison, the Earth mass is 6×10^{24} kg and its radius 6000 km.

[3]The acceleration of gravity at the surface of the Sun is 274m^2/s; on Earth's surface, it is 9.8m^2/s.

[4]Some heavier atoms such as oxygen, nitrogen and carbon also exist, although in very slow percentages ($< 0.1\%$).

[5]The physics of the Sun is very complex and, being ourselves fusion plasma physicists, it lies outside of our field of expertise. Therefore, it is not our intention here to provide a comprehensive description of the different physical processes that govern its most important dynamics. There are many books and reviews available that cover the Sun physics in detail, and that we strongly advise our readers to read [1–4]. Instead, we provide here an sketchy, but sound view of the main facts that will be of relevance to understand the discussion on whether the Sun behaves as a complex system or not. In particular, the first sections of this chapter borrow heavily from E.R. Priest's wonderful book, *Solar Magnetodydrodynamics* [2]. Priest's treatise is, in our opinion, very engaging and extremely clear, greatly succeeding at providing a vivid picture of the Sun dynamics.

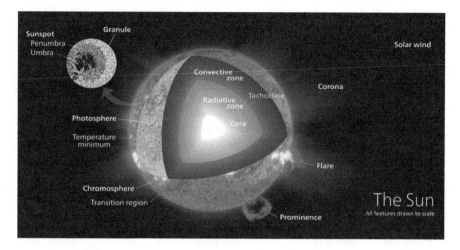

Fig. 7.1 Structure of our Sun. Credits: from Wikipedia.org (Kevin Ma, CC BY-SA 3.0 license)

to about 70% of the radius; and the **convection zone**, that runs all the way to the surface. The core is where 99% of all the energy is produced via fusion processes. It occupies about one fiftieth of its volume, but contains almost half of the total mass of the Sun. The core is surrounded by the radiative zone. Here, the energy is transferred towards the surface via *radiative diffusion*. By this name, it is meant that photons are the main carriers of the produced energy, being absorbed and emitted multiple times as they travel radially outwards.[6]

As one moves radially out from the core across the radiative region, the transport of energy via radiative diffusion eventually becomes less efficient than the transport carried by convection. As a result, huge convective cells (with linear sizes of the order hundreds of thousands of kilometres) populate what is known as the convective zone. Here, buoyancy is the main drive to push energy (together with mass and magnetic fields) towards the surface, very much as what happens in the case of water boiling in a pot. The boundary region between the radiative and convective zones is known as the **tachocline** [5, 6]. This is a particularly interesting region because it is a thin radial layer (its width is less that 4% of the Sun radius) where a very large latitudinally-differential rotation exists (with the equator rotating about a 30% faster than the regions close to the poles), in contrast to the solid rotation of the inner part of the Sun. The existence of this differential rotation makes the tachocline a very important player in the self-generation and sustainment of magnetic fields in the Sun: the so-called **solar dynamo** [7, 8]. Magnetic field lines get twisted, tensioned and deformed here; then, they are buoyantly pushed towards the surface by the huge convective cells that are present in the convective region.

---

[6]In fact, the absorption/emission processes undergone by these photons are so intensive that it would take on average a core photon almost ten million years to reach the surface of the Sun.

These magnetic fields, as they reach the surface, play an important part in much of the zoology that makes the Sun such an interesting system to study: active regions, sunspots, solar flares, prominences, etc. We will discuss many of them in Sect. 7.2.2.

The last part of the Sun is its atmosphere. It is a mostly ionized, low-density plasma that is often partitioned in three distinct regions according to the different physics that take place in them[1, 2]. The one closest to the surface is a very thin layer of plasma (a few hundred of kilometres wide) known as the **photosphere**. This is the part of the Sun that emits most of the solar radiation, being relatively dense and opaque, with a temperature of about 5000–6000°. It is also very granular in nature. This granulation is believed to represent the top of convective cells as they buoyantly ascend to the surface from the convection zone. The majority of these granules are of a size of about a thousand kilometres, and last for about 10 min or so. However, much larger granules—*supergranules*—are sometimes observed, with sizes up to tens of thousands in kilometres, being able to last for several days. The structure of the magnetic field in the photosphere is also very complicated, consisting of small magnetic domains that rapidly evolve and move around, but that are still able to organize themselves in large-scale structures. Among them, it is worth mentioning *sunspots*, that are regions with very large concentrations of magnetic flux and *unipolar regions*, that are very large, long-lived areas with linear dimensions of hundred of thousands of kilometres in which the magnetic has a dominant polarity (either radially inwards or outwards).

Above the photosphere lies the **chromosphere**, that extends for 2000–3000 km above the surface of the Sun. It is a region populated by the so-called *spicules*, which are plasma jets pointing radially outwards as they are ejected from the top of the convective cells that form the aforementioned supergranules, probably along the direction of the local magnetic field. Spicules can have lengths of a few thousand kilometres, and the ejected plasma reaches velocities of about 20 or 30 km/s. The chromosphere ends suddenly, in a narrow layer across which the plasma temperature increases suddenly by several orders of magnitude up to a few million degrees,[7] that provides the start of the so-called **solar corona**. The corona expands outwards from the Sun engulfing all the planets of the solar system and beyond. It is mostly invisible in white light, being originally observed only during eclipses. The plasma in the corona has a very low density, that further decreases as one moves away from the Sun. The corona is not uniform, and regions known as *coronal holes* are found in it. They form at those parts of the surface of the Sun where the magnetic field is unipolar and field lines are open (that is, these lines do not enter back into the Sun, as it happens in a solar prominence; for reference, one such prominence is shown in the upper, right frame of Fig. 7.2), thus allowing the plasma to flow outwards to form the **solar wind** (SW). In addition to particles and energy, the solar wind also takes

---

[7]This huge increase in temperature is one of the major, and still unresolved, problems in solar physics. Many explanations have been forwarded over the years (each championing for different players as responsible, including magnetic field reconnection, Alfven wave heating, plasma turbulence, and so on), but a definite answer is still lacking [2].

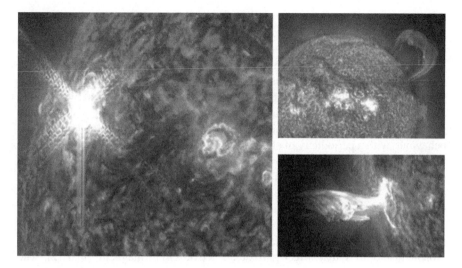

Fig. 7.2 Left: H_α image of a solar flare; Right: solar prominence with the form of an arc standing over the surface of the Sun (above) and coronal mass ejection or CME (below). Credits: all images (© ESA/NASA - SOHO/LASCO)

angular momentum, in sufficient amounts as to slow down the Sun significantly over its lifetime. It takes the solar wind about 5 days to reach the Earth, in comparison to the 8 min that solar light needs.

7.2.2 The Active Magnetic Sun

The Sun is a very active system. A closer look at its photosphere and overlying atmosphere reveals a plethora of activity. So-called **active regions** appear as bright regions in H_α photographs[8] (see Fig. 7.2), that are associated to moderate local concentrations of magnetic field (of the order of a hundred Gauss or so). Active regions form as emerging magnetic flux pops up from below the photosphere due to buoyancy.

Within the active regions, darker areas often appear that are associated to intense concentrations of magnetic fields. These darker areas are known as **sunspots**, and they are cooler than their surroundings. Sunspots often appear close to the equator,

---

[8]H_α images are photographs taken with an H_α filter. The H_α line is the first of six spectral lines of hydrogen in the visible part of the spectrum. It lies in the red part of it, at a wavelength of 656.28 nm. H_α filters work by rejecting all but the narrow sliver of $H-\alpha$ light. In this way, direct imaging of the lower part of the chromosphere—where the temperature is high enough (about 10,000°) so that the lower energy levels of hydrogen are often excited—, can be done while at the same time removing the excessive brightness coming from the underlying photosphere.

usually in pairs of opposite polarity (i.e., with a different sign of the component of the magnetic field normal to the surface of the Sun), and drift away from the equator at the highest activity periods. They are believed to mark the footprints of magnetic field flux tubes that are being brought out of the surface by buoyancy forces (see inset of Fig. 7.3), and that eventually may give rise to **filaments** and **prominences**, **solar flares** or **coronal mass ejections** (or CMEs). The state of activity of the Sun is often measured by counting the number of sunspots that are visible at any given time. It has been known since the late 1800s that this number has a cyclic behaviour, with a periodicity of about 11 years, although it is known to vary quite a bit, occasionally becoming as short as eight and as long as 17 years. However, the number eleven has stuck and the sunspot cycle is often referred to as the **11-year solar cycle**. There is also a wide variation in the number of sunspots appearing at shorter timescales than the 11-year cycle.

Prominences, flares and CMEs are different kinds of violent phenomena that release mass, energy and magnetic field towards the outer corona as a result of violent instabilities that affect part of an active-region magnetic field [2]. The distinction between them is, to a certain extent, phenomenological. Prominences are basically arcs of plasma that erupt from the surface of the Sun, but that keep their feet buried deep into the photosphere (see middle frame in Fig. 7.2; also the inset in Fig. 7.3). The plasma within the prominence is usually cooler and denser than typical coronal values. Prominences can sometimes last for months, being held above the Sun surface by strong magnetic fields that remain attached to the Sun main body. Solar flares and CMEs, on the other hand, differ from prominences in that, instead of being held above the surface of the Sun by the underlying magnetic fields, they manage to break free and be ejected away from the Sun, into open space (see lower frame in Fig. 7.2).

Solar flares are huge explosions that occur often near sunspots, usually along the boundaries that separate regions with different magnetic polarity, and that accelerate plasma particles to very high speeds. It is believed that they are the result of a huge release of energy caused by the reconnection of the magnetic field at or near a sunspot. The accelerated particles lead to a large X-ray emission (usually at energies larger than a few tens of KeV) as a result of their (Coulombian) interaction with the background plasma. Smaller flares are also possible (known respectively as micro- and nano-flares), that contribute to the total emissivity at lower frequencies. Solar flares can be identified as very bright flashes in H_α photographs (see left frame of Fig. 7.2).

Coronal mass ejections or CMEs, on the other hand, are huge, balloon-shaped bursts that are ejected by the Sun, carrying with them plasma with a mass in the order of tens to hundreds of billions of kilograms. As they follow the magnetic field of the Sun, they may heat up to tens of millions of degrees. It is also believed that CMEs, as solar flares, are the result of a reconnection event of the surface magnetic field within an active region, but one that is sufficiently powerful as to result in changes of the magnetic topology that liberate large packets of plasma from the main body of the Sun. The ejected mass is dragged away by its own inertia. Often, solar flares and CMEs are seen to happen together, near the same location.

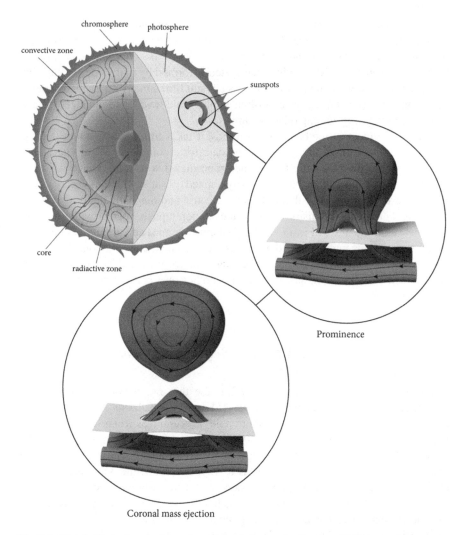

Fig. 7.3 Sketch illustrating the formation of a prominence leading to a CME event. A magnetic flux tube raises out of the photosphere due to buoyancy, forming the standing arc known as a prominence. The different polarity of the foots of the arc are marked by the different sign of the component of the magnetic field perpendicular to the surface of the Sun, where the sunspots would be. If the magnetic field is able to reconnect in such a way that the magnetic field in the arc region detaches itself from the main body of the Sun, the prominence ends up triggering a CME event. Credits: courtesy of Estefanía Cuevas

7.3 Is Our Sun a Complex System?

Our Sun is a very large system, and a very complicated one as well. The very brief description we just gave in Sect. 7.2 already shows some of the reasons that have led many to believe that the Sun might behave as a complex system. As discussed at length in Chap. 1, a complex system must exhibit self-organization and emergence. That is, *it must self-organize itself and permit the emergence of new complex behaviours and dynamics that allow it to carry out its goals most effectively.* Complex systems are also usually open systems, formed by many constituents that interact nonlinearly among themselves. As a result of the many interactions among these constituents, complex systems often exhibit scale-invariance and memory, and sometimes criticality, among other features.

With these ideas in the back of our minds, it takes little time to admit that the Sun indeed looks like a pretty good candidate for a complex system. Our star is an open system driven by the fusion energy produced at its core. Due to its large temperature, it is also a huge ball of (mostly ionized) plasma, with ions and electrons interacting nonlinearly with themselves via electric and magnetic fields. It sustains large radial gradients in density, pressure and temperature, that can provide abundant free energy for the onset of instabilities. To them, one must add gravity, rotation and an always evolving magnetic field. Finally, there are emergent behaviours in the Sun that appear via self-organization. Some examples are provided by the solar flares and coronal mass ejections just discussed. Another one, particularly interesting, is the *tachocline*.

7.3.1   The Tachocline: A Case of Self-Organization

The tachocline, as discussed in Sect. 7.2 is the thin radial boundary that exists between the radiative and convective zones [5]. In it, a large latitudinally differential rotation (i.e., from the North to the South pole) is present, as recently confirmed by helioseismic experimental evidence [9]. The origin of the tachocline is not clear, although there are theories that propose that it is of a turbulent nature[9] and the result of *self-organization* [12]. Be it as it may, the tachocline definitely plays an essential role within the solar global dynamo that must sustain the magnetic field of the Sun.[10] The differential rotation at the tachocline shears and twists any magnetic flux tubes that travel through it, strengthening the magnetic field at the cost of transferring kinetic and thermal energy from the plasma to it [8]. The strengthened magnetic field then keeps traveling radially outwards towards the photosphere and beyond, mostly carried by the buoyancy forces that push the surrounding plasma

---

[9]We discussed another region with a large differential rotation at the edge of tokamaks in Chap. 6. In tokamaks, they are known as (poloidal and toroidal) *zonal flows*. It is believed that they are self-driven by the plasma turbulence through the Reynolds stresses in order to dissipate energy more efficiently. As a result, these poloidal and toroidal zonal flows help to keep under control the radial transport that turbulence induces, improving plasma confinement in the process [10]. Regimes of operation with large zonal flows are in fact being considered as the standard operational configuration of the next-step ITER tokamak [11].

[10]In the absence of the solar dynamo, and given the magnetic resistivity of the Sun, any primordial magnetic field that might have been trapped during the formation of the Sun should be long gone.

to the surface. Once there, the magnetic field then plays a very active role in most (if not all) of the phenomena that take place in the solar atmosphere, and that we touched upon earlier: solar flares, CMEs, prominences, sunspots, etc. In the end, all of these phenomena provide avenues through which the energy that is produced at the core through fusion processes can leave the Sun, so that a stationary state can be maintained. The reason why these phenomena *emerge*, and not others, is probably because they happen to be the most efficient to transport energy given the state of the Sun in terms of how much fusion energy it produces, the magnitude of its gravitational forces, its temperature and so on.

7.3.2 Scale-Invariance of Solar Flare Data

Solar flares can also be considered to be an emergent phenomenon of a complex Sun.[11] There is some direct experimental evidence that supports such a claim. Over the years, the observation of solar flares, both from telescopes and satellites, has provided extensive catalogs that contain a plethora of information about these phenomena extending for more than 70 years. These catalogs contain information about the time when flares happened, their type, their duration, the energy that was released in them, and so on. Several interesting facts have been found while analysing data from many of these catalogs. We will not provide an extensive discussion of the results here, since some excellent reviews already exist [14, 15]. We will just mention that, if a composite of the flare frequency distribution (akin to the pdf we introduced in Chap. 2) is made as a function of the flare released energy using all the available data [16–20], an enormous power-law scaling is obtained that extends for more than *eight* decades in energy [15]. This is probably the longest we have seen for any system! Such a display of scale-invariance is indeed remarkable,[12] including flares with energies that range from the hard X-ray (HXR) part of the electromagnetic spectrum to the extreme-ultraviolet.

The exponent that best fits the scaling of the experimental solar flare data overall is about −1.8, although uncertainties exist due to the different origin of the data [15]. As we mentioned in Sect. 1.3.2, power-law scalings with exponents smaller than 2 are not only a signature of scale-invariance, but a telltale signal of an underlying lack of characteristic scales that might be related to criticality. Therefore, the observed scaling tells us that the size of solar flares lacks such a characteristic scale, and

---

[11]The appearance of self-organized shear flows such as the ones at the solar tachocline is not the only similarity between the active Sun and fusion plasmas confined in a tokamak. Explosive events also take place at the edge of tokamaks that are known as ELMs (see Chap. 6). In fact, some theoretical studies have proposed that the physical mechanism that drives ELMs, believed to be of the Rayleigh-Taylor type and that involves the interchange between neighbouring magnetic flux tubes, might not be all that different from what happens in the case of solar flares [13].

[12]It is also interesting to note that, when including only flares from a single active region instead, similar power-law scalings are also obtained [21].

that their size is not limited by the physical process that causes it (probably, local reconnections of the magnetic fields at the active regions in the photosphere), but by finite size effects. Or, in other words, that a larger Sun would yield even larger flares, together with a larger mesorange over which scale-invariance would be apparent. In the language we introduced in Chap. 1, one could say that the observed statistics of solar flare energy are not just scale-invariant, but that they suggest criticality as well.

Another interesting piece of experimental evidence has to do with the statistics of waiting-times between successive flares. As it would be remembered, we discussed in Sect. 4.4.5 that random triggering usually translates into exponential waiting-time pdfs, a characteristic feature of Poisson processes. The pdf of the waiting-times between solar flares appears to scale instead as $p(w) \sim w^{-(1.8-2.2)}$, as reported by many authors [22–24]. This scaling is consistent with the presence of some (long-term, scale-invariant, persistent) memory of the kind discussed at length in Chap. 4. However, it is fair to say that this is not the only possible interpretation of these pdfs. In particular, a double-exponential fit has also been proposed, that would imply a very different interpretation in which at least two different triggering processes are active, each with a different triggering rate. Another popular proposal is that of a time-dependent Poisson process, in which the triggering rate varies over time with power-law statistics [23, 25]. The controversy between these two views (time-varying Poisson vs. power-law) has been raging for the last two decades and it does not show any sign of receding just yet [23, 26, 27].

The last result about solar flares that we will mention has to do with their spatial scale-invariance, in the sense discussed in Chap. 2. Photographs of the solar corona provide views of the active areas on the surface of the Sun affected by individual solar flares, as defined by setting a minimum threshold for the local intensity. Box-counting algorithms can then be used to determine the spatial BC fractional dimension (Sect. 3.11) associated to these areas. One such exercise was carried out on images obtained by the TRACE satellite [28], yielding a mean BC dimension of 1.55. It has long been suggested that fractals appear in nature as a result of complex dynamics [29].

7.3.3 Lu-Hamilton SOC Flaring Model

In the case of our Sun, it is observations such as the ones just discussed that suggested that ideas such as self-organized criticality (see Sect. 1.3.2) might provide a possible dynamical paradigm to understand the Sun dynamics.

The power-law scaling of solar flare energies that we discussed earlier has been known for a long time. This knowledge was the main driver for a cellular automata model proposed by *Lu and Hamilton* (LH) in 1991 [30], in the wake of the self-organized criticality boom started by *P. Bak* a few years earlier [31]. The LU model was very simple. It considered a 3D lattice grid as a rough representation of the solar photosphere. At each point of that grid, labeled by the index i, a (scalar) magnetic field variable B_i was assigned representing its magnitude. A local gradient of this field was defined as:

$$\Delta B_i = B_i - \frac{1}{N} \sum_k B_{i+k}, \qquad (7.1)$$

where $N = 6$ was the number of nearest-neighbours considered, and k is the nearest-neighbour stencil (that corresponds to the nearest point to the left, to the right, above, below, in front and behind). This gradient was to be interpreted as a measure of the tension stored in the local curvature of the magnetic field. The dynamics of the automata were then set as follows. First, a critical threshold for this gradient, ΔB_c, was prescribed. At any cell in the lattice where $|\Delta B| > \Delta B_c$, a next-neighbour redistribution of magnetic field energy would follow. This was done according to[13]:

$$B_i \longrightarrow B_i - \frac{6}{7}\Delta B_c; \qquad B_{i+k} \longrightarrow B_{i+k} + \frac{1}{7}\Delta B_c, \ \forall k. \qquad (7.2)$$

The field at the nearby positions to i in the grid might then satisfy also the instability criterion, resulting in additional reconnection events, that would form an avalanche. The total energy released at site i would be given by:

$$E_i = \frac{6}{7}|\Delta B_c|^2; \qquad (7.3)$$

whilst the total energy released in the avalanche would be the sum of the energies released over all the sites affected by the relaxation.

The LH cellular automata exhibits power-law scalings for avalanche energies and their durations, among other things, behaving in way that is reminiscent to that of the running sandpile that we examined at length during each of the chapters of the first part of this book. The LU automaton provides in fact another popular realization of *self-organized criticality* (see Sect. 1.3.2). It certainly has all its basic ingredients: it is an open system that is slowly driven, has a local instability threshold and a fast relaxation rule. *Lu and Hamilton* argued that their cellular automata captured some of the bare essentials of the solar flare dynamics by invoking some well-known theories of that time. For instance, they suggested that the action of the convective motion coming to the photosphere from below would result in a sufficient random twisting of flux tubes as to guarantee a random, slow, local drive. They also justified the inclusion of a local instability threshold by enumerating several threshold-based instabilities that had been previously proposed by others as responsible for the triggering of local, explosive reconnections of the magnetic fields. In particular, they mention the proposal by *E.R. Parker* [32] of a minimum threshold value for the angle subtended by magnetic field vectors at opposite sites of any given current sheet that should be overcome for any explosive reconnection to take place [7], but other possibilities also exist [33].

---

[13]It is worth to point out that the redistribution rule is conservative in the field B, but not in the energy, that is proportional to B^2.

One thing that the LU cellular automata, as originally built, failed however to reproduce was a waiting-time pdf with a power-law scaling. Instead, the pdf of waiting-times between reconnections was exponential. At the time, this fact was used to argue against the validity of self-organized criticality as a relevant concept in the study of solar flare dynamics [34]. However, it was later shown with other SOC models that this is not the case. An exponential pdf for the waiting times is just a reflection of the random character of the drive, but the dynamics at the SOC state are pretty insensitive to the type of drive used as long as the system is not overdriven (see Sect. 1.3.2). It is in fact possible to change the pdf of the waiting-times to a power-law scaling. For instance, this can be done by using a correlated drive instead of random, without significantly changing the SOC dynamics [35]. Furthermore, if only avalanches of a sufficiently large size (as to be within the mesoscale are considered, the pdf of their waiting-times does follow power-law scalings in SOC systems. Given these facts, and considering the distance to the Sun and the finite resolution of the measuring equipment, the unavoidable effective thresholding present in the measured data might also determine the type of scaling seen [35, 36].

The impact of the *Lu and Hamilton* cellular automata on the solar physics community has been pretty remarkable.[14] It started a trend that encouraged to look at the Sun and the physics that govern it from a different perspective.[15] It has also lead to the import to solar and astrophysical problems of techniques that were used in many other complex systems and in widely different fields, as it is the case of the ones introduced in the first part of this book. We will illustrate the use of some of these techniques in the next section, although we will apply them to the analysis of CME data, instead of solar flares.

7.4   Case Study: Analysis of the SOHO-LASCO CME Database (1996–2016)

The dataset that we will examine in this section belongs to the SOHO-LASCO CME catalog [39]. It contains information about all coronal mass events (or CMEs) that could be manually identified[16] between January 1996 and May 2016 from within the data collected by the Large Angle and Spectrometric Coronagraph (LASCO) that is on board of the Solar and Heliospheric Observatory (SOHO) mission.

---

[14]The number of citations of the *Lu and Hamilton* 1992 paper are close to one thousand at the beginning of 2017.

[15]Several good reviews exist that discuss the impact of this paper and of the large number of contributions that followed [15, 19, 37, 38]. We strongly recommend any interested readers to use them as guides to dig deeper into this very interesting area of research.

[16]As stated by the maintainers of the database at NASA themselves, the list included in the catalog is necessarily incomplete because of the manual nature of the identification process.

There are several quantities of interest to us that are included in the LASCO catalog for each identified CME event, among many others. In particular, we will consider their **exit speed**, v_{exit}, as they move away from the Sun; their **ejected mass**, $m_{ejected}$; and their **ejected energy**, that is estimated from the other two using $E_{ejected} \simeq m_{ejected} v_{CME}^2 / 2$. Although there are three different speed values listed in the catalog, we will only analyze their *linear speed*, that is obtained by fitting a straight line to the height-time measurements. This linear value may not be the most physically representative, particularly for those CMEs that might experience a significant acceleration, but it provides an average speed that is convenient in order to group different CMEs together in a consistent manner. In regards to the ejected mass, there is a significant uncertainty in the numbers provided in the catalog. The estimation of the mass involves a number of unconfirmed assumptions, as mentioned in the catalog description, that advise to consider the values as more representative than realistic.

Before embarking in the analysis of the data, a last comment worth making is that a temporal gap exists in the dataset corresponding to the month of January of 1999. This gap might represent a problem for some of the analysis we will do to search for long-range time correlations and memory, such as by introducing artefacts. For that reason, we will restrict the analysis to the part of the CME dataset that extends from the period January 1999–May 2016 (see Fig. 7.4). There is a total of 17,051

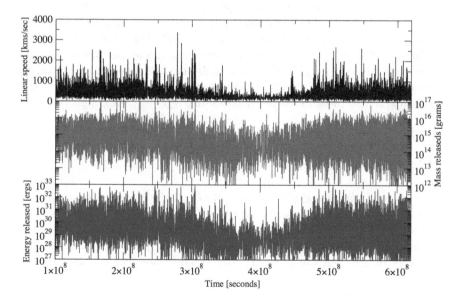

Fig. 7.4 Time series for the linear speed (above), ejected mass (middle) and ejected energy (below) of all CME events detected by the LASCO diagnostic between January 1999 and May 2016. The temporal axis measures the interval of time (in seconds) passed from the first CME event in the series (that happened on Jan. 1, 1996). The ejected energy is derived as $E_{ejected} \simeq m_{ejected} v_{CME}^2 / 2$ [For reference, 1 day = 8.64×10^4 s; 1 year = 3.15×10^7 s]

CMEs identified in the catalog within that period. The data spans a period of 17 years, which means that it will include roughly one 11-year cycle.[17] Thus, a very low frequency modulation is expected in the data (see Fig. 7.4, where it can be seen rather clearly), that must be considered when interpreting the results.

7.4.1 Waiting-Time Statistics

The first thing that we have calculated for the CME dataset is the pdf of the waiting-times in between successive events. The result, shown in Fig. 7.5 in both log-lin and log-log scalings, has been obtained using the constant-bin-size (CBS) method with 150 bins (see Sect. 2.4.1), and the constant-bin-content (CBC) method (see Sect. 2.4.2) with 25 events per bin. As mentioned in Sect. 2.3.4, the main feature of a Poisson process is an exponential pdf for its waiting-times, a direct consequence of its underlying random triggering process. However, the CME waiting-time pdf obtained here is not well fit by any exponential. Instead, it appears to have a power-law tail that scales as $p(w) \sim w^{-2.2}$, which is very reminiscent of what has been reported elsewhere for solar flares (see Sect. 7.3.2). The power-law is clearly defined over a little bit more than a decade, which is indeed suggestive of some complex dynamics.

The presence of a power-law, and the physical consequences regarding the role of complexity that it would lead to, must again be taken with caution. As happened in the case of solar flares (see discussion in Sect. 7.3.2), Fig. 7.5 also shows that a combination of (at least) two different exponentials, with average triggering rates that are respectively of the order of less than half a day and about 2 days, might also provide a reasonable fit. If this second interpretation was the correct one, it would suggest that at least two different mechanisms for triggering CMEs might coexist.[18]

7.4.2 Linear Speed Analysis

We turn now our attention to the analysis of the CME exit speed time series. Its pdf, as obtained by the methods described in Chap. 2, is shown in Fig. 7.6. For CBS, we use 150 bins; for CBC, we define a bin as one that includes 25 distinct events. At first inspection, it would appear that a power-law scaling $p(s) \sim v_{exit}^{-3.87}$ could exist for a range of exit speeds in between 600–2000 km/s. However, the range for this

---

[17]A modulation on the timescale of the length of the record can be indeed appreciated in the amplitudes shown in Fig. 7.4.

[18]This discussion parallels the one we already described for solar flares in Sect. 7.3.2, with some authors analyzing the solar flare waiting-time pdfs in terms of time-dependent Poisson processes. In fact, some authors have argued that the CMEs waiting-time data should be modelled in this manner as well [27, 40], instead of by means of power-law scalings.

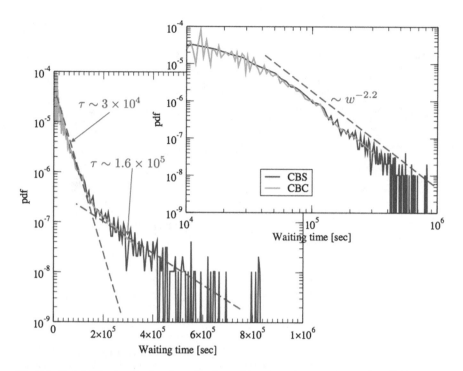

Fig. 7.5 Probability density functions for the waiting-times between successive CMEs in the LASCO catalog shown in log-linear (below) and log-log (above). The pdfs have been obtained using both the constant-bin-size (in blue) method described in Sect. 2.4.1 and the constant-bin-content method discussed in Sect. 2.4.2. The time is shown in seconds, as provided in the catalog [For reference, 1 day = 8.64×10^4 s; 1 year = 3.15×10^7 s]

scaling is too short and the exponent in that range too large to suggest any kind of complex behaviour.[19] Furthermore, the log-normal distribution that was discussed in Chap. 2 fits the exit speed pdf quite well (shown in the figure using a maroon dashed line). This fact might be suggestive of the existence of an underlying process of multiplicative nature that leads to the CME acceleration,[20] in principle devoid of any meaningful long-term memory (see discussion in Sect. 2.3.2 about the significance of log-normal distributions), as responsible for the explosive acceleration of the CME.

Although the CME exit speed pdf that we have obtained does not seem to point to any complex behaviour, we have other tools at our disposal. Therefore, we have also tested the speed data time series directly for any evidence of scale-free,

---

[19]The tail exponent should be smaller than 2 for a mean exit speed to be undefined, given that the exit speed is a positive definite quantity.

[20]In fact, some proposals in the literature suggest that cascades of reconnection processes could lead to solar flares and, possibly, to CMEs as well [41].

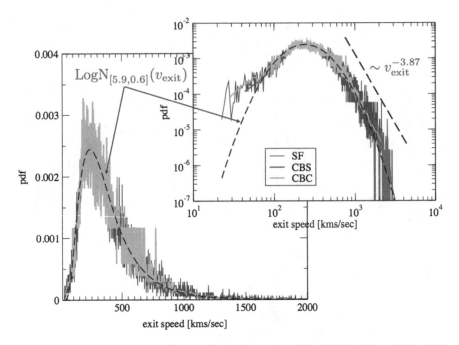

Fig. 7.6 Pdfs for CME exit velocities as obtained by means of the CBS (red; Sect. 2.4.1), CBC (green; Sect. 2.4.2) and SF (blue; Sect. 2.4.3) methods. In maroon dashed-line, the best lognormal fit to the data is also shown, corresponding to $\text{LogN}_{[5.9,0.6]}(x)$ (see Eq. 2.45)

long-term memory. The results of applying the R/S analysis (see Sect. 4.4.4) to the dataset results in the rescaled range shown in the upper frame of Fig. 7.7.[21] The related instantaneous Hurst exponent (see Eq. 6.12) is shown in the lower frame as a function of the elapsed time. This instantaneous Hurst exponent is very useful to determine scale-invariant regions, that should appear as flattish, horizontal regions in $H_{R/S}(\tau)$. As will be remembered, $H \sim 0.5$ is expected for any uncorrelated series with a finite variance (see Sect. 4.4.4). Instead, an extended range is found here, between a few days and a few years, where the exit velocity series exhibits strong positive correlations, since $H \sim 0.75$. Accordingly, exit speeds values above average tend to cluster together more often than not over that range of times, as also do exit speed values below average. Since this behaviour extends to timescales of the order

---

[21]It is worth mentioning that the R/S analysis must be done here a bit differently from what was explained in Sect. 4.4.4, since the CME data is not sampled at a uniform rate. Therefore, the R/S algorithm must be slightly modified to take this into account. It is not difficult to do. One simply needs to define blocks differently. Instead of by the number of data points they contain, as we did in Sect. 4.4.4, one must define them by their duration. In addition, all the sums that were previously calculated over blocks (to compute, say, means, variances or ranges) become now sums over all those data points that take place at times that are contained in each particular block. The rest of the analysis and its interpretation, remains unchanged.

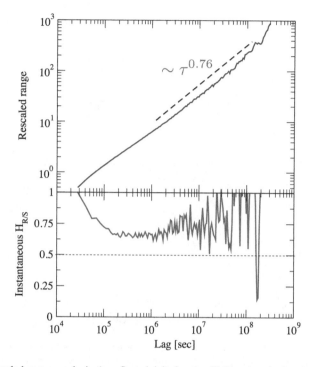

Fig. 7.7 Rescaled range analysis (see Sect. 4.4.4) for the CME exit velocity time series. The rescaled range (R/S) is shown in the upper frame as a function of the time lag (in seconds) measured from the first CME in the examined record; in the lower frame, the instantaneous scaling exponent $H_{[R/S]}(\tau)$ (see Eq. 6.12) is also shown [For reference, 1 day = 8.64×10^4 s; 1 year = 3.15×10^7 s]

of years, much longer than the typical duration of a CME, that is a symptom of complex behaviour. It is also worth mentioning that a clear flattening in the rescaled range is observed for $\tau \sim 7$ years (i.e. $\tau \sim 2 \times 10^8$ s). This flattening is due to the low-frequency modulation present in the data, most probably associated to the 11-year cycle. Due to the shortness of the available time series (roughly 17 years), the R/S analysis cannot say whether the detected persistence is maintained for even longer times or not.

7.4.3 Ejected Mass Analysis

Next, we turn our attention to the analysis of the CME ejected mass time series. Its pdf is shown in Fig. 7.8, this time calculated using only the CBC method (using 75 distinct events to define each bin) introduced in Chap. 2. Interestingly, a mesorange can be clearly identified in the plot. It spans the interval 5×10^{13}–2×10^{16} g, where the power-law scaling $\sim m^{-(1.2-1.4)}$ is apparent. The exponent value is sufficiently

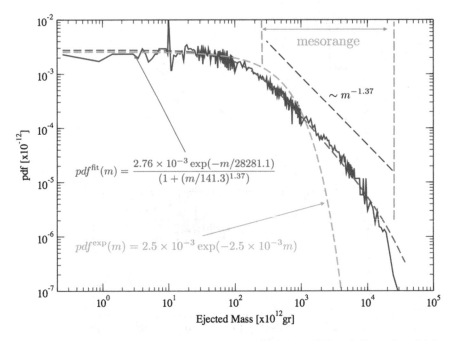

Fig. 7.8 Pdfs for CME ejected masses estimated by means of the CBC method (see Sect. 2.4.2). Two best-fits obtained by chi-square minimization techniques (see Sect. 2.5.3) are also included. One against Eq. 1.8 (shown in red), and a second using an exponential law (in green). The first fit is clearly much better (its chi-square value is more than an order of magnitude smaller), yielding a mesorange extending between $(0.1 - 30) \times 10^{15}$ g

small (less than 2, as it should be for a positive quantity) as to point not only to scale-invariance, but to a lack of characteristic scales as well. Or, in other words, it tells that the average ejected mass is not well defined in the sense that it would scale with the system (i.e., the Sun) size (see discussion in Sect. 1.3.2). This is an interesting finding, since it is rather different from the exponential pdf reported by other authors for CME data gathered by the SOLWIND satellite [42]. An exponential pdf would point to a (presumably random) Poisson process. Instead, the pdf obtained here is similar to the ones often reported for solar flares, that prompted several authors to propose the relevance of concepts such as self-organized criticality to the problem of solar flaring [14, 30].

We have also tested the mass time series for any evidence of long-term memory. The rescaled range that results from applying the R/S method to the data is shown in Fig. 7.9. A mesorange clearly exists in the data, extending again from a few days to a few years, over which $H_{[R/S]} \sim 0.65$. This value is again suggestive of the existence of long-term positive correlations. As in the case of the exit speeds, these correlations imply that larger[smaller]-than-average mass ejections are more frequently than not clustered together over time. Again, the lack of clear flattish regions in the R/S also points to a lack of strong periodicities. There is a very

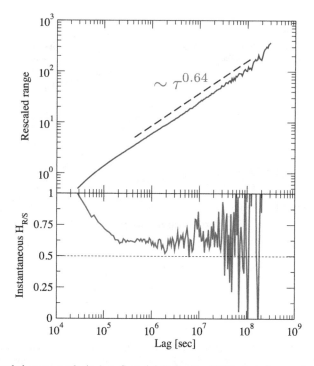

Fig. 7.9 Rescaled range analysis (see Sect. 4.4.4) for the CME ejected mass time series. The rescaled range (R/S) is shown in the upper frame as a function of the time lag (in seconds) measured from the first CME in the examined record; in the lower frame, the instantaneous scaling exponent $H_{[R/S]}(\tau)$ (see Eq. 6.12) is also shown [For reference, 1 day = 8.64×10^4 s; 1 year = 3.15×10^7 s]

narrow decrease of H around 26 days (i.e., $\tau \sim 2.2 \times 10^6$ s) that one might be tempted to associate to the rotation of the Sun around its own axis. Similarly, there is also a decrease at around 7 years (i.e. $\tau \sim 2 \times 10^8$ s), a consequence of the low-frequency modulation of the data that could be related to the 11-year solar cycle. None of these periodicities seems however to be sufficiently strong as to affect the rescale range significantly (see discussion in Sect. 4.4.4). Also, as in the case of the exit velocities, the shortness of the record prevents us from analyzing whether the observed persistence may extend for a period longer than a few years.

Since both the ejected mass pdf and its R/S analysis have revealed mesoranges across which scale-invariance appears to exist, we have gone ahead an tested the level of monofractality of the CME mass time series. We have used the multifractal analysis presented in Chap. 4.[22] The family of multifractal exponents $\xi(q)$ (see

---

[22] Again, due to the fact that the CME data is not uniformly distributed in time, the prescriptions given in Chap. 4 must be modified. Analogously to what we did in the case of the R/S analysis, a block must be defined again by its temporal duration, not by the number of data points it contains.

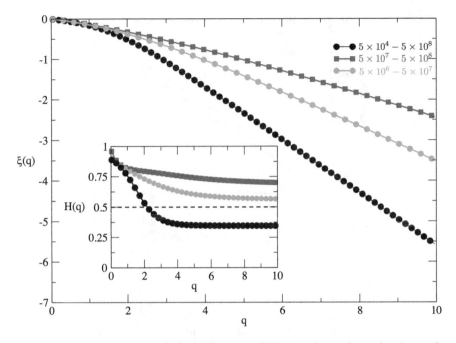

Fig. 7.10 Multifractal exponents $\xi(q)$ obtained for the CME ejected mass time series. As seen in the inset, the generalized Hurst exponent $H(q) = 1 + \xi(q)/q$ has different behaviours depending on the timescales of the blocks included in the analysis. When boxes larger than 5×10^6 s are used, a scale-invariant behaviour seems to appear for $q > 2$, with a Hurst exponent value close to $H \sim 0.6$

Eq. 3.97) that is obtained for the CME ejected mass time series when including blocks of all sizes is shown (in black) in Fig. 7.10. As discussed in Sect. 3.3.7, one would theoretically expect that the generalized Hurst exponent (Eq. 3.98) satisfies $H(q) = 1 + \xi(q)/q \sim H$, $\forall q > -1$ if the time series is a monofractal with self-similarity exponent H. Although $H(q)$ is never really constant for any finite time series, the generalized Hurst exponent for the CME masses varies over a much wider range than what is obtained for fBm[23] when blocks of all possible sizes are included in the calculation. However, if we limit the analysis to the range of block sizes that lies within the mesorange previously identified in the R/S analysis (namely, for durations between 5×10^6–5×10^8 s or, equivalently, for blocks larger than 2

In addition, all sums and operations should be computed including only those data points whose time of occurrence lies within the block of interest.

[23]For instance, one could compare Fig. 7.10 with Fig. 3.30, that shows the family multifractal exponents and the related generalized Hurst exponent for a numerical realization of fBm with self-similarity exponent $H = 0.8$.

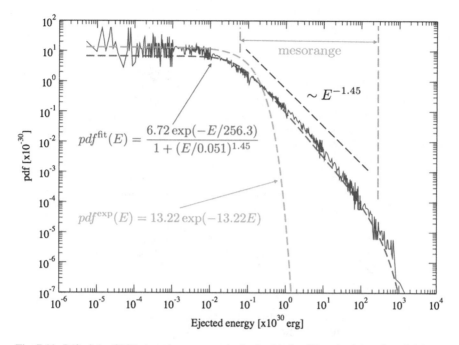

Fig. 7.11 Pdf of the CME ejected energy as obtained with the SF method (see Sect. 2.4.3). In addition, best-fits to an exponential law (in green) and to Eq. 1.8 (in red), obtained with the minimum chi-square method (see Sect. 2.5.3), are included. The second fit is much better (its chi-square value is two orders of magnitude smaller), providing an estimate for the self-similar mesorange within the interval $(0.05 - 300) \times 10^{30}$ erg

months), the spread of generalized Hurst exponent becomes much narrower. In fact, $H(q)$ is confined within the range $H \sim 0.6 - 0.7$, as shown in the same figure.[24]

7.4.4 Ejected Energy Analysis

We complete the analysis of the CME data in the LASCO catalog by examining the time series of the CME ejected energies. The ejected energy is a derived quantity, calculated using $E_{ejected} = (1/2)m_{ejected} \times v_{ejected}^2$. The pdf of the energies, obtained this time by applying the survival function method (see Sect. 2.4.3), is shown in Fig. 7.11. The result is rather remarkable, since it exhibits a scale-invariant mesorange that extends for almost four decades, approximately over $(0.05 - 300) \times 10^{30}$ erg. The scaling over this mesorange is $p(E) \sim E^{-1.45}$, again with an exponent

---

[24]In a sense, these results suggest that R/S is often more reliable than multifractal analysis to identify the presence of a mesorange in real data.

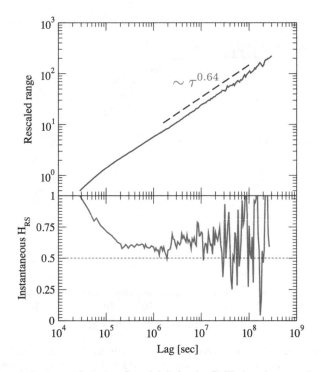

Fig. 7.12 Rescaled range analysis (see Sect. 4.4.4) for the CME ejected energy time series. The rescaled range (R/S) is shown in the upper frame as a function of the time lag in years, as measured from the first CME in the examined record; in the lower frame, the instantaneous scaling exponent $H_{[R/S]}(\tau)$ (see Eq. 6.12) is also shown

smaller than two, as needed to impede a well-defined mean energy. This scaling behaviour suggests once more that complex dynamics might be at play here (see Sect. 1.3.2), leading to the observed lack of characteristic scales.[25]

We proceed now to look for evidence of long-term correlations by carrying out the R/S analysis of the ejected energy time series. The obtained rescaled range is shown in the upper frame of Fig. 7.12, with the instantaneous $H_{[R/S]}$ exponent (Eq. 6.12) in the lower frame. Again, an extended range is found—for roughly the same timescales as those found for the ejected mass time record—across which a power-law scaling with a Hurst exponent significantly larger than 0.5. This is a consequence of the persistent correlations that were found in both the speed and time series, that contribute to the establishment of similar positive correlations between the successive CME energy ejections. The value of the Hurst exponent is however a tad smaller than for the other two series, with $H \sim 0.62 - 0.64$. This might be

---

[25]This result, although similar to that found for solar flares (see Sect. 7.3.2), is however at odds with other reports that have found CME energy pdfs much closer to exponentials [43].

a consequence of the fact that the energy is a derived quantity, as explained earlier, which introduces decorrelations and reduces the overall value of H.

7.5   Conclusions

In this chapter we have discussed the possibility of our Sun behaving a complex system. In particular, we have discussed how complex features may be detected in dynamical processes such as those related to solar flaring, or to the ejection of CMEs. The main experimental piece of evidence found in favour of complex behaviour is probably the enormous power-law scaling that is seen in the frequency distribution of flares, spanning more than seven orders of magnitude in energy. Also, the value of the scaling exponent itself, that suggest that concepts such as criticality might be relevant to the understanding of the dominant dynamics (see discussion in Sect. 1.3.2). From the analysis of the CME data gathered from the SOHO satellite, we have found that a similar scaling is also exhibited by the pdf of the CME ejected energies, that extends for more than four orders of magnitude with a critical scaling exponent. In addition, the R/S analysis has showed clear evidence of long-term, persistent correlations in the CME records, confirming what other authors have reported in the literature for solar flares. These correlations appear to extend from a few days to at least few years, although they might extend for much longer.[26] These timescales are much longer scales than the typical duration of a solar flare or a CME, that are usually of the order of tens of minutes. The presence of these long-term correlations thus suggests that long-term memory is somehow stored inside the solar corona, perhaps in the intricate magnetic field patterns that result from the reconnection events that lead to flaring and/or CME events. Finally, the analysis carried out has also illustrated the importance of having a deep knowledge of the underlying processes and the physics of the system. Otherwise, it becomes very difficult to make any sense of the results provided by many of the methods use to characterize complex behaviour.

References

1. Mullan, D.J.: Physics of the Sun: A First Course. Chapman and Hall, New York (2009)
2. Priest, E.R.: Solar Magnetohydrodynamics. D. Reidel Publishing, London (1982); Priest, E.R.: Magnetodydrodynamics of the Sun. Cambridge University Press, Cambridge (2014)
3. Golub, L., Pasachoff, J.M.: The Solar Corona. Cambridge University Press, Cambridge (2009)
4. Zirin, H.: Astrophysics of the Sun. Cambridge University Press, Cambridge (1988)
5. Spiegel, E.A., Zahn, J.P.: The Solar Tachochline. Astron. Astrophys. 265, 106 (1992)

---

[26]This is clearly impossible to say without examining much longer time series.

6. Hughes, D.W., Rosner, R., Weiss, N.O.: The Solar Tachocline. Cambridge University Press, Cambridge (2007)
7. Parker, E.N.: Cosmic Magnetic Fields. Clarendon Press, Oxford (1979)
8. Moffat, H.K.: Magnetic Field Generation in Electrically Conducting Fluids. Cambridge University Press, Cambridge (1978)
9. Charbonneau, P., Christensen-Dalsgaard, J., Henning, R., Larsen, R.M., Schou, J., Thompson, M.J., Tomczyk, S.: Helioseismic Constraints on the Structure of the Solar Tachocline. Astrophys. J. 527, 445 (1999)
10. Diamond, P.H.: Zonal Flows in Plasmas - A Review. Plasma Phys. Controll. Fusion 47, R35 (2005)
11. Shimada, M., et al.: Progress in the ITER Physics Basis - Chapter 1: Overview and Summary. Nucl. Fusion 47, S1 (2007)
12. Gough, D.O., McIntyre, M.E.: The Inevitability of a Magnetic Field in the Sun's Radiative Interior. Nature 394, 755 (1998)
13. Cowley, S.C., Wilson, H., Hurricane, O., Fong, G.: Explosive Instabilities: From Solar Flares to Edge Localized Modes in Tokamaks. Plasma Phys. Controll. Fusion 45, A31 (2003)
14. Aschwanden, M.J.: Self-Organized Criticality in Astrophysics. Springer, New York (2014)
15. Aschwanden, M.J.: 25 Years of Self-Organized Criticality: Solar and Astrophysics. Space Sci. Rev. 198, 47 (2017)
16. Crosby, N.B., Aschwanden, M.J., Dennis, B.R.: Frequency Distributions and Correlations of Solar X-ray Flare Parameters. Sol. Phys. 143, 275 (1993)
17. Shimizu, T.: Energetics and Occurrence Rate of Active-Region Transient Brightenings and Implications for the Heating of the Active-Region Corona. Proc. Astron. Soc. Jpn. 47, 251 (1995)
18. Krucker, S., Benz, A.O.: Energy Distribution of Heating Processes in the Quiet Solar Corona. Astrophys. J. 501, L213 (1998)
19. Aschwanden, M.J., Tarbell, T., Nightingale, R., Schrijver, C.J., Title, A., Kankelborg, C.C., Martens, P.C.H., Warren, H.P.: Time Variability of the Quiet Sun Observed with TRACE: II. Physical Parameters, Temperature Evolution, and Energetics of EUV Nanoflares. Astrophys. J. 535, 1047 (2000)
20. Parnell, C.E., Jupp, P.E.: Statistical Analysis of the Energy Distribution of Nanoflares in the Quiet Sun. Astrophys. J. 529, 554 (2000)
21. Wheatland, M.S.: Flare Frequency-Size Distributions for Individual Active Regions. Astrophys. J. 532, 1209 (2000)
22. Grigolini, P., Leddon, D., Scafetta, N.: Diffusion Entropy and Waiting Time Statistics of Hard X-ray Solar Flares. Phys. Rev. E 65, 046203 (2002)
23. Wheatland, M.S., Litvinenko, Y.E.: Understanding Solar Flare Waiting-Time Distributions. Sol. Phys. 211, 255 (2002)
24. Aschwanden, M.J., McTiernan, J.M.: Reconciliation of Waiting Time Statistics of Solar Flares Observed in Hard X-rays. Astrophys. J. 717, 683 (2010)
25. Wheatland, M.S., Sturrock, P.A., McTiernan, J.M.: The Waiting-Time Distribution of Solar Flare Hard X-ray Bursts. Astrophys. J. 509, 448 (1998)
26. Lepreti, F., Carbone, V., Veltri, P.: Solar Flare Waiting Time Distribution: Varying-Rate Poisson or Lévy Function? Astrophys. J. 555, L133 (2001)
27. Telloni, D., Carbone, V., Lepreti, F., Antonucci, E.: Stochasticity and Persitence of Solar Coronal Mass Ejections. Astrophysi. J. 781, 1 (2014)
28. Aschwanden, M.J., Aschwanden, P.D.: Solar Flare Geometries: I. The Area Fractal Dimension. Astrophys. J. 574, 530 (2008); Solar Flare Geometries: II. The Volume Fractal Dimension. Astrophys. J. 574, 544 (2008)
29. Bak, P.: How Nature Works: The Science of Self-Organized Criticality. Springer, New York (1996)
30. Lu, E.T., Hamilton, R.J.: Avalanches and the Distribution of Solar Flares. Astrophys. J. 380, L89 (1991)

31. Bak, P., Tang, C., Wiesenfeld, K.: Self-Organized Criticality: An Explanation of the 1/f Noise. Phys. Rev. Lett. 59, 381 (1987)
32. Parker, E.N.: Nanoflares and the Solar X-ray Corona. Astrophys. J. 330, 474 (1988)
33. Priest, E.R., Forbes, T.: Magnetic Reconnection. Cambridge University Press, Cambridge (2000)
34. Boffetta, G., Carbone, V., Guiliani, P., Veltri, P., Vulpiani, A.: Power Laws in Solar Flares: Self-Organized Criticality or Turbulence? Phys. Rev. Lett. 83, 4662 (1999)
35. Sanchez, R., Newman, D.E., Carreras, B.A.: Waiting Time Statistics of Self-Organized-Criticality Systems. Phys. Rev. Lett. 88, 068302 (2002)
36. Paczuski, M., Boettcher, S., Baiesi, M.: Interoccurrence Times in the Bak-Tang-Wiesenfeld Sandpile Model: A Comparison with the Observed Statistics of Solar Flares. Phys. Rev. Lett. 95, 181102 (2005)
37. Charbonneau, P., McIntosh, S.W., Liu, H.-L., Bogdan, T.J.: Avalanche Models for Solar Flares. Sol. Phys. 203, 321 (2001)
38. Crosby, N.B.: Frequency Distributions: From the Sun to the Earth. Nonlinear Process. Geophys. 18, 791 (2011)
39. Gopalswamy, N., Yashiro, S., Michalek, G., Stenborg, G., Vourlidas, A., Freeland, S., Howard, R.: The SOHO/LASCO CME Catalog. Earth Moon Planet. 104, 295 (2009)
40. Wheatland, M.S.: The Coronal Mass Ejection Waiting-Time Distribution. Sol. Phys. 214, 361 (2003)
41. Hughes, D., Paczuski, M., Dendy, R.O., Helander, P., McClements, K.G.: Solar Flares as Cascades of Reconnecting Magnetic Loops. Phys. Rev. Lett. 90, 131101 (2003)
42. Jackson, B.V., Howard, R.A.: A CME Mass Distribution Derived from SOLWIND Coronograph Observations. Sol. Phys. 148, 359 (1993)
43. Jackson, B.V.: Magnetic Storms. Geophysical Monograph, 98, p. 59. American Geophysical Union, Washington (1997)

Chapter 8
Planetary Plasmas: Complex Dynamics in the Magnetosphere of the Earth

8.1 Introduction

It was around 1600 when William Gilbert, then physician to Queen Elisabeth I of England, first proposed that the Earth acted as a permanent magnet with the form of a sphere. Since then, the study of the structure of the **magnetic field of the Earth**, its origin, how it is sustained and the ways it interacts with the Sun (mainly, by its reaction to the solar wind; see Chap. 7) have captivated the interest of many researchers. Not just because of scientific curiosity, but also because its mere existence plays a crucial role in our own. Indeed, life on Earth would have been very different (if it had appeared at all) in the absence of a magnetic field that could shield it from the solar wind (SW) and other highly-energetic cosmic radiations.

The **aurora** has also fascinated humans for centuries. This natural display of light and colour that is often visible close to the Earth poles, but sometimes also at much lower latitudes, is mentioned in works from as far back in time as ancient Greece and Rome or the Middle Ages, often related to religious sentiments. The aurora was first associated to the magnetic field of the Earth at the beginning of the eighteenth century, probably by Edmund Halley [1]. Its intimate relationship with solar phenomena was not realized until the 1850s, though. The relation was established after noticing the coincidence between periods of strong solar activity and the appearance of powerful auroras. However, the physical reasons behind this connection were not clear at the time.[1] It was not until the end of the 19th and the beginning of the twentieth centuries that the idea of a **magnetic storm** was clearly formulated for the first time. It was then theorized that a "flying cloud of charged particles" was emitted from sunspots after a solar flare had taken place, leading to a sudden increase of the geomagnetic field and to powerful auroras in the sky a few

---

[1] In fact, even the famous Lord Kelvin asserted in an important speech to the Royal Society that such correlation was just a lucky coincidence [2].

© Springer Science+Business Media B.V. 2018
R. Sánchez, D. Newman, *A Primer on Complex Systems*,
Lecture Notes in Physics 943, https://doi.org/10.1007/978-94-024-1229-1_8

days later [3]. Magnetic storms could last from a few hours to a few days. Although it was at first thought that these particles were mainly ions, it was soon realized that only a quasi-neutral plasma could avoid its dispersion (a consequence of its mutual Coulombian repulsion) during their trip to Earth [4].[2]

In this chapter we will provide a brief introduction to the magnetosphere of the Earth and to some of its main dynamics, including how it is shaped by its interaction with the solar wind. As it has been the case for all chapters in the second part of this book, the discussion does not intend to be comprehensive,[3] but aims instead at providing the fundamental materials needed for the understanding of the conditions that may lead to complex dynamics. These conditions stem, on the one hand, from the interaction with the solar wind[4] and, on the other hand, from the turbulent motions of conducting media that take place within the Earth core and that sustain the geomagnetic field. As a result, the magnetosphere of the Earth can be considered as an open, driven system—both by the solar wind and by the turbulent Earth core— with a large number of degrees of freedom that interact nonlinearly. Therefore, it contains many of the ingredients required to develop complex dynamics (see discussion in Chap. 1). Those readers already familiar with the inner workings of the magnetosphere of the Earth may decide to jump directly to Sect. 8.3, where we will discuss several possible complex behaviours that might be happening in this system, focusing mainly on the phenomena known as **magnetic storms** and **substorms**. We will conclude the chapter, as always, by analyzing various datasets of interest in Sect. 8.4 by means of some of the tools presented in the first part of this book.

8.2   The Magnetosphere of the Earth

The magnetic field that exists around our planet and that we referred to under the general name of **magnetosphere** is formed by the addition of two main components: (1) the magnetic field generated by the turbulent motion of the conducting fluid metals present at the Earth core, also known as the **geomagnetic field**, and (2) the magnetic field generated in the Sun by the processes discussed in Chap. 7 and that

---

[2]In fact, many people identifies this suggestion as the start of modern space plasma physics [5].

[3]Some great texts do exist that provide a much more thorough description of this area of research, including both the geomagnetic field and the magnetosphere [5–11]. We are particularly fond of [5] and [8]. We have followed [11] closely in certain areas such as the storm/substorm dynamics, that are covered in it with great clarity. Those readers interested in a more in-depth look at any of these topics, either from a more theoretical or observational perspective, are strongly encouraged to browse through them.

[4]Interestingly, the solar wind is itself a result of the complex dynamics that could be taking place in the Sun (see Sect. 7.3). This fact suggests the interesting possibility of studying a system, the magnetosphere of the Earth, that might self-generate its own complex behaviour while, in addition, being driven by an external source that might also possess its own complex features. In Sect. 8.4 we will analyze several datasets, using the tools presented in the first part of this book, to try to discern which one of these components, if any, might be dominant.

is transported to us by the solar wind. This magnetic field is usually referred to as the **interplanetary magnetic field** or, simply, the **IMF**.

8.2.1 The Geomagnetic Field

The magnetic field of the Earth is believed to be generated and sustained by the motion of the electrically conductive fluid material that exists at its core [10]. The core is the region of the Earth extending radially outwards from its centre to approximately 3000–3200 km.[5] With a temperature of about 6000 K, the iron alloys that form the Earth core are in a fluid state. The process by which a magnetic field is generated from fluid (or plasma) motion is known as a **dynamo**. It all starts with the motion of parcels of conductive material at the core, mainly driven by rotation and buoyancy, in the presence of a seed magnetic field.[6] When a conductor moves in a magnetic field it drags with it the surrounding magnetic field, resulting in an opposing (Lorenz) force to its motion [12]. In this process, the field can get twisted, tensioned and reinforced; the process is more efficient, the more complicated the motion pattern becomes. In this way, given a seed magnetic field, the dynamo process just sketched could reinforce and sustain a magnetic field for as long as the mass motion continues, assuming that it becomes sufficiently turbulent. The specific details of this process are however very complicated, and intense theoretical and modelling work is still carried out to try to understand all its intricacies [9, 13].

As a result of rotation and its associated Coriolis forces, mass motion at the core has a dominant component in the plane perpendicular to the rotation axis of Earth.[7] For this reason, the geomagnetic field becomes endowed with a strong dipole-like structure similar to the one illustrated in Fig. 8.1, in which a few of its magnetic surfaces[8] are shown. Magnetic field lines are thus basically parallel to the surface of the Earth at the Equator, enter into the crust in the neighbourhood of the North pole, and exit around the South pole.[9] Also, magnetic surfaces will have a larger

---

[5] As a reference, the radius of the Earth is about 6400 km.

[6] This seed could have been perhaps provided by any remnant magnetic field that might have got trapped in the process of formation of the Earth.

[7] Convective, rotating structures are formed that are strongly elongated in the direction of rotation of Earth. They are often known as **Taylor columns**.

[8] **Magnetic surfaces** are surfaces to which the magnetic field is tangent everywhere. **Magnetic field lines**, on the other hand, are those curves to which the field is always tangent. Magnetic field lines are contained within magnetic surfaces, assuming that the latter can be defined. If not, magnetic field lines may ergodically fill a volume, or escape to infinity (see Chap. 6 for a more detailed discussion).

[9] This is how the geomagnetic field is directed nowadays. However, the geomagnetic field is known to reverse its orientation intermittently, although over a timescale of hundred of thousands of years, as has been found in ferromagnetic sediments from the ocean floor. In fact, the geomagnetic field is far from quiet, even at smaller timescales. For instance, scientists have found that the North magnetic pole is also drifting northwards at a rate of a few tens of miles per year [14].

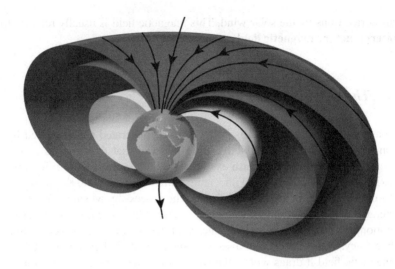

Fig. 8.1 Sketch illustrating the dipole-like geomagnetic field in the close neighbourhood of Earth. A few magnetic surfaces (in yellow, orange and red) and some magnetic field lines (in black) are also shown, including the magnetic axis [Credits: courtesy of Estefanía Cuevas]

poloidal cross section as the point of entrance[exit] of the magnetic field lines they contain approaches the North[South] pole. Eventually, magnetic surfaces degenerate into a single curve known as the *magnetic axis* that crosses the Earth from North to South, as shown in the figure. The magnetic axis does not coincide exactly with the rotation axis of the Earth, though, being tilted about ten degrees from it. The strength of the geomagnetic field is small, being about 20–60 μT at the surface.[10] As we mentioned, its direction is mostly horizontal (i.e., parallel to the surface) at the Equator, and vertical (i.e., perpendicular to the surface) at the poles.

8.2.2   Structure of the Magnetosphere of the Earth

The dipole shape of the geomagnetic field is not a bad approximation close to the surface of the Earth. However, at a distance of just a few times the radius of the Earth, the actual magnetic field becomes rather different due to the ever-present contribution of the interplanetary magnetic field (IMF). That is, of the solar magnetic field that is carried to us by the solar wind (SW). This solar stream, that brings to us both a quasi-neutral energetic plasma and the magnetic field trapped with it, is continuously present, although its strength may change considerably over time, particularly after a solar flare or a CME reaches the Earth.

---

[10]The geomagnetic field is still sufficiently strong as to have measurable effects, as sailors from all ages have learnt to appreciate while using compasses for navigation.

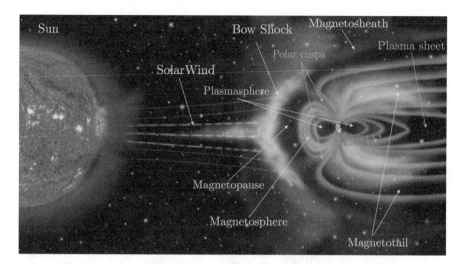

Fig. 8.2 Main regions of the magnetosphere of the Earth [Credits: public domain image created by NASA; labelling has been added by the authors]

The constant presence of the solar IMF strongly modifies the structure of the magnetic field around the Earth. The general layout of the magnetosphere is illustrated in Fig. 8.2. Several different parts can be distinguished in it. The **bow shock** is the outermost layer of the magnetosphere. At the bow, a shock wave forms due to the fact that the SW speed (of the order of 400 km/s) suddenly drops from supersonic to subsonic as a result of the encounter with the geomagnetic field. The bow shock is located at about 90,000 km from the planet, on its dayside, and its thickness is about 100 km. At the bow shock, the solar wind is not only slowed down, but compressed and heated as well. The bow shock is stationary and the processes taking place inside are mostly collisionless. That is, they occur in such a low-density medium that collisions among SW particles become negligible, with most of the energy and momentum transfer happening via wave-particle interactions.[11]

The first region downstream from the bow shock, mostly occupied by the "shocked" solar wind, is known as the **magnetosheath**. The magnetic field becomes very irregular in this region, that contains a very turbulent plasma of a density that is much lower than in the bow shock. The magnetosheath is essentially the transition region between the inner magnetosphere and the solar wind. As such, it controls the influx of particles from the solar wind that make it into the magnetosphere.

The **magnetopause** is the first region within the magnetosphere. It is an abrupt region where the pressure of the geomagnetic field manages to balance the kinetic

---

[11]Some SW particles are reflected back from the shock, populating a region known as the **foreshock**. The foreshock is extremely active and turbulent, with many waves and instabilities being generated there that often propagate downstream along the magnetosheath.

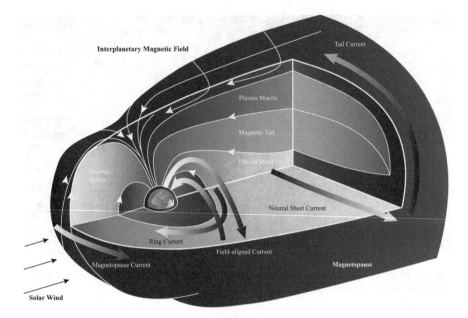

Fig. 8.3 Layout of the magnetosphere of the Earth that shows the main currents that flow in the different regions, both parallel and perpendicular to the magnetic field [Credits: courtesy of Estefanía Cuevas]

pressure of the solar wind.[12] It effectively sets the boundary between solar wind and magnetosphere, determined dynamically by this balance. As a result, the position and shape of the magnetopause changes as the pressure of the incoming solar wind changes over time.[13] This boundary does not insulate the inner magnetosphere completely, though. When large blobs of solar plasma travel along the magnetopause they can give rise to instabilities that lead to the breaking and reconnection of local magnetic field lines, opening paths through which SW particles can traverse into the inner magnetosphere.

The region in between the magnetopause and the Earth is known as the **magnetosphere**.[14] In the part of the Earth that faces the solar wind (i.e., the dayside), it retains a structure similar to that of (a compressed) dipole-like field. On

---

[12]These two contributions are the dominant ones since the magnetic and thermal pressures of the solar wind are negligible compared with its kinetic pressure [11].

[13]The width of the magnetopause is not that small, extending for a few thousand kilometers. The reason for this width is that charged particles within the solar wind do penetrate to a certain depth before being deflected black by Lorenz forces. Since the Lorenz force has a different sign for positive and negative charge, this deflection gives rise to a current in the magnetopause that is known as the **Chapman-Ferraro current** (see Fig. 8.3).

[14]Although many authors, as we have done, also use this name to refer to the whole magnetic field around the Earth.

the downstream part, however, a **magnetotail** develops, where the magnetosphere extends far beyond the Earth (the length of the magnetotail can easily exceed a few millions of km). Further downstream the tail breaks into two lobes (referred to as the **Northern** and **Southern lobes**). Magnetic field lines in the upper lobe point towards the Earth, whilst lines in the lower lobe point away from it. They are separated by a **plasma sheet**, where the magnetic field is much weaker,[15] that collects all the solar plasma that comes down the magnetotail, simultaneously drifting in the direction perpendicular to it. As a result, the plasma sheet builds up a relatively large density. Maxwell equations also predict that a current must flow within the plasma sheet, that is known as the **neutral sheet current**. This current is perpendicular to the magnetic field in the magnetotail. The resulting current loop is closed by two other currents, known as the **tail currents**, that are also shown in Fig. 8.3, one around the Northern lobe, the other around the Southern one.

Another important part of the magnetosphere are the two so-called **polar cusps** that form on the dayside part of the magnetosphere, one on the Northern hemisphere, the other on the Southern one (see Figs. 8.2 and 8.3). The polar cusps correspond to the two singularities that form in the region where the closed field lines on the dayside separate from open field lines that are swept to the magnetotail in the nightside. Since the magnetic field vanishes in the cusps, particles and plasma can penetrate freely into the magnetosphere through them, crossing the lower atmosphere and reaching the surface of the Earth. The cusps are however not connected with the dipole axis, but at lower latitudes of around 80^0.

The inner part of the magnetosphere is known as the **plasmasphere**, shown in violet in Figs. 8.2 and 8.4. It is located just above the **ionosphere**, the last layer of the upper atmosphere that owes its name to the fact that is ionized by solar radiation. In the plasmasphere, a very low density plasma exists, and the dipole-like structure of the geomagnetic field is the dominating one.[16] The location of the outer boundary of the plasmasphere varies over time, depending on the strength of the solar wind above all, but it is usually within 30,000–50,000 km from the Earth surface. It mostly corotates with Earth. Its role is very important, particularly for us and for any other form of life on Earth, since it provides shielding against cosmic and solar highly energetic particles.

Two other important regions exist in the magnetosphere, known as the **Van Allen radiation belts**, that are shown (in red) in Fig. 8.4. These regions are tyre-shaped belts where highly energetic particles (of the order of a few MeVs) are trapped by the geomagnetic field.[17] The Van Allen belts overlap partially with the plasmasphere,

---

[15]In fact, there are regions in the **plasma sheet** where the magnetic field will vanish. These regions, whose location will vary over time as the whole magnetosphere changes, provide the seeds where the reconnection processes that lead to magnetic substorms happen, as we will discuss later.

[16]As a result, the plasmasphere has a doughnut shape that is very reminiscent of the geometry of the tokamaks used in MCF and that we described lengthily in Chap. 6.

[17]Trapped particles are a very common occurrence in tokamak plasmas as well. In both cases, particle trapping happens because the magnetic field is inhomogeneous, and because particle gyromotion is much faster than any other process [15]. In those conditions, it can be shown that

Fig. 8.4 Positions of the inner (shown in yellow) and outer (in orange) Van Allen radiation belts with respect to the plasmasphere of the Earth (in red) [Credits: courtesy of Estefanía Cuevas]

with the inner belt located at about 6000–10,000 km from the Earth surface, and the outer one at about 30,000–45,000 km. Depending on the size of the plasmasphere, as determined by the solar activity at any specific time, the outer Van Allen belt can be fully inside (as in Fig. 8.4), partially outside, or even fully outside of it. In addition to forming the Van Allen belts, particles trapped within them also drift perpendicularly to the magnetic field, giving rise to the so called **ring current** that flows within the belts as shown in Fig. 8.3.

In addition to the ring current, the tail currents, the Chapman-Ferraro current and the neutral sheet current, all of which are perpendicular to the local magnetic field, there are also some currents in the magnetosphere that flow parallel to the local magnetic field. These field-aligned currents, known as **Birkeland current**, flow at high latitudes (see Fig. 8.3, in yellow), forming a closed circuit with parts in which (mostly electrons) flow from the ionosphere to the magnetosphere and reversely.[18]

---

the magnetic moment of a particle, defined as $\mu = mv_\perp^2/2B$, remains essentially constant as the particle moves along the magnetic field (here, m is the particle mass, B the magnetic field strength and $v_\perp$ is the part of the particle velocity perpendicular to the magnetic field). Since the particle energy, $E = (m/2)(v_\parallel^2 + v_\perp^2)$ must also be conserved, the velocity of the particle along the magnetic field must be given by $v_\parallel = \sqrt{2E/m - \mu B}$. Therefore, if a particle moves in a direction in which the magnetic field strength B increases, it may happen that it reaches a point where $v_\parallel$ vanishes, and where the particle has to reverse its motion. When all other drifts are considered, in addition to parallel motion, the resulting trajectory has a poloidal cross-section that looks like a banana. This is the reason for the tyre-shape of the Van Allen belts. It is also the reason why these type of orbits are known under the general name of banana orbits in tokamaks [16].

[18]The evolution and distribution of Birkeland currents is closely related to auroras at quiet periods of solar activity, that are usually observed at those locations where the electrons stream down to the ionosphere (i.e., where the Birkeland current flows upwards).

8.2.3   Dynamics of the Magnetosphere of the Earth

The magnetosphere isolates the surface of the Earth from the majority of ultra-energetic particles, either of solar or cosmic origin. However, a fraction of the charged particles does penetrate into the magnetosphere, more abundantly during periods of high solar activity. Some do it by funnelling down the polar cusps that were described earlier. Others will spiral down magnetic field lines[19] and permeate into the lower atmosphere, usually near the Northern and Southern regions.

The more interesting dynamics happen, however, when a reconnection of magnetic field lines driven by local instabilities takes place near a point with zero-magnetic field (or X-point), and results in plasma transfer, among other things. The general process is sketched in Fig. 8.5.[20] For the discussion, we will follow closely [11]. The process starts with some IMF (line 1 in Fig. 8.5) that is brought by the solar wind is assumed to have a southward direction.[21] Since this orientation is opposite to that of the geomagnetic field in the dayside magnetosphere, an X point will eventually form and reconnection follows. The reconnected line (line 2) thus mixes together the IMF and Earthly fields. These magnetic lines are then convected tailwards with the solar wind, eventually becoming part of the magnetotail (lines 3–5). During the convection, plasma from the magnetosheath will be convected into the tail and the neutral plasma sheet. At the same time, magnetic flux is also being transferred from within the region with closed field lines (the plasmasphere of the Earth) into the magnetotail. Clearly, this magnetic flux must somehow make it back to the magnetosphere, since the latter manages to maintain its overall structure over time. The way in which this happens takes place at the nightside magnetosphere, deep in the plasma neutral sheet. An X point will eventually form at some location where field lines from the Northern and Southern lobes of the magnetotail meet (line 6). When the proper conditions are met, reconnection will take place there, forming a closed geomagnetic field line (line 7) and a purely IMF field line further down the magnetotail (line 8). As magnetic tension relaxes, the reconnected geomagnetic field line will move sunwards carrying the lost magnetic flux back within the plasmasphere (line 9). Plasma from the plasma sheet will also be convected, making it down the closed, dipole-like magnetic field lines into the lower atmosphere, and causing strong auroral displays at high latitudes. At the same time, the tension in the

---

[19]Any charged particle in the presence of a constant magnetic field will tend to follow a helical trajectory around the magnetic field line, in a direction determined by its parallel velocity to the magnetic field. In addition, if an electric field exists, it will also be subject to a drift perpendicular to the magnetic field and the electric field. If the field is not uniform, additional drifts exist perpendicular to the magnetic field and its gradient and curvature. These drifts are, in fact, the reason for the formation of the Van Allen belts, as was discussed earlier.

[20]The figure has been put together inspired by Fig. 8.27 of [11], which is itself a reprint of a figure from [17].

[21]Interesting things also happen when the direction of the solar magnetic field points in the northward direction, but its effects are much weaker [11].

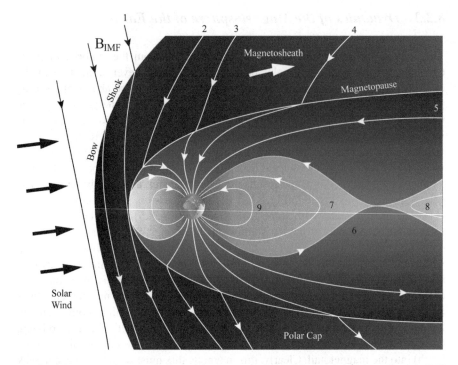

Fig. 8.5 Illustration of the interaction of the magnetosphere of the Earth with a southwards IMF that gives rise to a magnetic storm [Credits: courtesy of Estefanía Cuevas]

IMF line leads to its shortening, and its subsequent pulling beyond Earth along the magnetotail.

8.2.3.1   Magnetic Storms

The process just sketched is the main ingredient of what is referred to as a **geomagnetic storm**. These storms are usually associated to a coronal mass ejection (CME) from the Sun (see Chap. 7). Therefore, magnetic storms take place more frequently during the maximum of the solar 11-year cycle.[22] A few days after the CME took place in the Sun, the large amount of mass and the embodied magnetic field ejected with it will reach the Earth, compressing the geomagnetic field at the equator and increasing its value at the surface. The effect is more dramatic if the direction of the IMF is southwards, as we mentioned earlier. It is possible to

---

[22]Magnetic storms can also happen around the solar minimum, though. In this case, they are associated to high speed solar streams that are intermittently ejected from within regions of open field lines that often appear in between sunspots, and that are known as *coronal holes* (see Chap. 7).

detect geomagnetic storms[23] by measuring with magnetometers the variation on the horizontal component of the geomagnetic field at the Equator. The time series of this variation (expressed in nanoTeslas) is known as the D_{st} **index**[24] [18, 19]. The typical pattern that is observed during a storm is a sudden increase of the D_{st} index followed by a rapid and irregular decrease below its usual value.[25] Then, a much slower recovering back to the stationary value takes place. Typical durations of each phase are a few hours for the initial increase, maybe up to a day for the decrease, and several days for the recovery. In total, magnetic storms typically last up to a week [19]. The typical range of change of the horizontal magnetic field is between (10–1000) nT, being the smaller values naturally much more frequent.

8.2.3.2   Magnetic Substorms

The action of the solar wind on the magnetosphere of the Earth is however incessant, even if at a much lower level than that taking place during geomagnetic storms. As a result, particles are continuously convected into the plasma sheet within the magnetotail. A so-called **plasmoid** can thus form and grow over time within the magnetotail (see Fig. 8.6[26]), fed by the particles and energy convected by the solar wind that manage to make it into the neutral plasma sheet. As the plasmoid grows in size and density, an X-point (i. e, a point where the magnetic field vanishes) will eventually form between the plasmoid and Earth (see frames 3–5 in Fig. 8.6, where the X-point location is marked by the left arrow). Eventually reconnection will take place at the X-point, pushing the plasmoid tailwards while simultaneously accelerating a smaller amount of plasma towards the Earth. Disturbances in the geomagnetic field will grow as a result of this process, leading to the formation of the aurora at the high-latitude nightside of the Earth (since that is the direction from where the plasma is coming). This process is usually referred to as a **magnetic substorm**. After the substorm passes, the emptied magnetotail will start again to be slowly filled by the solar wind and the whole process repeats itself. It must be noted, however, that magnetic substorms are not periodic or quasi-periodic, but happen instead intermittently. In addition, it is also appears that substorms are not triggered randomly with a well-defined mean triggering rate (i.e., as in the typical Poisson process discussed in Sect. 4.4.5), but they appear to be instead a persistent process

---

[23]In addition, magnetic storms lead to the observation of auroras at latitudes lower than usual. They also may have an strong impact on global communications, power grids or satellites.

[24]D_{st} is an abbreviation for Disturbance Storm Time.

[25]Physically, the increase phase is related to the initial compression of the geomagnetic field by the CME; the decrease is related to the growth of the ring current that follows the injection of particles from the plasma sheet into the radiation belts. The larger ring current creates a magnetic field that opposes the geomagnetic field, causing the rapid reduction of the D_{st} index.

[26]The figure has been put together inspired by Fig.11.6 of [11] which is itself a reprint of a previous figure from [20].

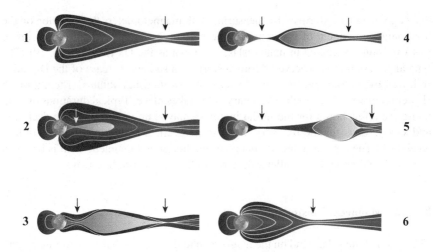

Fig. 8.6 Sketch illustrating the development of a magnetic substorm: formation of a plasmoid via convection to the plasma sheet (1–3); formation of X-point and reconnection of the field (4); ejection of the plasmoid (5) aurora formation and return to the initial state (6) [Credits: courtesy of Estefanía Cuevas]

that, to a certain extent, lacks characteristic scales. We will come back to discuss these peculiarities soon, in Sect. 8.3.

Experimentally, substorms can be tracked by means of the so-called **auroral electrojet (AE) index** [21]. Similarly to the D_{st} index, the AE index also measures the excursion of the horizontal component of the magnetic field as compared with stationary values.[27] *But in contrast to the D_{st} index, the AE index is measured at much higher latitudes.* In particular, at several positions around the auroral oval where the effect of substorms is felt much more strongly, as discussed previously. The geomagnetic disturbances associated to magnetic substorms are usually in the range of (200–2000) nT, typically lasting between 1 and 3 h.[28]

8.2.3.3 Auroras

We have mentioned the aurora several times already in this chapter. It is a phenomenon that is typical of high latitudes where, under suitable conditions (no full moon, clear sky), it can be often observed. It goes under several names, such as **aurora** (Boreal or Austral, depending on the hemisphere), **Northern lights** or **polar lights**. It usually appears as a coloured arc or band that extends in the sky

---

[27]Physically, these excursions of the magnetic field at the surface of the Earth are caused by the electrical currents that appear associated to the auroras. This is the reason for the index name.

[28]Any temporal variation with a timescale much shorter than any of these is usually related to other phenomena, such as magnetospheric waves or pulsations [11].

from east to west, and that evolves in time exhibiting various, highly-structured patterns that may change rather quickly. Auroral patterns are usually named after their appearance [22]: draperies, rays, coronas, etc. Their colour also varies, ranging from green to red, depending on different factors that we will enumerate soon. Auroras are closely related to solar activity, and one could in fact use the frequency of auroral activity at mid-latitudes as a proxy for solar activity with amazing success.

The physics behind the aurora are as follows. As charged particles are accelerated in the plasma sheet within the magnetotail (say, by the occurrence of a magnetic substorm), they move towards the Earth and penetrate into the atmosphere down the magnetic field lines, entering predominantly at high latitudes. From space, auroras look like bright coronas (see Fig. 8.7) that become more intense on the dayside (if caused by a magnetic storm, that pushes the field mainly from the direction of the sun) or nightside (by a magnetic substorm, that pushes particles from that tail towards the nightside of the Earth) depending on their cause. While energetic particles move along the field towards the surface of the Earth, they excite the atoms and molecules present in the atmosphere, which in turn emit the light seen as aurora. The colour of the aurora then depends on which is the dominant element excited (say, nitrogen and oxygen, either in atomic or molecular form). Since the composition of the atmosphere changes with height, the different colours that we see simply tell us how deep into the atmosphere the incoming particles made it, which ultimately depends on their initial energy.[29] Thus, a red aurora is usually related to

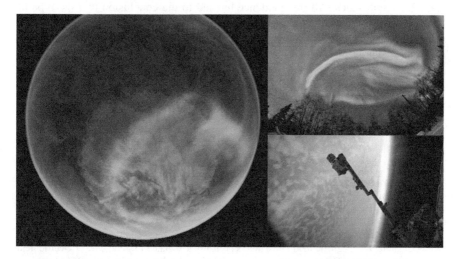

Fig. 8.7 Left: auroral corona as seen from space by the IMAGE spacecraft. Right: green (above) and red (below) auroral displays [Credits: coronal space view (© NASA—IMAGE); green aurora (courtesy of José M. Reynolds); red aurora (© ESA/NASA—ISSS, courtesy of Scott Kelly)]

---

[29] Although this is the general picture of the relevant physics, there are still some unsolved problems with it. Since the magnetic field has a dipole structure, the magnitude of field gets stronger

the excitation of atomic nitrogen (N_2) at about 70–90 km of height; green is usually a signal of atomic oxygen (O) being excited at about 100–150 km[30]; violet-purple is usually emitted by ionized nitrogen (N^+) that get excited at heights > 1000 km.

8.3   Is the Magnetosphere of the Earth a Complex System?

The picture we just drew in the last section, put together thanks to the hard work and dedication of many researchers over the last 100 years, provides a great insight regarding the nature of the magnetosphere of the Earth but it is, undoubtedly, too simplistic. The magnetosphere is a system that is dynamically alive, highly structured over many scales and with such a degree of complexity that any description based solely on the paradigms of reconnection and convection—under the external drive of the solar wind—is unable to fully capture. The magnetotail and its neutral plasma sheet have been experimentally probed over the years by various satellites (among others, the Geotail, AMPTE and THEMIS missions lead by NASA, or the CLUSTER mission lead by the European Space Agency), that have shown that their dynamics are turbulent, anisotropic, multiscale and strongly intermittent [24, 25]. The strong turbulence present in the magnetotail [26] leads to complicated spatial structures in which plasma vortices and bursty flows,[31] current sheets and local magnetic fields coexist while being strongly modified over many different scales. All this evidence has led to the conclusion that the type of slow-convective-loading, fast-release-via-reconnection cycle previously described as conducent to a magnetic substorm is not a smooth process, but the result of many incremental processes spread over a wide variety of scales that involve the intermittent generation, growth, drift and merging of coherent plasma structures and their embodied magnetic field. Eventually, the accumulation of the consequences of such processes build up to the global, non-linear instability that becomes a magnetic substorm.

---

as one moves towards the Earth. Having a finite energy, and since their magnetic moment μ needs to be conserved, particles will eventually be deflected back towards the magnetosphere way before reaching the surface. In fact, considering their typical energy spectrum, the majority of the incoming particles should be reflected back at about a 1000 km, which is too high for any significant light emission to occur. It is therefore believed that some additional acceleration must take place along the magnetic field lines, that is probably associated to some kind of electric field that develops naturally. The origin of this electric field is still not well understood [23].

[30]Red auroras can also be due to the excitation of atomic oxygen, but the process then happens at higher altitudes (> 200 km) where less energetic particles may excite a different transition.

[31]In particular, the so called **bursty bulk flows** or (BBFs) [27], that are intermittent high-speed flows that predominantly travel earthwards, perpendicularly to the local magnetic field, with durations of up to a few minutes. Some authors have linked BBFs to local reconnections of the magnetic field [28, 29], although there are also others that report cases in which these flows are detected with no apparent evidence of any reconnection of the nearby current sheets [27].

It was in the late 80s and early 90s that the dynamics of the magnetotail started to be looked at in the light of modern chaos theory. Using various indices—among them, the AE index previously described—as surrogates for substorm activity, it was established that magnetospheric substorm dynamics exhibited properties characteristic of low-dimensional chaotic systems such as self-similarity and fractality [30–32]. These properties are also characteristic of high-dimensional complex systems that exhibit dynamics such as self-organized criticality (see Sect. 1.3.2). The situation in the mid 90s was thus ripe for the proposal of new paradigms based on those ideas to try to understand, at least at a qualitative level, what was really going on in the magnetotail.

8.3.1   Chang's SOC Substorming Model

The plasma sheet in the magnetotail is an open system, driven by the solar wind, in which a large separation of timescales exist between loading (slow) and release (fast). In addition, the plasma sheet is in a highly turbulent state, interacting nonlinearly with itself via electric and magnetic fields, in a landscape in which local instability thresholds possibly abound. A separation of spatial scales over many decades thus exists, extending from the relevant microscopic scales (mainly, the ion gyroradius and the ion inertial length, which are of the order of a few tens or hundreds of kilometers) to the magnetotail global size (of the order of several millions of kilometers). Therefore, it seems clear that the plasma sheet is a system in which many of the ingredients that are required for the onset of complex dynamics (see Chap. 1) appear to be present. Therefore, it is no wonder that the first proposal to understand the magnetotail dynamics based on the concept of self-organized criticality (SOC; see Sect. 1.3.2) was proposed by Tom Chang in the early 1990s [33], just a few years after Per Bak's seminal paper on SOC [34] was published.

The main ingredient of Chang's proposal was the spontaneous emergence of coherent structures in the plasma sheet, in the form of magnetic flux tubes, at locations where the energy carried by existent MHD waves (such as Alfven waves [28, 33] or whistler waves [35], among other possibilities) would accumulate due to the fact that their propagation vector, $\mathbf{k}$ became perpendicular to the local, fluctuating magnetic field $\mathbf{B}$, which impedes its propagation. In this way, a large number of flux tubes, the majority of which would be aligned with the direction of the dominant neutral-sheet current, would appear effortlessly (see Fig. 8.8). Each flux tube could then interact with other tubes and, depending on the respective polarity of their magnetic fields and the magnitude of their enclosed currents, they could either merge and grow in size,[32] or break apart. The process would in principle

---

[32]The merging would happen, under certain conditions, it the two flux tubes with the same magnetic field polarity approach. In this case, the region of the two structures facing each other would be similar to an X point, where reconnection might eventually take place. *Chang* went as far

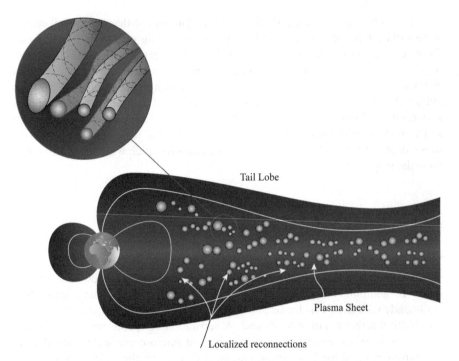

Tail Lobe

Plasma Sheet

Localized reconnections

Fig. 8.8 Sketch illustrating the formation of multiscale coherent structures at the magnetotail according to Chang's SOC model, inspired by Figs. 1 and 2 of [28] [Credits: courtesy of Estefanía Cuevas]

remain the same over structures of very different scales, resulting in a turbulent landscape which is intrinsically anisotropic, inhomogeneous and multiscale. The conclusion was that, continuously fed by the solar wind drive, this landscape would evolve into large structures that would trigger the global instability that causes the substorm in the magnetotail. If one assumes that, in addition, the growth process lacks any characteristic scales,[33] it is easy to envision how substorms would also lack any characteristic scales. Furthermore, if the modifications to the turbulent landscape caused by previous substorms could survive for sufficiently long times, they would also provide the needed mechanism to store memory in the magnetotail that might result in strong, scale-free long-term correlations among substorms.

Chang's original proposal, however, is somewhat different from other popular SOC models such as the paradigmatic running sandpile or the SOC model for tokamak radial transport discussed in Sect. 6.4.3. The most important difference probably is, as stated by the authors themselves, that a local threshold condition

---

as to suggest that the result of these reconnection, if successful, was precisely the excitation of the bursty bulk flows (BBFs) that various satellites have detected inside the magnetotail [33].

[33] In this case, the growth of a particular flux tube would somewhat resemble a percolation process.

is not required to control the local merging between structures, in contrast to what happens in these two other examples.[34] In the absence of a local threshold condition, a well-defined global SOC state with average profiles close to threshold (see Sect. 1.3) cannot be defined. Instead, the plasma sheet would evolve through a time-dependent global state determined by the temporal variations of the external drive provided by the solar wind [36]. Chang argued that, if the timescale over which the solar wind varies significantly is long compared with the timescales in which the coherent structures change and grow, it should still be possible for the system to store memory in the disturbances of the turbulent landscape carved by previous substorms. Such a dynamical state has been termed *forced criticality* by some authors [35, 36], in order to distinguish it from more standard SOC models. Studies towards elucidating the main features of forced criticality are still being carried out to this day, either by means of cellular automata [37] or using simplified MHD simulations [35].

It is however not clear to us that local thresholds for reconnection are necessarily absent in the flux tube merging process that could be taking place at the magnetotail. Recent kinetic simulations [38] of island (i.e., the cross section of a flux tube) coalescence for a range of system sizes greatly exceeding the kinetic scales show that the coalescence process is driven by the attraction of the currents associated with two neighbouring flux tubes. As they approach each other, a current sheet is formed between them that may become susceptible to reconnection. If reconnection does not take place, the magnetic field in the region between the tubes would pileup until the repulsive force due to the magnetic field gradient balances the attraction of the currents, which causes the flux tubes to repel each other and bounce. Thus, the local threshold is probably related to a maximum amount of flux that may pile up between the flux tubes beyond which reconnection does not happen. If this was indeed the case, a model closer in spirit to standard SOC might perhaps be more adequate for the magnetotail [39, 40].

In addition, there is another factor that must also be taken into account regarding the fact that the solar wind that intermittently drives magnetic substorms is also a product of another complex system, our Sun (see Chap. 7). The solar wind thus exhibits some complex features on its own (see also Sects. 8.4.3 and 8.4.4, where some SW datasets will be analyzed), that might also influence the magnetospheric dynamics. The important question then arises regarding whether the complex features of the substorm cycle are really due to just the internal magnetotail

---

[34]In the running sandpile, local relaxations are controlled by Z_c; the instantaneous profile, when compared with Z_c, provides the memory storage. In tokamaks, on the other hand, the local threshold condition is that of the dominant tokamak instability, that has been experimentally seen to be related to a temperature or pressure gradient.

dynamics or if they might be a mere reflection of those in the SW drive.[35] We will
have more to say about this question in the next few sections.

8.3.2 Evidence of Critical Dynamics in the Magnetotail

In recent years, many works have been published that try to substantiate, using
experimental data, the claim that criticality (either forced or standard, or even
some other variant) is relevant to the understanding of substorm dynamics.[36] A
large number of these studies have consisted of statistical analysis of indices of
various types such as the AE index previously discussed, as well as of satellite data
regarding the detection of bursty bulk flows (BBFs). For instance, it has been shown
by several authors that the AE index exhibit self-similar properties that extend for
several decades under various conditions [37, 46–48]. Similar results have also
been obtained when analyzing the statistics of BBFs [49]. In regards to evidence
of temporal self-similarity, long-term persistence has been found in the analysis of
AE data using the R/S method [48], as well as in the form of power-law statistics
for the waiting-times between AE events above a certain threshold [36, 42]. The
aforementioned question of which is the origin of these complex features—whether
they are due to internal magnetospheric dynamics or have a solar origin through
the SW drive—remains however open and is an active field of investigation. For
example, the connection of some AE features with those of the solar wind has also
been addressed, suggesting that at least their waiting-time distribution are certainly
related [42, 43].

8.4 Case Study: Magnetospheric and Solar Wind Indices

In this section we will first apply some of the analysis tools described in the first
part of this book to the two indices that were previously mentioned in this chapter:
the D_{st} index, that provides a surrogate to study storm dynamics, and the AE index,
that plays a similar role to investigate magnetic substorm dynamics. In the analysis

---

[35]In the context of the SOC running sandpile, it is known that a correlated drive, as long as
its strength is not sufficiently large as to overdrive the system, does not essentially change the
underlying SOC dynamics. Its main effect seems to be to change the avalanche waiting-time
distribution, that ceases to be exponential and reflects instead the character of the drive [41]. For
instance, by developing power-law tails. Several studies of satellite SW data seem to suggest that
the substorm waiting-time distribution might actually be dictated by the nature of the solar wind
[42, 43], being thus consistent with this view.

[36]We do not intend to be comprehensive here, since we are not experts in this area. A much more
detailed description of the state-of-the-art in this area can be found in any of the several excellent
reviews that have recently appeared on the subject [44, 45].

of the D_{st} index, one should expect to find complex features that are a reflection of the dynamics of solar CMEs, these being the main reason for the storm triggering. The AE index should be a different story, though, since substorm dynamics could be the result of complex dynamics intrinsic to the Earth magnetosphere under the intermittent, but incessant action of the solar wind drive. The analysis carried out here will be useful to illustrate how to make apparent complex features in this type of data, that requires a considerable amount of interpretation. In addition, we will also examine a couple of indices that are often used to characterize the properties of the solar wind (SW) that drives the magnetosphere. We expect that, by comparing the results obtained for the AE and the SW datasets, we might shed some light on the possible origin of the substorm complex features.

8.4.1 Analysis of the D_{st} Index (1957–2008)

As previously discussed, the D_{st} index is derived from hourly readings of low-latitude (i.e., near the equator) variations of the horizontal component of the geomagnetic field. The index reflects any changes in the globally symmetrical westward equatorial ring current, that is the main culprit of the depression of the horizontal component of the magnetic field during large magnetic storms. For that reason, it is so often used as a surrogate to investigate storm dynamics. The data, as provided by NOAA/NGDC, provides the daily average of the index after being treated to remove secular trends, including the latitude dependence of the various measuring sites.

The time series that we will analyze is shown in Fig. 8.9, with the temporal axis showing the elapsed time (in days) from the first value present in the record. In the inset of the figure, a zoom of the time series exposes the typical shape of a magnetic storm as detected by the D_{st} index (see Sect. 8.2.3.1): a rapid initial increase, followed by a slightly slower decrease and a much longer recovery (see Sect. 8.2.3.1). The typical duration of a storm is usually stated to be between 7 to 10 days,[37] thus giving a first estimate for the typical time that solar CMEs need to overcome the obstacle that Earth sets in their paths. However, as we will argue later, the existence of a well-defined characteristic storm duration may not be completely supported by the data.

8.4.1.1 Statistics

The pdf of the D_{st} signal is shown in Fig. 8.10, calculated by using both the CBC and CBS methods introduced at the end of Chap. 6. At first inspection, it might appear that a power-law scaling $p(d) \sim d^{-3.6}$ could exist for a range of values

---

[37] For that reason, we will express the elapsed time (from the first point in the series) also in days.

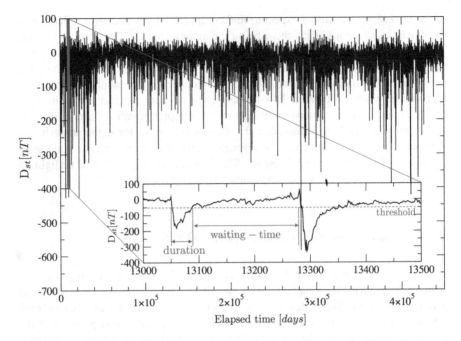

Fig. 8.9 Time series of the D_{st} index in the period Jan. 1990–Dec. 2008. [Source: Solar and Terrestrial Physics Division, NOAA/NGDC; public domain.] In the inset, the meaning of storm duration, waiting-time and threshold is illustrated using a zoom out of the signal

between 100–300 nT. However, the range is too short and the exponent too large to suggest any kind of complex behaviour, at least from such a simple analysis. However, the statistics of the D_{st} signal by itself carries little physical meaning. It is much more meaningful to estimate the statistics of the storms themselves. We can identify them by choosing a suitable threshold to separate the different events. We have used −50 nT as the threshold value (see inset of Fig. 8.9, where it is marked with a red horizontal line on a zoomed-out part of the dataset), so that periods of time when the signal is below the threshold (i.e., more negative) are considered as magnetic storms, and periods when the signal is above as waiting times. Using his value of the threshold, 3548 distinct magnetic storms can be identified within the dataset.[38]

---

[38]The value used for the threshold is, naturally, rather subjective. We have used −50 nT because, after a quick visual inspection of the dataset, it worked very well and introduced very few artefacts. One could have chosen even a more negative threshold, but the number of storms then diminishes quickly. For instance, the number of distinct events is 627 for −100 nT. Power-law scalings are still apparent in the statistics using −100 nT, though, but they are naturally much noisier. Various practical criteria have been proposed in the literature to choose a threshold value in situations like this one. Among them, it is rather popular to calculate first the mean μ and standard deviation of the signal, σ, and then choose the threshold to be μ plus a few times σ.

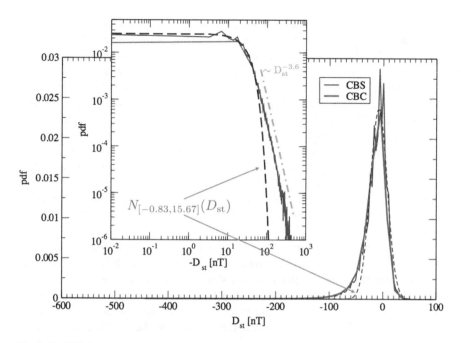

Fig. 8.10 PDF for the D_{st} data estimated by means of the CBC (using 25 events to define each bin; see Sect. 2.4.2) and CBS (200 bins; Sect. 2.4.1) methods shown in lin-lin and log-log scales. The best fit to a normal distribution (Eq. 2.30) is also included (in dashed brown)

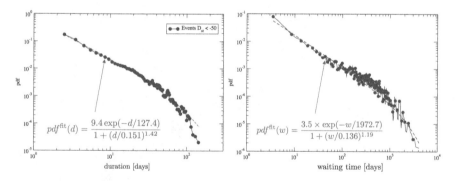

Fig. 8.11 PDFs for the magnetic storm durations (left) and the waiting-time in between successive storms (right) calculated using the SF method (see Sect. 2.4.3). The best fit to Eq. 1.8 is also included (in dashed red)

The pdfs obtained (applying the SF method, Sect. 2.4.3) for the durations and the waiting-times between storms larger than the selected threshold are shown in Fig. 8.11. The results are now much more interesting. For the storm durations, we find an extended decaying power-law scaling roughly as $p(d) \sim d^{-1.42}$ that extends for almost two decades, from one to a hundred days. The value of the exponent is sufficiently small (i.e., between 1 and 2) to suggest that a characteristic scale for storm duration is absent, consistent with critical behaviour (see Sect. 1.3.2). The fall-off at the end of the duration power-law suggests that storm durations are probably limited by finite-size effects (see discussion in the next paragraph). The pdf of the waiting-time between magnetic storms sufficiently large also exhibits a decaying power-law scaling $p(w) \sim w^{-1.19}$ that extends for almost three decades, from 1 day to a few years. The fact that the waiting-times do not follow an exponential power law implies that their triggering is not random, and the small exponent again points to critical behaviour (see Sect. 4.4.5). We have also defined an "effective storm size" as the area subtended by the signal over the duration of each event. The size pdf is shown in Fig. 8.12 in log-log scale. A very clear power-law scaling $p(s) \sim s^{-1.21}$ is then apparent for more than two decades, which points once more towards a divergent size scale in the sense discussed in Sect. 1.3.2 since the exponent is less than two.

Since each magnetic storm is triggered by a large solar event reaching the Earth, be it a CME or a massive ejection from within a coronal hole, it is thus reasonable to expect that the scaling behaviour just shown should somehow be a reflection of the solar dynamics. As we discussed in Chap. 7, solar CMEs appear to lack a well-defined characteristic size scale in the sense that their size is only limited by finite-size effects. Therefore, the lapse of time required by these CMEs to pass by the

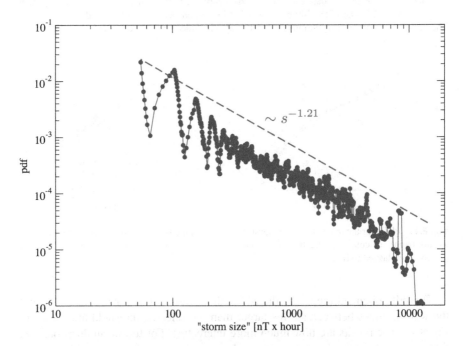

Fig. 8.12 PDF for the 'storm sizes' (defined in the text) estimated by means of the SF method (see Sect. 2.4.3) shown in log-log scale

Earth would also lack a characteristic scale, which might be the reason for the scale-free nature found in the storm durations. Similarly, it was also shown in Sect. 7.4 that solar CMEs appear to exhibit long-term persistent correlations that extended, at least, from a few days to tens of years. Accordingly, the pdf of waiting-times between magnetic storms, that should reflect this correlation, is not exponential and exhibits instead power-law scalings (see discussion in Sect. 4.4.5).

8.4.1.2   Power Spectrum

Next, we have calculated the power spectrum of the D_{st} signal by means of Eq. 4.43. The result, shown in Fig. 8.13, is rather interesting. The first thing to notice is that there are two distinct scaling regions, one scaling as $f^{-1.8}$ for timescales shorter than roughly a hundred days, and a second one scaling as $f^{-0.6}$ for longer timescales that extends for as long as data is available (circa 50 years). As we discussed at length in Sect. 4.2.2, scaling exponents between 0 and -1 are characteristic of self-similar, persistent dynamics. The breakpoint between the two scalings happens at about 15–20 weeks, separating a region dominated by storm self-correlation at the higher frequencies from a region with long-term persistence at the smaller frequencies. As discussed previously, this persistence is probably just a reflection of correlations within the solar drive (i.e., among CMEs) and that have their origin in the Sun. We will have more to say on this later in this section.

It is also interesting to note that there are several periodicities that appear very clearly in the power-spectrum and that have been marked with vertical arrows in Fig. 8.13. The easiest ones to interpret are the one appearing at about 26 days, that probably corresponds to the *solar synodic rotation*, and the one showing up at about 12 years, that is very possibly related 11-*year cycle* discovered by Schwabe (see Chap. 7). Another one, that appears at about 700 days, is probably related to the so-called *quasi-biennial oscillations* [50] that are found in many other records of solar magnetism, such as the number of solar flares, the solar magnetic field index and the sunspot areas. It is however not clear what the other peaks in the spectrum correspond to, although we suspect that some of them are just higher harmonics of the already mentioned ones.

8.4.1.3   *R/S* Analysis

We have also applied Hurst's rescaled range analysis to the signal in order to look for further evidence of long-term correlations. The rescaled range R/S of the D_{st} record is shown, as a function of time lag (measured in days) in Fig. 8.14. Two distinct scaling regions are clearly appreciated in the plot, particularly in the inset where the instantaneous Hurst exponent (Eq. 6.12) is shown. A first region, across which $R/S \sim \tau^{0.92}$, extends up to time lags of approximately 100–200 days, consistent with the scaling region found in the power spectrum at the highest frequencies.

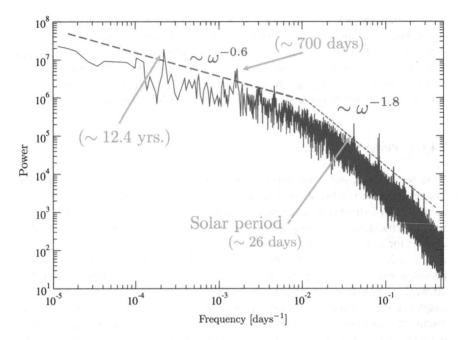

Fig. 8.13 Power spectrum of the D_{st} data. Peaks corresponding to the terrestrial and solar periods (and some of their harmonics) are identified with arrows

More interestingly, a second region is found, for time lags $\tau > 1000$ days, where $R/S \sim \tau^{0.81}$ that also seems consistent with the persistent scaling region found in the power spectrum at the lowest frequencies. However, the latter extended all the way to time lags of the order of a few hundred. In the case of the rescaled range, the intermediate region extending between (200–900) days lacks a clear scaling due to a periodicity that clearly shows up at about 700 days, causing the expected flattening in the rescaled range around it (see discussion in Sect. 4.4.4). As we already mentioned, this periodicity is probably related to the *solar quasi-biennial oscillations*. The amplitude of this perturbation must be significant, since it distorts the R/S scaling considerably over this range. The remaining periodicities identified in the power spectrum are also apparent in the instantaneous H exponent. The first one, corresponding to the *solar synodic rotation* (~ 26 days) is probably of rather small amplitude since it has almost no effect on the rescaled range. The second periodicity, related to the *11-year cycle* (~ 12.4 years), has a larger effect than the solar rotation, but still not too dramatic. Although this is probably due to the fact that the record is just 50 years long, so that it just contains three or four oscillations.

In any case, we can conclude that the results of the R/S analysis and the power spectra are pretty consistent with each other, and combining them, one can conclude that persistent, self-similar memory is indeed present starting at about $\tau > (100–200)$ days, and extending for at least as long as the record allows (about 50 years). The location of this break-point is probably related to the self-correlation

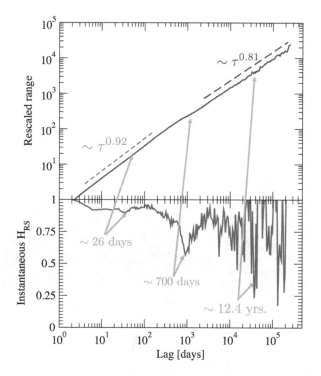

Fig. 8.14 Rescaled range analysis (see Sect. 4.4.4) for the D_{st} record. The rescaled range (R/S) is shown in the upper frame as a function of time lag (with respect to the first point in the record) in days; in the lower frame, the corresponding instantaneous Hurst exponent (see Eq. 6.12) is also shown. Some periodicities present in the record are identified with arrows

of magnetic storms. Its value suggests that they appear to last much longer than the usually stated 7–10 days, being closer to a few hundred days (and probably limited only by finite-size effects, as we discussed earlier). This finding is not necessarily inconsistent with previous results, being just related to how one defines the duration of storm. For example, the inset of Fig. 8.9 shows a couple of events in the D_{st} signal that take times closer to a hundred days to get back to stationary values, although the largest changes associated to the storm certainly take place during the first ten days or so. These long duration values are not unusual for magnetic storms due to the slow power-law decay of their statistics, as revealed by the storm duration pdf that was calculated earlier (see Fig. 8.11, left frame).

8.4.2   Analysis of the AE Index (1990–2008)

The record of the AE index that we will examine next corresponds to the 18-year long period between January 1, 1990 and December 31, 2008 (see Fig. 8.15). The

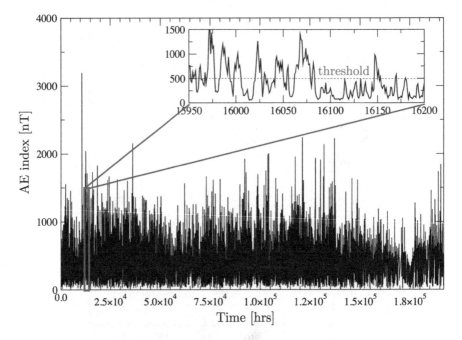

Fig. 8.15 Time series of the auroral electrojet (AE) index in the period Jan. 1990–Dec. 2008. [Source: Solar and Terrestrial Physics Division, NOAA/NGDC; public domain.] The temporal axis gives the time lag in hours, as measured from the beginning of the record

index is constructed by estimating the separation between the upper and lower envelopes of the superposed plots of the horizontal component of the magnetic field measured hourly at various auroral-zone magnetic observatories.[39] Scientists have relied for decades on the AE index as a measure of the level of geomagnetic disturbance that results from the auroral electrojets and hence, by proxy, of the state of the magnetosphere [21]. Since the magnetometers used to calculate it are at higher latitudes, the AE index is more sensitive to distortions due to magnetic substorms, as previously discussed. Magnetic substorms last typically just a few hours, in contrast to storms that lasted from several days to a few weeks. In addition, events in the AE index do not seem to have a characteristic temporal pattern, being much more irregular in shape, amplitude and duration. The lack of a characteristic pattern makes it more difficult to reliably separate individual substorms, in contrast to what happened for magnetic storms.

---

[39]In fact, the upper and lower envelopes are used to define two other indices, the so-called AU and AL indices. Clearly, AE = AU − AL.

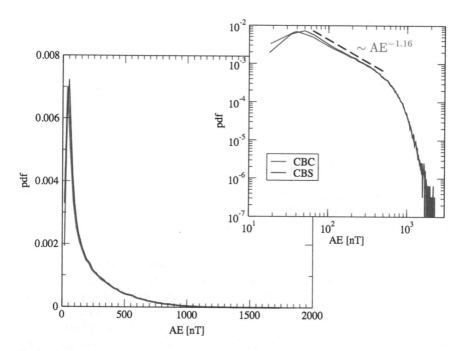

Fig. 8.16 PDF for the AE data estimated by means of the CBC (using 25 events to define each bin; see Sect. 2.4.2) and CBS (200 bins) methods shown in lin-lin and log-log scales

8.4.2.1 Statistics

We start by looking at the pdf of the AE record. The result is shown in Fig. 8.16, as calculated by using both the CBC and CBS methods introduced at the end Chap. 6. In contrast to the D_{st} index that we discussed previously, one can see here a clear power-law scaling $p(AE) \sim AE^{-1.16}$ that extends for more than a decade, between 50–700 nT. The value of the scaling exponent is sufficiently small (between 1 and 2, as discussed in Sect. 1.3.2) as to suggest that a typical AE amplitude may not exist, at least across that mesorange, which points to some kind of critical behaviour.[40]

As we did for magnetic storms, we have also tried to determine the statistics of individual magnetic substorms by means of a threshold (marked in red in Fig. 8.15). We have found adequate to use now 500 nT as the threshold value, that results

---

[40]The reason why the pdf of the AE data exhibits critical behaviour whilst the pdf of the D_{st} dataset did not deserves some comments. It is probably related to the fact that the sampling rate of the AE signal analyzed, in which data is provided every hour, is relatively close to the typical magnetic substorm duration (a few hours). In contrast, in the case of the D_{st} index, the sampling rate was one per day while magnetic storm often extended to almost a hundred days. For that reason, it appears that, in the case of the AE index, the statistics of the amplitude of the signal probably reflect better the underlying scaling of the substorm durations.

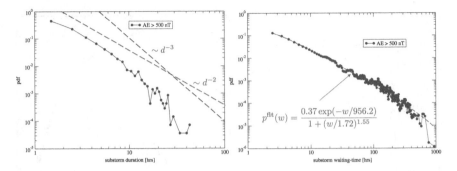

Fig. 8.17 PDFs for the magnetic substorm durations (left) and the waiting-time in between successive substorms (right) calculated using the SF method (see Sect. 2.4.3). The best fit to Eq. 1.8 is also included for the waiting-time pdf (in dashed red)

in 6988 distinct 'substorm events'. The pdfs for the resulting substorm durations and waiting-times are shown in Fig. 8.17. Clearly, their waiting-time pdf exhibits a fat power-law tail, scaling as $\sim w^{1.55}$. This value of the exponent implies that the substorm triggering process is correlated. The case of the substorm duration pdf, though, is more complicated to interpret. The reason is, as we mentioned before, that the large sampling rate of the signal (every hour) is very close to the substorm usual durations (a few hours). As a result, the duration values span just a decade, and it is very difficult to claim any power-law scaling with such a short range (we have included the scalings d^{-2} and d^{-3} in Fig. 8.17 to help guide the eye). We have defined also an "effective substorm size", using again the area subtended by the signal over the duration of each event, where this scaling is much clearer. The size pdf is shown in Fig. 8.18 in log-log scale. A very clear power-law scaling $p(s) \sim s^{-1.9}$ is then apparent for more than a decade, which points to a divergent size scale in the sense discussed in Sect. 1.3.2.

8.4.2.2 Power Spectrum

Next, we have calculated the power spectrum of the AE signal. The result, shown in Fig. 8.19, brings another piece of evidence that suggests critical behaviour. A power-law scaling region, $p(\omega) \sim \omega^{-0.9}$, is present in the range that goes between 10 and 400 h. The value of the exponent, between -1 and 0, is suggestive of the presence of self-similar, persistent correlations.[41]

For timescales longer than 400 h, the power spectrum become more flattish, although it is difficult to say due to the appearance of several strong periodicities. In

---

[41]It is worth noting that this range of timescales is well below the one across which self-similar, persistent memory was found for magnetic storms while analyzing the D_{st} index, that started at about a hundred days and extended to much longer times.

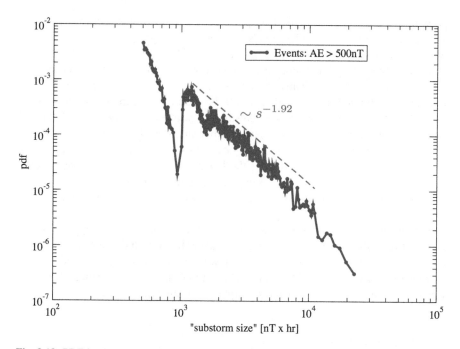

Fig. 8.18 PDF for the 'substorm sizes'(defined in the text) estimated by means of the CBC method (using 20 events to define each bin; see Sect. 2.4.2) shown in log-log scale

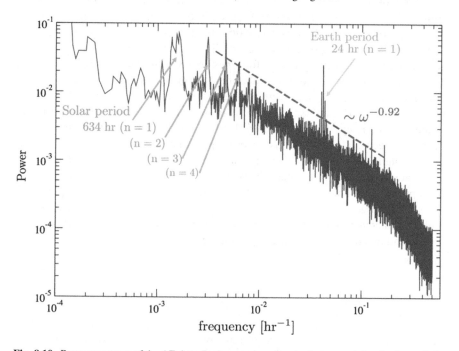

Fig. 8.19 Power spectrum of the AE data. Peaks corresponding to the terrestrial and solar periods (and some of their harmonics) are identified with arrows

particular, there is a periodicity that appears at about 640 h that seems to be pretty strong (several of its harmonics are also present in the signal), that could be related with the *solar synodic period* (approximately, $\sim$ 26 days). That periodicity might be responsible for the flattening, since the power spectrum seems to increase again at even lower frequencies. Thus, the persistence range might extend to even lower frequencies, although a longer data would be required to explore this possibility using the power spectrum. It is interesting to note that, in the case of the D_{st} index, the impact of this periodicity was rather small.[42] The explanation probably is that, whilst storms are triggered mostly by CMEs, substorms are triggered by the incessant action of the solar wind. If the latter is being fed primarily by specific active areas on the surface of the Sun, they should be strongly modulated by the solar rotation that would effectively switch them on[off] as the Earth enters[exits] their line-of-view.

Finally, at the largest frequencies, a second breakpoint appears at approximately 2–3 h. This breakpoint marks the temporal scale below which the self-correlation of individual substorms dominates. In fact, a few hours is what is usually stated for the average substorm duration. This finite value, in the case in which critical dynamics were indeed operative, would be a consequence of finite-size effects.

8.4.2.3   *R/S* Analysis

We have looked for additional evidence of persistence by means of the R/S analysis of the AE data, using the prescription given in Sect. 4.4.4. The resulting rescaled range is shown in Fig. 8.20 as a function of the temporal lapse (in hours) with respect to the first datapoint in the record. The rescaled range shows two distinct scaling regions, one scaling as $\tau^{0.83}$ for times between 2–3 and 400–500 h; a second one, scaling as $\tau^{0.9}$, exists for timescales longer than 1000 h. The first scaling region coincides quite well in range with the persistent scaling region we found in the power spectrum. The influence of the solar synodic rotation is again felt strongly in the rescaled range. It is revealed by the deep flattening that appears around that value, between 500–1000 h. This becomes particularly clear in the time trace of the instantaneous Hurst exponent, shown in the inset, that goes down to about 0.3 in the neighbourhood of the periodicity. Regarding the second scaling region, the one appearing at times longer than 1000 h, the fact that the scaling exponent is just slightly larger than the one seen at lower timescales suggests that we might be looking at a continuation of the same type of dynamics that dominated the previous region, and that the power spectrum just hinted at.

---

[42]Additionally, the periodicity associated to rotation of the Earth (that is, 24 h) also appears in the spectrum, although its effect on the AE spectrum seems to be rather small.

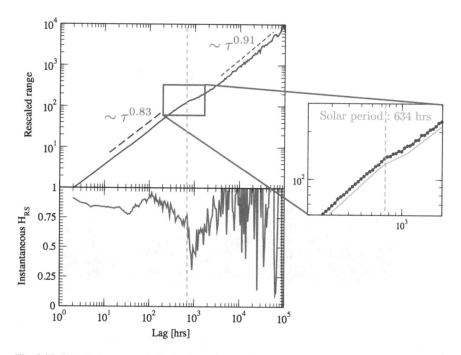

Fig. 8.20 Rescaled range analysis (see Sect. 4.4.4) for the AE data. The rescaled range (R/S) is shown in the upper frame as a function of time lag (with respect to Jan 1, 1990) in hours; in the lower frame, the instantaneous Hurst exponent (see Eq. 6.12) is also shown

8.4.3   Analysis of the Scalar B in the Solar Wind (1963–2017)

The analysis of the AE data has revealed self-similar and persistent features in the magnetospheric substorm cycle. Whether these features are the result of (complex) dynamics intrinsic to the magnetosphere (as suggested, for instance, by Chang's SOC model [33] and many others [26, 39, 40, 44, 45]) or if they are a mere reflection of the properties of the solar wind drive, it remains to be seen. In order to investigate this question a bit, we will analyze now two different data records that can be used as proxies of the solar wind dynamics. These records provide values of the plasma density and magnetic field magnitude (among many others) in the solar wind (SW) as measured over the last 50 years by a number of geostationary satellites around the $L1$ Lagrange point that is located in front of the Earth, at about 225 times its radius. This database has been put together as part of the OMNI initiative at NASA [https://omniweb.gsfc.nasa.gov].

We will examine first the SW scalar magnetic field B. The complete data record is shown in Fig. 8.21, with the horizontal axis corresponding to the lapse of time (in hours) passed from the beginning of the record (November 28, 1963) and the vertical one giving the magnetic field magnitude in nanoteslas. The record provides

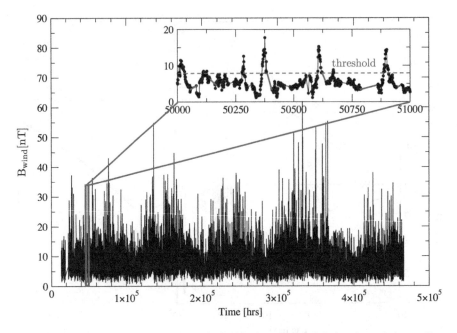

Fig. 8.21 Time series of the scalar magnetic field in the solar wind during the period extending from Nov. 1963 to Jul. 2017. [Source: OMNI initiative at NASA; public domain.] The temporal axis gives the time lag in hours, as measured from the beginning of the record

hourly values for B, although there is a significant percentage of entries are missing (a few are marked in red in the inset of the figure for illustration purposes).

8.4.3.1   Statistics

We start by looking at the pdf of the SW magnetic field record. The result is shown in Fig. 8.22, as calculated by using both the CBC and CBS methods introduced at the end Chap. 6. In contrast to the power-law scaling previously found in the pdf of AE index, there is no meaningful power-law scaling here. Of if there is, it corresponds to exponent values too large to provide any evidence of critical dynamics.

Again, we will define a "magnetic burst" by introducing a meaningful threshold in the SW magnetic field series, in a analogous way as how we treated magnetic storms and substorms previously. After various considerations, we have chosen the value 7.75 nT as the threshold (see Fig. 8.21), that allows to differentiate 8893 distinct SW magnetic bursts. Pdfs for the burst durations, the waiting-time between successive bursts and their effective size (again, defined as the area subtended by the signal over the duration of a burst) are respectively shown in Figs. 8.23 and 8.24.

From the figures, it is once more apparent that the SW magnetic field dataset exhibits self-similar scalings for at least two decades in durations and sizes, with

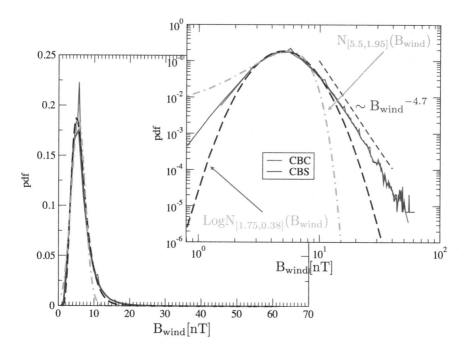

Fig. 8.22 PDF for the SW magnetic field data estimated by means of the CBC (using 25 events to define each bin; see Sect. 2.4.2) and CBS (200 bins) methods shown in lin-lin and log-log scales. Best fits to the normal (Eq. 2.30) and log-normal (Eq. 2.45) pdfs are also included

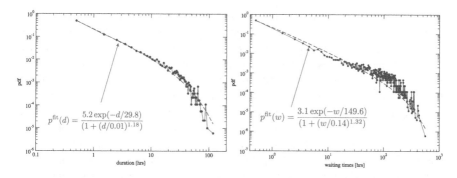

Fig. 8.23 PDFs for the SW magnetic field bursts (left) and the waiting-time in between successive bursts (right) calculated using the SF method (see Sect. 2.4.3). The best fit to Eq. 1.8 is also included for the waiting-time pdf (in dashed red)

values of the scaling exponents between 1 and 2, thus pointing to the action of some kind of divergent, critical dynamics (see Sect. 1.3.2). In the case of the durations, this scaling extends up to about 10–30 h. In addition, the pdf of the waiting-times between magnetic bursts also exhibits a divergent power law that extends at least until a couple hundred hours, suggesting that magnetic bursts are not triggered randomly, but in a correlated way.

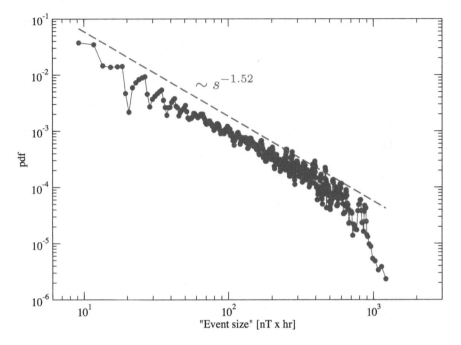

Fig. 8.24 PDF for the SW magnetic field 'burst sizes' (defined in the text) estimated by means of the CBC method (using 20 events to define each bin; see Sect. 2.4.2) shown in log-log scale

These results are rather interesting and deserve some comments. If the SW magnetic field data had lacked any self-similar properties, one could easily conclude that the presence of those properties in the AE data suggests that some process internal to the magnetosphere is responsible for the observed dynamics. The fact that the SW data exhibits self-similar properties does not mean, on the other hand, that the AE features are a reflection of them since there are many examples of systems driven by a correlated drive (for instance, the running sandpile [41]) that are able to establish their own internal dynamics at sufficiently long time scales. However, an interesting observation comes from the fact that the fall-off of the waiting-times or the SW data happens at about 100–200 h, whilst the one for the "substorm events" we introduced earlier happens at about 1000 h (see Fig. 8.17). This gap suggests that the magnetosphere might be loading up (via convection and reconnection process) for much longer timescales than those of the drive before unloading via the triggering of a substorm, which points to a possible important role for the internal magnetospheric processes.

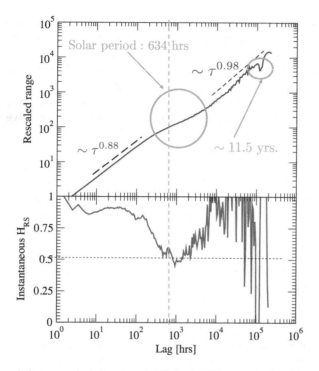

Fig. 8.25 Rescaled range analysis (see Sect. 4.4.4) for the SW magnetic data. The rescaled range (R/S) is shown in the upper frame as a function of time lag (with respect to Nov 28, 1967) in hours; in the lower frame, the instantaneous Hurst exponent (see Eq. 6.12) is also shown

8.4.3.2 R/S Analysis

The R/S analysis of the SW magnetic field data is shown in Fig. 8.25 as a function of the temporal lapse (in hours) with respect to the first point in the record.[43] This rescaled range is very interesting, since it clearly shows two distinct scaling regions, one scaling as $\tau^{0.88}$ for time lags between 1 and 300–400 h; a second one, scaling

---

[43] As we already mentioned in Chap. 8, while analyzing CME data with the R/S technique, the analysis must be done here a bit differently from what was explained in Sect. 4.4.4. The reason is that the SW data, although sampled in principle at a uniform rate, has many missing entries. The R/S algorithm must be slightly modified to take this into account or, otherwise, temporal scales and possibly exponents could be estimated wrongly. The changes are not difficult to do. One simply needs to define the averaging blocks differently. Instead of by the number of data points they contain, as we did in Sect. 4.4.4, one must define them by their real temporal duration. In addition, all the sums that were previously calculated over blocks (to compute, say, means, variances or ranges) become now sums over all those data points that take place at times that are contained in each particular block. The rest of the analysis and its interpretation, remains unchanged. We will have a chance to use this modified approach again when analyzing the CME data in Chap. 7. In that case, the uneven sampling is real, not an artefact of the measuring process.

as $\tau^{0.98}$, that exists for timescales longer than 1000–2000 h. In between these two regions there is a large flattening of the rescaled range, extending between 400–1000 h, that reveals the strong effect of a periodicity. In this case, it is probably the solar synodic rotation, for the same reasons that were explained when discussing the AE results (the reader should compare Fig. 8.20 with Fig. 8.25). If nothing else, this similarity suggests that some of the AE features (at the very least, the impact of various solar periodicities) might indeed be a reflection of the SW drive.

At the largest time lags, the rescaled range of the SW magnetic field shows that a second periodicity is also felt, although much more weakly that the synodic rotation. In fact, its the famous 11-year cycle, that can also be appreciated by inspecting the SW magnetic data directly (see Fig. 8.21). It is probable that the strong influence of the synodic rotation, combined with the 11-year periodicity, is what causes the scaling exponent to go above 1 across the second range.

8.4.4   Analysis of the Proton Density in the Solar Wind (1963–2017)

The second record from the OMNI catalog that we will analyze is that of the SW proton density, n, that provides information of the plasma component of the solar wind. The complete data record is shown in Fig. 8.26, with the horizontal axis corresponding to the lapse of time (in hours) passed from the beginning of the record (November 28, 1963) and the vertical one giving the proton density in particles per cubic centimetre. The record provides hourly values for n, when available. In fact, the percentage of missing entries (a few are marked in red in the inset of the figure for illustration purposes) is larger than in the case of the SW magnetic field dataset.

8.4.4.1   Statistics

As we did previously, we start once again by looking at the pdf of the SW proton density record. The result is shown in Fig. 8.27, as calculated by using both the CBC and CBS methods introduced at the end Chap. 6. As with the SW magnetic field, there is no meaningful power-law scaling here. Of if there is, it corresponds again to exponent values too large to provide any evidence of critical dynamics.

Once more we will define a "plasma burst" in a similar way to how we introduced the "magnetic burst" previously. The threshold has been set, in this case, to 15 part/cm^{-3} (see Fig. 8.26), that allows to identify 5242 distinct SW plasma bursts. The resulting pdfs for the burst durations, the waiting-time between successive bursts and their effective size (again, defined as the area subtended by the signal over the duration of a burst) are respectively shown in Figs. 8.28 and 8.29.

From the figures, it is apparent that the SW plasma density dataset also exhibits self-similar scalings for at least one decade in durations and two decades in sizes

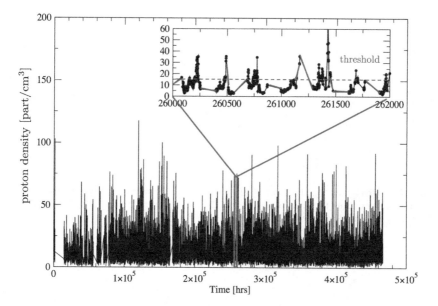

Fig. 8.26 Time series of the proton density in the solar wind during the period extending from Nov. 1963 to Jul. 2017. [Source: OMNI initiative at NASA; public domain.] The temporal axis gives the time lag in hours, as measured from the beginning of the record

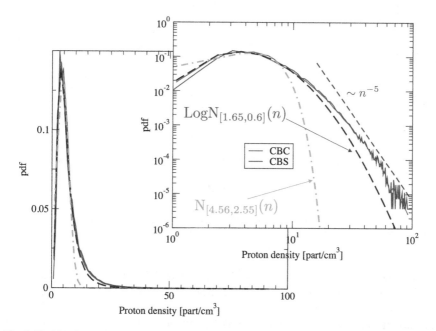

Fig. 8.27 PDF for the SW magnetic field data estimated by means of the CBC (using 25 events to define each bin; see Sect. 2.4.2) and CBS (200 bins) methods shown in lin-lin and log-log scales. Best fits to the normal (Eq. 2.30) and log-normal (Eq. 2.45) pdfs are also included

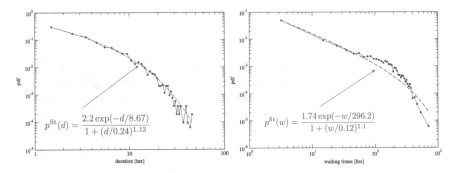

Fig. 8.28 PDFs for the SW plasma bursts (left) and the waiting-time in between successive bursts (right) calculated using the SF method (see Sect. 2.4.3). The best fit to Eq. 1.8 is also included for the waiting-time pdf (in dashed red)

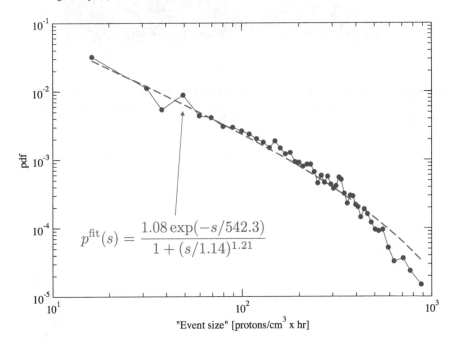

Fig. 8.29 PDF for the SW plasma 'burst sizes' (defined in the text) estimated by means of the CBC method (using 20 events to define each bin; see Sect. 2.4.2) shown in log-log scale

with values of the scaling exponents between 1 and 2, thus pointing once more to the action of some kind of divergent, critical dynamics (see Sect. 1.3.2). In the case of the durations, the power-law scaling extends up to about 8–10 h, that is a bit shorter that what we found for the SW magnetic field.[44] In addition, the pdf of

---

[44]This may not be all that significant, since it depends a bit on the threshold values used.

the waiting-times between plasma bursts also exhibits a divergent power law that extends at least until a couple hundred hours suggesting, as was also found in the case of the SW magnetic bursts. This value is much shorter than the few thousand hours needed for the AE waiting-time pdf to exhibit its fall-off. This difference again suggests to us that several plasma bursts (and magnetic bursts, as we saw earlier) may take place during the magnetospheric loading process that leads to a magnetic substorm, thus pointing again to a possible important role for the internal magnetospheric dynamics in the substorming process.

8.4.4.2 *R/S* Analysis

The *R/S* analysis of the SW proton density data is shown in Fig. 8.30 as a function of the temporal lapse (in hours) with respect to the first point in the record. The result is quite similar to that of the SW magnetic field examined earlier. As then, the density rescaled range shows two distinct scaling regions, one scaling as $\tau^{0.85}$ for time lags between 1 and 300–400h; a second one, scaling as $\tau^{0.91}$, that exists for all timescales longer than 1000–2000h. In between these two regions one finds

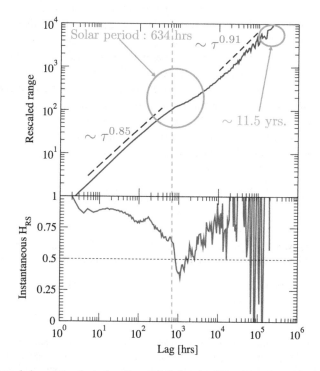

Fig. 8.30 Rescaled range analysis (see Sect. 4.4.4) for the SW proton data. The rescaled range (*R/S*) is shown in the upper frame as a function of time lag (with respect to Nov 28, 1967) in hours; in the lower frame, the instantaneous Hurst exponent (see Eq. 6.12) is also shown

the same large flattened region that was found in the rescaled range for the SW magnetic data, that extends between 400–1000 h revealing the strong influence of the solar synodic rotation. At the largest time lags, the plasma SW rescaled range also shows a second minor flattening that is probably a consequence of the 11-year cycle.

8.5   Conclusions

In this chapter we have discussed the main physics of the magnetosphere of the Earth, that can be envisioned as a complex open system driven by the (itself complex) Sun, be it via CMEs, coronal hole ejections or the incessant action of the intermittent solar wind. In particular, we have explored the possibility that complex dynamics internal to the magnetosphere are responsible for the magnetic substorm cycle, as suggested by several authors (see Sect. 8.3.1). We have also examined several datasets of relevance in this context by means of some of the tools that were presented in the first part of this book. These datasets have included various proxies for the magnetic storm (i.e., the D_{st} index) and substorm (the AE index) activities, as well as for the strength of the action of the solar wind (via local SW magnetic field amplitude and proton density records). From these analysis, it seems clear that all these records exhibit self-similar, divergent properties and long, persistent correlations. In addition, the comparative analysis of the AE index and the SW records have revealed some evidence (in the form of rather different cutoff values in their waiting-time pdfs) regarding the fact that internal magnetospheric dynamics are important in the substorming process, as suspected by many authors. However, it is clear that there is still much more to be done in this area. We hope that some of the tools that were presented in the first part of this book might be helpful in this task, assuming that any of our readers is up for the challenge.[45]

References

1. Halley, E.: An Account of the Late Surprising Appearance of the Light Seen in the Air on the Six of March Last; with an Attempt to Explain the Principal Phenomena. Philos. Trans. R. Soc. 55, 406 (1716)
2. Kelvin, W.T.: Address to the Royal Society at Their Anniversary Meeting, Nov 30, 1892. Proc. R. Soc. Lond. A 52, 300 (1892)

---

[45]Once again, we encourage all interested readers to take a look at some of the very good recent reviews published in this area [44, 45], where they can find a more complete description of activities, results and efforts carried out in the investigation of complex dynamics in the magnetosphere during the last 30 years. A more extensive list of relevant publications will also be found inside those reviews, thus providing a more complete view on this topic that what we manage to do in this introductory chapter.

3. Fitzgerald, G.F.: Sunspots and Magnetic Storms. The Electrician 30, 48 (1892)
4. Lindemann, F.A.: Note on the Theory of Magnetic Storms. Philos. Mag. 38, 669 (1919)
5. Kennel, C.F.: Convection and Substorms. Oxford University Press, Oxford (1995)
6. Parker, E.N.: Interplanetary Dynamical Processes. Wiley, New York (1963)
7. Lemaire, J.F., Gringauz, K.I.: The Earth's Plasmaphere. Cambridge University Press, Cambridge (1998)
8. Kivelson, M., Russell, C.T.: Introduction to Space Physics. Cambridge University Press, Cambridge (1995)
9. Merrill, R.T., McElhinny, M.W., McFadden, P.L.: The Magnetic Field of the Earth. Academic, San Diego (1998)
10. Cambell, W.H.: Introduction to Geomagnetism. Cambridge University Press, Cambridge (2003)
11. Kallenrode, M.B.: Space Physics. Springer, Heidelberg (2013)
12. Davidson, P.A.: An Introduction to Magnetohydrodynamics. Cambridge University Press, Cambridge (2001)
13. Moffatt, K.H.: Magnetic Field Generation in Electrically Conducting Fluids. Cambridge University Press, Cambridge (1978)
14. Jacobs, J.A.: Reversals of the Earth's Magnetic Field. Cambridge University Press, Cambridge (1995)
15. Roederer, J.: Dynamics of Geomagnetically Trapped Particles. Springer, Berlin (1970)
16. Wesson, J.: Tokamaks. Oxford University Press, Oxford (2004)
17. Hughes, W.J.: Introduction to Space Physics. Cambridge University Press, Cambridge (1995)
18. Chapman, S.: The Earth's Magnetism. Wiley, Hoboken (1951)
19. Sugiura, M.: Hourly Values of Equatorial D_{st} for the IGY. In: Annals of the International Geophysical Year 35. Pergamon Press, Oxford (1964)
20. Hones, E.W.: Magnetic Reconnection. American Geophysical Union, Washington, DC (1984)
21. Davis, T.N., Sugiura, M.: Auroral Electrojet Activity Index AE and Its Universal Time Variations. J. Geophys. Res. 71, 785 (1966)
22. Davis, T.N.: The Aurora Watcher's Handbook. University of Alaska Press, Fairbanks (1992)
23. Haerendel, G.: Auroral Acceleration in Astrophysical Plasmas. Phys. Plasmas 8, 2365 (2001)
24. Russel, C.T.: Noise in the Geomagnetic Tail. Planet. Space Sci. 20, 1541 (1972)
25. Baker, D.N., Pulkkinen, T.I., Buchner, J., Kilmas, A.J.: Substorms: A Global Instability of the Magnetosphere-Ionosphere System. J. Geophys. Res. 104, 601 (1999)
26. Zimbardo, G., Greco, A., Veltri, P., Voros, Z., Taktakishvili, A.L.: Magnetic Turbulence in and Around the Earth's Magnetosphere. Astrophys. Space Sci. Trans. 4, 35 (2008)
27. Angelopoulos, V., Baumjohann, W., Kennel, C.F., Coroniti, F.V., Kivelson, M.G., Pellat R, Walker, R.J., Luhr, F., Pashmann, G.: Bursty Bulk Flows in the Inner Central Plasma Sheet. J. Geophys. Res. 97, 4027 (1992)
28. Chang, T.: Self-Organized Criticality, Multi-Fractal Spectra, Sporadic Localized Reconnections and Intermittent Turbulence in the Magnetotail. Phys. Plasmas 6, 4137 (1999)
29. Pandey, B.P., Lakhina, G.S.: Driven Reconnection and Bursty Bulk Flows. Ann. Geophys. 19, 681 (2001)
30. Baker, D.N., Kilmas, A.J., McPherron, R.L., Buchner, J.: The Evolution from Weak to Strong Geomagnetic Activity: An Interpretation in Terms of Deterministic Chaos. Geophys. Res. Lett. 17, 41 (1990)
31. Vassiliadis, D.V., Sharma, A.S., Eastman, T.E., Papadopoulos, K.: Low-Dimensional Chaos in Magnetospheric Activity from AE Series. Geophys. Res. Lett. 17, 1841 (1990)
32. Takalo J, Timonen, J., Koskinen, H.: Properties of AE Data and Bicolored Noise. Geophys. Res. Lett. 20, 1527 (1993)
33. Chang, T.: Low-Dimensional Behavior and Symmetry Breaking of Stochastic Systems Near Criticality-Can these Effects be Observed in Space and in the Laboratory? IEEE Trans. Plasma Sci. 20, 691 (1992)
34. Bak, P., Tang, C., Wiesenfeld, K.: Self-Organized Criticality: An Explanation of the $1/f$ Noise. Phys. Rev. Lett. 59, 381 (1987)

35. Chang, T.: Complexity and Anomalous Transport in Space Plasmas. Phys. Plasmas 9, 3679 (2002)
36. Consolini, G., Chang, T.: Magnetic Field Topology and Criticality in Geotail Dynamics: Relevance to Substorm Phenomena. Space Sci. Rev. 95, 309 (2001)
37. Consolini, G.: Sandpile Cellular Automata and the Magnetospheric Dynamics. In: Cosmic Physics in the Year 2000, Proceedings of 8th GIFCO Conference, Bologna (1997)
38. Karimabadi, H., Dorelli, J., Roytershteyn, V., Daughton, W., Chacon, L.: Flux Pileup in Collisionless Magnetic Reconnection: Bursty Interaction of Large Flux Ropes. Phys. Rev. Lett. 107, 025002 (2011)
39. Chapman, S.C., Watkins, N.W., Dendy, R.O., Helander, P., Rowlands, B.: A Simple Avalanche Model as an Analogue for the Magnetospheric Activity. Geophys. Res. Lett. 25, 2397 (1998)
40. Lui, A.T.Y., Chapman, S.C., Liou, K., Newell, P.T., Meng, C.I., Brittnacher, M., Parks, G.K.: Is the Dynamic Magnetosphere an Avalanching System? Geophys. Res. Lett. 27, 911 (2000)
41. Sanchez, R., Newman, D.E., Carreras, B.A.: Waiting Time Statistics of Self-Organized-Criticality Systems. Phys. Rev. Lett. 88, 068302 (2002)
42. Freeman, M.P., Watkins, N.W., Riley, D.J.: Evidence for a Solar Wind Origin of the Power Law Burst Lifetime Distribution of the AE Indices. Geophys. Res. Lett. 27, 1087 (2000)
43. Freeman, M.P., Watkins, N.W., Riley, D.J.: A Minimal Substorm Model that Explains the Observed Statistical Distribution of Times Between Substorms. Geophys. Res. Lett. 31, L12807 (2004)
44. Valdivia, J.A., Rogan, J., Muoz, V., Gomberoff, L., Klimas, A., Vassiliadis, D., Uritsky, V., Sharma, S., Toledo, B., Wastavino, L.: The Magnetosphere as a Complex System. Adv. Space Res. 35, 961 (2005)
45. Sharma, A.S., Aschwanden, M.J., Crosby, N.B., Klimas, A.J., Milovanov, A.C., Morales, L., Sanchez, R., Uritsky, V.: 25 Years of Self-Organized-Criticality: Space and Laboratory Plasmas. Space Sci. Rev. 198, 167 (2016)
46. Uritsky, V.M., Pudovkin, M.I.: Low Frequency $1/f$-Like Fluctuations of the AE-Index as a Possible Manifestation of Self-Organized Criticality in the Magnetosphere Ann. Geophys. 16, 1580 (1998)
47. Dobias, P., Wanliss, J.A.: Intermittency of Storms and Substorms: Is it Related to the Critical Behaviour? Ann. Geophys. 27, 2011 (2009)
48. Price, S.P., Newman, D.E.: Using the R/S Statistic to Analyze AE Data. J. Atmos. Sol. Terr. Phys. 63, 1387 (2001)
49. Angelopoulos, V., Mukai, T., Kokubun, S.: Evidence for Intermittency in Earth's Plasma Sheet and Implications for Self-Organized Criticality. Phys. Plasmas 6, 4161 (1999)
50. Stenflo, J.O., Gudel, M.: Evolution of Solar Magnetic Fields: Modal Structure. Astron. Astrophys. 191, 137 (1988)

Chapter 9
Laboratory Plasmas: Dynamics of Transport Across Sheared Flows

9.1 Introduction

In this last chapter, we will go back to the same laboratory plasmas discussed in Chap. 6 to look for our last example of complex behaviour in plasmas. In particular, we will discuss the dynamics of **turbulent transport across sheared flows**. **Sheared** (or **shear**) **flow** is a label used to refer to any kind of flow that varies in space. Naturally, small-scale sheared flows are always present in any turbulent medium. For instance, any turbulent vortex (or "eddy") is a region where the local flow (mostly directed around the center of the eddy) varies quickly as one moves away from the center. The action of turbulence also tends to generate local patterns of flow that vary quickly in space and time to facilitate the dissipation of energy at the (usually very small) viscous scales. These are not the kind of sheared flows that interest us, though. We will be concerned about **large-scale sheared flows** instead. That is, flows that maintain their coherence over scales much larger and longer than any local turbulent scales. Large-scale sheared flows are somewhat rare in nature because shear very often drives instabilities of the Kelvin-Helmholtz (KH) type [1]. It is only in special conditions that shear flows can be kept stable, at least for sufficiently long times as to have an strong impact in the system dynamics. Shear flows are stabilized, usually, by either differential rotation or magnetic fields. For that reason, large-scale sheared flows often appear in stars, planetary atmospheres and oceans or in magnetized fusion toroidal plasmas, to name just a few (see Fig. 9.1). In the context of magnetic confinement fusion (MCF), they are usually referred to as **zonal flows**.[1]

---

[1]In MCF, it is sometimes usual to distinguish between mean (over time) and fluctuating shear flows; in that case, the term "zonal flow" is used to denote the fluctuating part of the shear flow. In this chapter, however, we will not make that distinction.

© Springer Science+Business Media B.V. 2018
R. Sánchez, D. Newman, *A Primer on Complex Systems*,
Lecture Notes in Physics 943, https://doi.org/10.1007/978-94-024-1229-1_9

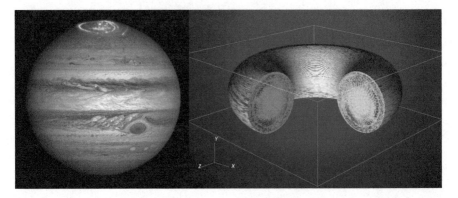

Fig. 9.1 Left: latitudinal shear flow patterns around the northern pole of Jupiter's atmosphere. **Right**: isocontours of turbulent electrostatic potential from a tokamak Ion-Temperature-Gradient (ITG) turbulence simulation showing the presence of a strong, radially-shear poloidal flow. Credits: Jupiter image (© ESA/NASA - Hubble)

We will start the chapter by providing a brief introduction to turbulent sheared flows in Sect. 9.2, with a particular focus on how and why they appear in tokamaks. We will then discuss one of the main consequences of the presence of zonal flows in any system: the *reduction of turbulent transport across them.*[2] For us, this process is of interest because it turns out that turbulent transport across shear flows often becomes endowed with non-Guassian and subdiffusive features, in contrast to the more traditional understanding of how shear flows acted on transport. We will briefly sketch the probable physical origin of both subdiffusion and non-Gaussianity in Sect. 9.2.3. As a result of these features, Fick's law no longer provides a good phenomenological description of transport across shear flows and other paradigms are needed. We will illustrate, as in every other chapter, this phenomenology by analyzing data from gyrokinetic simulations of tokamak turbulence (see right frame in Fig. 9.1) in Sect. 9.4. In this case we will employ some of the tools that were introduced in Chap. 5, particularly those related to the propagator of the fractional transport equation (Eq. 5.3.3).

9.2 Stable Sheared Flows

It is often the case in hydrodynamics that sheared flows are unstable, generating turbulence usually via the excitation of Kelvin-Helmholz (KH) instabilities [1]. Some examples of this behaviour are provided, for instance, by jets, flows past obstacles or flows close to walls. Turbulence is generated in all these cases and any

---

[2] As it is often the case, in some systems this reduction might be bad news; but in the case of fusion toroidal plasmas, this reduction of radial transport becomes extremely handy!

large-scale sheared flow pattern that might be present quickly disappears [2]. There are however situations in which sheared flows can be stabilized, maintaining their coherence for very long times. This usually happens in the presence of differential rotation or space-varying magnetic fields [3].

9.2.1   Differential Rotation and Magnetic Fields

Hydrodynamic shear flows are sometimes stabilized by some kind of **differential rotation** [3]. The simplest example is possibly that of a two-dimensional circular vortex, whose own vorticity profile makes it stable.[3] Other examples found in nature are any of the many shear flows that exist in planetary atmospheres, that are stable at least most of the time [4]. In the Earth's atmosphere, for instance, one has the so called **jet streams** that have such an enormous impact on air travel. In other planets, such as Jupiter, latitudinal banded zonal flows are clearly visible from afar in its atmosphere (see left frame of Fig. 9.1). Stable sheared flows stabilized by differential rotation also exist in the ocean, as it is the case of the **Antarctic circumpolar current**.

Strong, space-varying magnetic fields can also stabilize shear flows under certain conditions. In confined toroidal fusion plasmas, such as in tokamaks or stellarators, stable shear flows (known as **zonal flows**) may also develop, either spontaneously or driven externally. Zonal flows are usually established over narrow radial regions, known as **transport barriers** (see right frame in Fig. 9.1), with mass motion taking place within a magnetic surface.[4] The reason for using the word "transport barrier" will be clear soon.

---

[3] The vorticity of a flow is defined as the curl of the local velocity vector, $\mathbf{w} \equiv \nabla \times \mathbf{v}$. Thus, the vorticity is zero everywhere for a uniform flow. For a solid-rotation in two-dimensions, though, it can be shown that its vorticity vector $\mathbf{w} \propto \Omega$, being Ω the angular velocity vector that satisfies $\mathbf{v} = \mathbf{r} \times \Omega$. In a more general two-dimensional flow, the relation between vorticity and angular velocity changes, but regions where the vorticity is large are still indicative of an intense local rotation. In particular, turbulent eddies can be seen as regions where vorticity accumulates, with the orientation of the local rotation (clockwise or counterclockwise) being consistent by the sign of the local vorticity.

[4] It should be remembered that the tokamak and stellarator magnetic topology can be approximated by a family of nested, toroidal magnetic surfaces to which the magnetic field is tangent. The radial direction is the one perpendicular to these magnetic surfaces. The two directions on the surface are the poloidal and toroidal ones. The magnetic field has, therefore, both a poloidal and toroidal component, being the latter much larger. The flow motion that constitutes a zonal flow in tokamaks appears to be of electrostatic origin, thus flowing in the direction of the $\mathbf{E} \times \mathbf{B}$ drift, that is contained in the magnetic surface but perpendicular to the magnetic field.

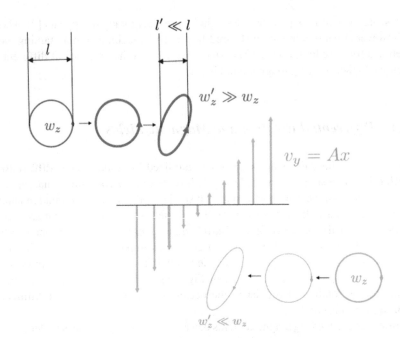

Fig. 9.2 Action of a linear shear flow (in blue), $v_y(x) = Ax$, $A > 0$, on eddies containing vorticity of different sign: positive (in red) and negative (in red). The shear flow will stretch both type of eddies along the shear direction, strengthening the one with vorticity of the same sign as that of the shear flow ($w_z^{\text{shear}} = A$) and weakening the other one

9.2.2 Turbulence Suppression by Sheared Flows

The presence of a stable shear flow affects turbulence in different ways. The most evident one is that it distorts eddies in such a way that their characteristic scales (in space, time and amplitude) are changed [5]. The way in which this happens is best understood by means of the sketch shown in Fig. 9.2. In it, the action of a flow (along y) with a spatial dependence (on x) on a two dimensional eddy is shown. The eddy is stretched and distorted because its different parts would travel at different velocities, as dictated by the shear flow. The characteristic size of the eddy across the flow could thus be significantly reduced, as shown in the figure. Other things also happen. First, depending on the sign of the vorticity of the eddy, its local rotation may be enhanced (in red) or suppressed (in green). In addition, because the eddy is part of a turbulent flow, its lifetime may also be reduced. The reason is that the life of any eddy is determined by how long it takes for its neighbouring eddies to shear it apart. This time could be reduced by the shear flow because the stretching might bring the eddy faster into the domain of influence of its neighbours; the more, the further it becomes stretched. Finally, the average fluctuating energy in the turbulence could also be reduced, due to the aforementioned shortening of the eddy lifetime.

After considering all these facts, it should come as no surprise that, if the conditions in which the shear flow is acting on turbulence are appropriate, the effective turbulent transport across the shear flow could be significantly reduced. Indeed, by invoking plain diffusive arguments, one could estimate the across-the-shear-flow eddy diffusivity by means of a CTRW-like argument (see discussion on turbulent diffusion in Chap. 5; see also Fig. 6.6), leading to an effective eddy diffusivity of the order of: $D \sim l^2/\tau \sim lv \sim v^2\tau$, being l the typical eddy size along the direction of transport, v the mean amplitude of the turbulent velocity fluctuations and τ the typical eddy lifetime. Clearly, if the eddy lifetime, its size and the turbulent amplitude are all reduced by the action of the shear flow, the resulting eddy diffusivity for across-the-shear transport would also be much smaller.[5]

Which are the conditions that a shear flow must satisfy to effectively suppress turbulence and reduce transport according to the picture we just draw? Naturally, the shear flow must be stable to begin with. This implies that strong turbulence suppression would more often than not be observed in the presence of either strong differential rotation or space-varying magnetic fields. In addition, several other conditions must also be met. First, if the eddy stayed at rest with respect to the shear flow, the characteristic size of the eddy l should be of the order (or larger) of the length over which the sheared flow varies significantly, $L_v \simeq v(dv/dx)^{-1}$. However, eddies are often not static, being pushed around by the background flow. Therefore, this condition must be replaced by a temporal condition that requires that τ_S, the time that the shear flow needs to act effectively on the eddy (i.e., its rate of differential advection), be shorter than τ_v, the time that the eddy needs to cross the sheared region. A second condition that must be fulfilled is that the time the shear flow needs to act effectively must also be shorter than the eddy lifetime, τ_l. Otherwise, the eddy would disappear before the shear can act on it. Or, in other words,

$$\tau_S \ll \tau_v, \tau_l. \tag{9.1}$$

It is worth commenting now that there are cases in which a shear flow might affect transport across it even if some of the aforementioned conditions are not met. One example is the kind of situation discussed in Chap. 6, where we considered transport in regimes of near-marginal turbulence.[6] In these cases, turbulent transport does not resemble the type of CTRW-like motion around single eddies that appears in the diffusive view of turbulent transport (see Chap. 5). Instead, in near-marginal conditions turbulent transport is dominated by avalanche processes that result from the successive concatenation of instabilities excited at nearby locations along a

---

[5]This type of argument, that heavily relies on a diffusive description of the transport process, has however been challenged recently, as we will discuss in Sect. 9.2.3.

[6]In these cases, the transport dynamics were heavily reminiscent of those of the running sandpile, the poster child of self-organized criticality (see Chap. 1).

certain direction.[7] In this type of situations, any shear flow with a characteristic scale *much longer than that of a typical eddy* could still reduce turbulent transport by decorrelating the avalanching process, whose scale length diverges with the system size. In a certain sense, the shear flow acts effectively in order to short-circuit any long-range radial propagation [6]. We will however not deal with these type of situations in this chapter.[8]

9.2.3   Zonal Flows in Tokamaks

Shear flows are of particular importance in fusion toroidal plasmas because they are associated with the formation of the so-called **edge transport barriers** (ETB). These barriers were first seen in tokamaks in the late 1980s, forming spontaneously very close to the plasma edge after a certain heating-power threshold was overcome [10] (see Fig. 9.3; see also the discussion on tokamak confinement regimes in Sect. 6.4.1). This operational regime is usually known as the *H*-**mode**, being the standard operation regime for next step tokamaks, such as the ITER experiment [11]. The *H*-mode is remarkably universal. Almost every tokamak with access to sufficient external power[9] has been able to reproduce *H* modes [12].[10] The general properties of tokamak *H*-modes were already outlined in Sect. 6.4.1, but we will repeat the three most important ones here. First, there is a minimum external power requirement that must be overcome, known as the *H*-**mode power threshold**. Secondly, temperature and pressure profiles become rather steep across the transport barrier, a consequence of having to balance the same local flux with a much lower local transport. And third, there is a very strong radially sheared (poloidal and toroidal) flow at the ETB, that seems to have a clear electrostatic nature, as has repeatedly been shown by experiments [3]. By electrostatic nature it is meant that the flow observed is very close to the one estimated by computing the $\mathbf{E} \times \mathbf{B}$ drift using the measured mean electric field.

In the case of the *H*-mode, the shear flow appears spontaneously, probably self-generated by turbulence via the Reynolds stress tensor [14], thanks to the anisotropy and inhomogeneity that is inherent to any toroidal fusion plasma.[11] In contrast to what often happens in fluid turbulence—where any flow shear typically drives

---

[7]In the case of tokamaks, it is the radial direction; for a sandpile, it is down the slope.

[8]The interested reader can find additional information about the action of shear flows on near-marginal transport in several papers recently published in the literature [7–9].

[9]In addition, access to the *H* mode also requires a *divertor configuration*, as shown in Fig. 9.3, that permits to insulate the plasma from the wall sufficiently well.

[10]Other MCF toroidal devices, such as stellarators, have also produced *H*-modes and ETBs [13].

[11]ETBs have also been externally induced in tokamak plasmas, mostly via plasma biasing [15]. In this technique, a potential difference is forced between the plasma and the edge, that results in a radial electric field and an associated $\mathbf{E} \times \mathbf{B}$ flow shear that acts on edge turbulence, reducing its levels and associated transport, and causing a steepening of the plasma profiles.

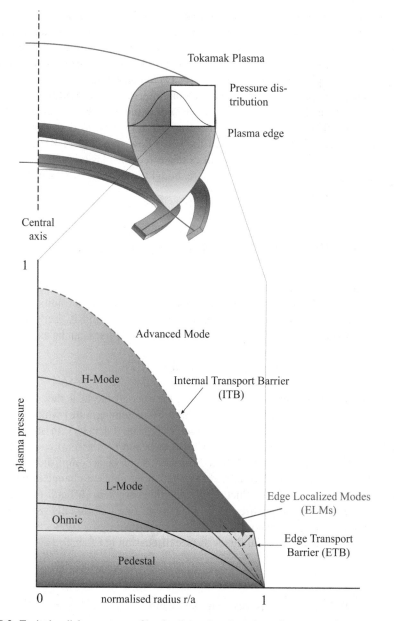

Fig. 9.3 Typical radial pressure profiles for the main tokamak confinement regimes reached as external power is increased: ohmic (black), L-mode (blue), H-mode (red, solid) and advanced modes (red, dashed). The transition from *L*-mode to *H*-mode takes place when the edge transport barrier develops, above a certain power threshold. The transition to advanced confinement modes happens when an internal transport barrier is formed at some intermediate radial location. Credits: courtesy of Estefanía Cuevas

turbulence via KH instabilities, leading to stronger turbulence—, the edge shear flow is stable and reduces local fluctuations and turbulence as we discussed previously. Since tokamak plasmas are driven systems, the reduced levels of turbulence lead to a reduced effective transport, which subsequently leads to a further steepening of the gradients (or plasma temperature or density) to balance the external drive. As a result, energy and plasma confinement are both improved. The conditions that limit how large a shear flow can become are still not completely clear, being probably situation-dependent [14]. Among them, one probably have to consider drag, the excitation of secondary instabilities of the shear flow (possibly of KH nature), or the excitation of other types of instabilities that affect the plasma gradients (such as the pressure-gradient-driven modes that are thought responsible for the excitation of ELMs, as we discussed in Chap. 6).

9.3 Non-diffusive Transport Across Sheared Flows

The traditional understanding of how the suppression of radial transport by the kind of radially sheared zonal flow that takes place in a tokamak has traditionally been based on the same diffusive arguments we introduced previously. Let's discuss in detail how it goes. In turbulence theory, turbulent radial fluxes can be expressed as:

$$\tilde{\Gamma}_n \simeq \langle \tilde{n}\tilde{v}_r \rangle \quad \text{and} \quad \tilde{q} \simeq n_0 \langle \tilde{T}\tilde{v}_r \rangle, \tag{9.2}$$

where the first one is the particle flux, and the second the heat flux. In them, $\tilde{n}$ and $\tilde{T}$ represent the advected fields (say, density or temperature) whilst $\tilde{v}_r$ is the fluctuating radial velocity (n_0 is the average background density). In all cases, the bracket stands for ensemble average. The suppression of transport caused by the shear flow leads to a reduction of these fluxes. This can be accomplished either via a reduction in the amplitude of any of the fluctuating fields, a change in the phase between them or both. These fluctuating quantities are very difficult to calculate, even for the simplest geometries. For that reason, one traditionally assumes that some effective (eddy) transport coefficient (i.e., diffusivity or conductivity) could be used to estimate these radial fluxes as (see discussion in Chap. 5):

$$\tilde{\Gamma}_n \sim -D_{\text{eff}}\nabla n \quad \text{and} \quad \tilde{q} \sim -n\chi_{\text{eff}}\nabla T. \tag{9.3}$$

Using CTRW-like arguments, one can then assume $D_{\text{eff}}, \chi_{\text{eff}} \sim v^2\tau \sim l^2/\tau \sim vl$ where l, τ and v the (space, time or amplitude) represent characteristic scales of the turbulence. These typical scales must then be determined either theoretically, experimentally or from numerical simulations.

If this interpretation is accepted as essentially valid, it is then natural to reduce the action of the shear flow to changes in either l, v or τ, according to the principles that we briefly sketched in Sect. 9.2.2. The key point here is to note that the whole argument rests on the hypothesis (usually assumed implicitly) that the nature of radial transport must be and remains to be diffusive. Or, more precisely, that the underlying transport dynamics must be and remain (near)-Gaussian and (near)-Markovian, thus guaranteeing that l, τ and v are physically meaningful quantities (see discussion in Chap. 5).

Recently, several works made within our research group have investigated this question and expressed some doubts about the validity of this paradigm [16–18]. In the context of simulations of a turbulent toroidal plasma in a tokamak geometry, these studies have tested whether the nature of transport across a shear flow remains diffusive or if it becomes something different instead. The results appear to suggest that, in contrast to the aforementioned diffusive picture, *radial turbulent transport across poloidal shear flows become endowed with marked subdiffusive and non-Gaussian features*. The characteristics become more increasingly non-diffusive as the shear in the flow becomes stronger.

Possible physical mechanisms responsible for the emergence of both subdiffusion and non-Gaussianity have also been identified [18]. Subdiffusion appears to be a consequence of the differential effects that the shear flow has on eddies with positive or negative vorticity (see Fig. 9.2), as we discussed earlier. Since only eddies of the same vorticity as that of the shear flow are enhanced, while the others are suppressed, the resulting vorticity landscape in which transport across the flow takes place is dominated by a single local rotation. This situation leads to subdiffusive transport across the flow due to the fact that reversing the direction of transport becomes then more probable than staying on course. The reasons are more clearly visualized in two-dimensional turbulent simulations than in three-dimensional toroidal runs. For instance, Fig. 9.4 shows the vorticity of a drift-wave turbulence [19] simulation carried out on a period two-dimensional slab in which the magnetic field is perpendicular to the plane using the BETA code [8, 20]. Being two-dimensional, the local vorticity of the flow (akin to local rotation, as we mentioned earlier) always points in the perpendicular direction to the plane of the simulation. In the figure, red is used for positive vorticity, blue for negative. The right frame shows a situation without any significant shear flow. In it, vortices of both sign are clearly seen, with a rather uniform distribution in space (and time).

Turbulent convection, in the neighbourhood of any pair of vortices, can reverse its direction or continue its course depending on the relative sign of their vorticity (as shown by means of sketches at the right side of Fig. 9.4). In the absence of shear flow, this leads to an approximate CTRW with a step-size of the order of the eddy size and a timescale of the order of the eddy lifetime. Diffusive behaviour is then expected. In the presence of a shear flow (we use a sinusoidal profile along x, as shown at the top of the same figure), however, the landscape changes significantly, with the majority of space occupied with areas of the same sign of vorticity as the shear flow. Thus, convection is more probable to reverse than to continue on its current course, which is the seed for subdiffusion. The same general picture can be used to understand subdiffusive in toroidal simulations, although some modifications must be made to account for the more complex geometry [18].

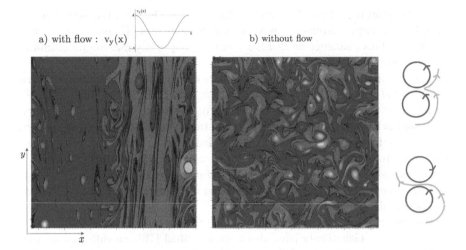

Fig. 9.4 Typical vorticity landscape in two simulations of drift-wave plasma turbulence in a two-dimensional periodic slab. In the absence of a shear flow (right), positive (in red) and negative (blue) vorticity are distributed intermittently in space and time, without any preferred pattern. However, in the presence of the shear flow $v_y(x) = A\cos(x)$, $A > 0$, the resulting vorticity is dominantly negative (blue) on the left half of the slab, and positive (right) on the right half due to the action of the shear flow on the background turbulence. As a result, transport across the shear flow (i.e., along x) tends to reverse its direction more often compared with the case in which there is not a preferential vorticity sign, as shown with the two sketches shown at the right side of the figures. The successive accumulation of these reversals leads to subdiffusion along x [18]. In a three-dimensional toroidal plasma, a similar argument can be made using the component of the vorticity parallel to the magnetic field

It is interesting to remark that, in the case just described, the observed subdiffusive behaviour across the shear flow is not, in itself, a complex behaviour.[12] Instead, it is much more similar to the kind of transport that takes place in a fractal landscape, as might be the case of water percolating through a bed of porous rocks. Subdiffusion is, in these cases, a mere reflection of the structure of the underlying landscape and not a dynamical response.[13] Indeed, transport across externally-driven shear flows have also been shown to behave subdiffusively [16].

The reasons for transport across shear flows to exhibit non-Gaussian features are, however, quite different. In the context of the same tokamak plasma simulations, its origin could be traced to a complex, nonlinear interaction between local turbulent fluctuations and local fluctuations of the shear flow amplitude [18]. The dynamics between them was found to be very reminiscent of a predator-prey type

---

[12]At least, not in the sense of cases such as the near-marginal transport discussed in Chap. 6, where transport was a consequence of the complex dynamics taking place in the system.

[13]Whether the origin of the fractal structure or the shear flow is due to complex dynamics or not, is another matter. In the case of the self-generated edge transport barrier that appears in tokamaks, it certainly is!

of interaction, with the shear flow being the predator and the local fluctuations, the prey. The situation is more or less, as follows: whenever there is a location where fluctuations become larger (or smaller) than the average, they will tend to drive[inhibit] the local shear flow via the Reynolds stresses. The action of this reinforced[weakened] local shear flow on the local fluctuations will shear and stretch them, whilst at the same time pushing them to nearby radial locations. As a result, the local levels of turbulence at these nearby locations may increase[decrease], and the whole process repeats itself, propagating radially in avalanche-like manner. The radial propagation of the leading fluctuations and pursuing shear flows were found to lack any characteristic spatial scale, which leads to the observed non-Gaussian (i.e., Lévy-like) features.

9.4 Case Study: Transport Across Self-Consistent Zonal Flows in Ion-Temperature-Gradient (ITG) Tokamak Turbulence

We will investigate the nature of transport across shear flows in simulations of (ITG) tokamak turbulence by means of (some of) the analysis tools introduced in Chap. 5. In particular, we will illustrate the use of *propagators* as tools to characterize the nature of transport (see Sect. 5.4.1). All numerical simulations have been carried out using the UCAN2 gyrokinetic code [21, 22] in a simplified tokamak geometry, with circular toroidal cross-sections, with typical tokamak parameters (details can be found in [16]).

UCAN2 is a so-called particle-in-cell code (or PIC) that solves a (gyro-averaged) kinetic equation [23] for the plasma ion distribution function[14] coupled to a simplified Poisson equation. Electrons, on the other hand, are considered to be adiabatic.[15] Once the ion distribution function is known, the plasma density, flow velocity and temperature can be easily obtained in the usual way, by taking velocity moments [25]. The Poisson equation, on the other hand, provides the temporal evolution of the electrostatic potential consistent with the changing charge distributions that result from the ion and electron motions. An illustration of the output of a UCAN2 simulation is provided in Fig. 9.5, that shows four different instants of the evolution

---

[14]To be more precise, UCAN2 is a so-called global δf code, that assumes closeness to an equilibrium distribution f_0 that contains the plasma equilibrium profiles (i.e., plasma density, flow and temperature), so that only the deviations from f_0 are followed in time. These type of setups are very useful to evolve turbulence over a large domain as long as the variations of the profiles are not too large. For that reason, they would not be good for turbulence simulations in near-marginal conditions, where intense profile evolution is expected at the fluctuation scales. The approach is however adequate for the kind of study carried out here.

[15]This means that electrons are considered to be extremely mobile, being able to move along the field lines to provide any force balance required. This approximation importantly simplifies the simulations, since the electron distribution function does not need to be computed [24].

Fig. 9.5 Successive snapshots of electrostatic potential fluctuations in the UCAN2 toroidal simulations showing the development of a strong poloidal shear flow (second frame from above), that breaks the linear ITG structures (top frame), and shears eddies reducing their typical size, as shown in the last two frames. The last frame corresponds to the starting time considered for the propagator studies carried out in this chapter to determine the nature of radial transport, that is already well inside the nonlinearly saturated phase

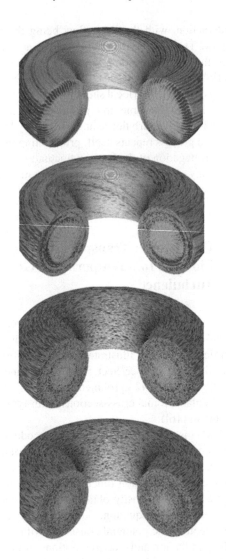

time

(time increases from top to bottom) of the electrostatic potential for a run in which the dominant instability was the so-called ion-temperature-gradient (ITG) mode.[16] The sequence of frames selected is particularly interesting since it shows, in its second frame, that an intense poloidal shear flow is being self-driven by the turbulence and that it shears apart the linear turbulent structures present in the top frame. The final result, at saturation, is a turbulent state with much smaller turbulent structures (see last frame).

---

[16]ITG instabilities are thought to be responsible for a large fraction of the radial ion heat transport in tokamaks [19].

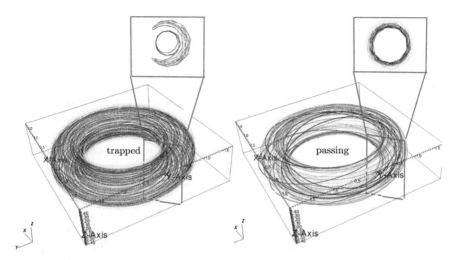

Fig. 9.6 Typical gyro-averaged ion trajectories from the UCAN2 toroidal simulations studied here. **Left:** trajectory of a trapped ion gyrocenter in 3D (below), and its intersection (above) with the toroidal plane $\phi = 0$. **Right:** trajectory of a passing ion gyrocenter in 3D (below) and intersection with the $\phi = 0$ plane (above). Passing particles are identified by circular intersections, whilst trapped particles exhibit banana-like shapes

The fact that UCAN2 is a PIC code is particularly useful for us. The reason is that PIC codes solve the gyro-averaged ion kinetic equation by means of *marker particles* [26]. That is, they follow the trajectories of selected ion gyrocenters in phase space,[17] from which the gyro-averaged ion distribution function can be easily built, instead of discretizing the phase space on which the gyro-kinetic equation must be solved using a discrete mesh. The spatial part of these trajectories correspond to the Lagrangian (gyro-averaged) trajectories that the gyrocenters of actual ions would follow in the toroidal domain of the simulation (see Fig. 9.6). Two main types of gyrocenter trajectories appear in these simulations: **passing** and **trapped**. The distinction is due to the fact that the magnetic field is not uniform in strength, but has a strong (decaying) dependence with the major radius of the torus

---

[17]The usual phase space in which any kinetic equation is solved has six dimensions, three spatial ones corresponding to position and another three that correspond to the velocity vector. In gyro-kinetics, however, the averaging over the gyro-motion eliminates one velocity dimension [23]. The velocity part of phase space is thus reduced to two dimensions, one for $v_\perp$ and another for $v_\parallel$, respectively the perpendicular and parallel components of the velocity with respect to the local magnetic field. $v_\perp$ is no longer a vector since the gyro-phase, that gives its orientation, is gone after the averaging.

(i.e., $B \propto 1/R$). Since the gyro-averaging process makes the magnetic moment $\mu = mv_\perp^2/2B$ of each ion to be conserved [23] and since its energy, $E = m(v_\parallel^2 + v_\perp^2)/2$, is also conserved in the motion,[18] it follows that the parallel velocity of the ion gyrocenter with respect to the magnetic field is given by $v_\parallel^2 = 2E/m - \mu B$. As a result, if an ion gyrocenter moves in a region where B is increasing, it may reach a position where $\mu B > 2E/m$, forcing the particle to reverse its parallel direction.[19] This usually happens, for ions with sufficiently small energy, in the inside part of the torus as illustrated in Fig. 9.6. This distinction will be important in what follows.

PIC codes have advantages and disadvantages, as any other numerical method does. For us, however, the fact that they employ marker particles is great news since we can easily use their trajectories to build the kind of Lagrangian diagnostics discussed in Chap. 5 to characterize the nature of turbulent transport in the simulations. In particular, we will use them to build **radial propagators** (discussed in Sect. 5.4.1). The case we will study corresponds to a UCAN2 simulation in which the dominant instability is the ion-temperature-gradient mode. The geometry is a torus of major radius $R = 1.7$ m and minor radius $a = 0.4$, with a magnetic field value at the magnetic axis $B_0 = 1.87$ T. The actual profiles of the rotational transform, density and temperature profiles considered can be found elsewhere [16], since we will consider here the same simulation data used in that paper.[20] For the purpose of this discussion, it suffices to say that the plasma profiles are such that they are well above the local instability threshold for ion-temperature-gradient modes, thus excluding the possibility of any near-marginal dynamics. These choices allow us to focus exclusively on the effect of the shear flow on radial transport. The simulations have been run for a long time (of the order of a few milliseconds, that must be compared with the local turbulent timescales, of the order of a few microseconds), sufficient for shear flows to develop and saturate with the turbulence (see Fig. 9.7). It is only for times beyond the saturation time (roughly given by t_0 in Fig. 9.7) that the nature of radial turbulent transport is examined. It is also worth mentioning that the poloidal zonal flows that are established at the steady state of this simulation extend over the whole radius, as shown in the right inset of Fig. 9.7, which gives us ample radial space to run the diagnostics.

---

[18]We use the common practice of denoting $v_\perp$ and $v_\parallel$ as the components of the ion velocity perpendicular and parallel to the local magnetic field.

[19]When all other drifts are considered, in addition to parallel motion, the projection of the trajectory of any trapped particle on any toroidal cross section has a crescent moon or banana shape. For that reason, trapped ion orbits are usually known as banana orbits.

[20]The same goes for the numerical details of the simulation, such as the number of cells considered, the number of particles per cell included or the time resolution used.

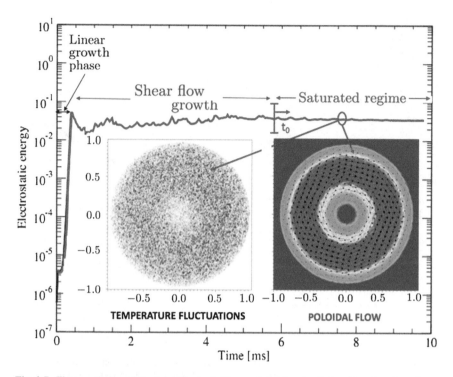

Fig. 9.7 Time evolution of the total electrostatic energy of the simulation (namely, the volume integral of $|\tilde{\phi}|^2$) showing three distinct phases: (a) the linear ITG growth phase (up to 0.3–0.4 ms); (b) the slower growth phase that corresponds to the build-up of the shear flow and its saturation with the background turbulence (from 0.4 to 4.5 ms); (c) the quasi-steady phase (for times larger than $t_0 \simeq 5.7$ms). The characterization of transport is done over the quasi-steady phase. A snapshot of the potential fluctuations and the poloidal flow at the $\phi = 0$ toroidal cross-section are shown for the instant marked with a circle

9.4.1 Radial Propagator Analysis

A propagator, $G(x, t|x_0, t_0)$, is just the probability of finding a particle at position x and time t if it was previously at position x_0 at time t_0. We talked about propagators lengthily in Chaps. 3 and 5. In Sect. 5.4.1, we also discussed how propagators can be used to characterize the nature of transport. In the simplest implementation of these methods, one simply took advantage of the fact that their asymptotic behaviour is known for model transport equations of special significance, such as the classical diffusive equation (Eq. 5.3) or the fractional transport equation (Eq. 5.66), and use this knowledg to estimate the values of the fractional transport exponents (using, for instance, Eqs. 5.89 and 5.82).

However, in order to illustrate the full power of using propagators, we will fit the evolution of the full numerical ion gyrocenter propagator to that of the propagator of the fractional transport equation (Eq. 5.60) using the chi-square minimization

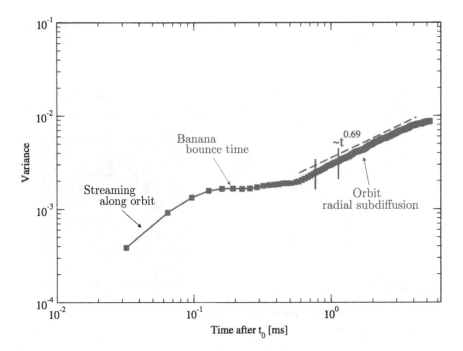

Fig. 9.8 Variance of the ion radial propagators as a function of the elapsed time, as measured from t_0. The times corresponding to the propagators in Fig. 9.9 are marked with vertical red lines

procedure described in Sect. 5.4.1. The outcome of the minimization will yield the values of the two fractional exponents, α and β, that define the fractional transport equation that provides the best effective description for the process. The numerical propagator is obtained by considering the evolution in time of a sub-population of gyrocenters that is initially localized around the radial location $r/a \simeq 0.4$, sufficiently close to the center of the simulation box to minimize finite-size effects. In order to improve the statistics, all gyrocenters initially localized in the range $r/a \simeq 0.39$–0.41 were used, amounting to approximately 75,000 of the 4 million particles used in the UCAN2 simulation. The propagators are then constructed, at every time step up to a maximum elapsed time of 5 ms, by calculating the pdf of their radial displacement with respect to their initial condition. The initial time was chosen to be $t_0 = 5.7$ ms to ensure that the turbulence was already well within the saturated area.

There is however an issue that must be dealt with before the minimization that will yield the exponents can be carried out. Figure 9.8 illustrates the problem by showing the variance of the propagator as a function of the time lag measured from t_0. Clearly, a subdiffusive scaling (the variance would scale linearly for a diffusive process) is exhibited at time lags longer than half a millisecond, which is longer than the turbulent characteristic timescale (of the order of tens of microseconds). At shorter times, however, motion is not just ballistic as one would expect from self-correlation. Instead, an almost flat scaling region is found at around 0.1–0.2 ms, that suggests that the gyrocenters barely move radially over that timescale. It turns out that his timescale coincides with the **average banana bouncing time**, the time that ion gyrocenters need to complete one banana orbit when they are trapped (see left frame of Fig. 9.6). Since the gyrocenter of any trapped ion comes back to its initial radial location after exploring a radial extension of the order of the banana width, the fact that flattening is observed suggests that the trapped population is significant in these simulations. Since it is long-term radial transport dynamics that we are interested in, we need to adapt the minimization process accordingly to avoid trapping effects.

We proceed as follows to determine the fractional exponents α and β. First, we build a target function according to Eq. 5.86, that includes the difference between the numerical and the analytical propagator as a function of the exponents. In order to circumvent the trapping issue just mentioned, we only considered propagators for $\Delta t > 0.4$ ms in order to calculate the target function. Snapshots of the propagator are shown in Fig. 9.9 for two different values of the elapsed time. A Levenberg-Marquardt minimization of the target function then leads to a value of the spatial fractional exponent $\alpha \sim 1.42$ and a value of the temporal fractional exponent $\beta \sim 0.43$ as the ones that provide the best fit. Therefore, the transport dynamics across the shear flow present in the UCAN2 simulation are strongly non-Gaussian and non-Markovian. The resulting transport exponent is then $H = \beta/\alpha \sim 0.30$, that shows that radial transport across a strong poloidal shear flow is clearly subdiffusive.

9.4.2   Other Considerations

The analysis just described shows that, by means of the tools introduced in Chap. 5, it is possible to show that radial transport across the poloidal shear flow generated by the ITG turbulence in the UCAN2 run is both subdiffusive and non-Gaussian, at least for timescales much longer than the local turbulent decorrelation times (and the banana bouncing time). It does not prove, however, that the poloidal shear flow is the cause of the observed subdiffusion and non-Gaussianity, nor it provides an explanation of why this behaviour takes place. These questions can however be addressed by applying similar techniques. In particular, the role played by the shear flow can be characterized by repeating the same analysis used here on another UCAN2 simulation, with the same parameters, but with the shear flow *artificially zeroed-out at each iteration*. This was done within our group and we found that the resulting radial transport became Gaussian and diffusive. This different behaviour pointed to the shear flow as the main culprit for the observed change in the nature of transport[16]. We also carried out additional UCAN2 simulations that included either stronger or weaker turbulence drives, resulting in shear flows with different shearing capabilities. By means of the propagator analysis of these runs, it was also shown that subdiffusion became more pronounced (i.e., H gets smaller) as the shearing capability of the flow becomes larger [17]. We also found that transport

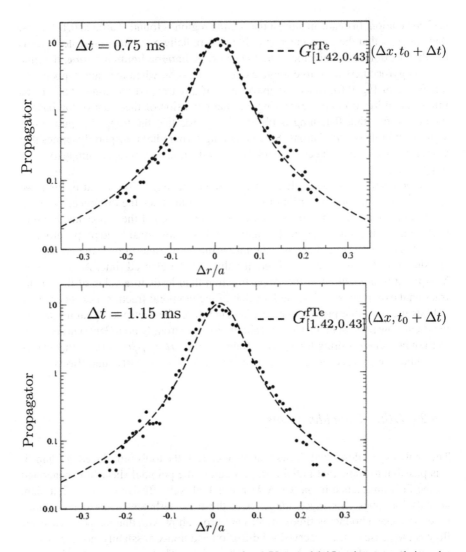

Fig. 9.9 Numerical radial propagators obtained after 0.75 ms and 1.15 ms have passed since the initial time at $t_0 = 5.7$ ms. The best fit found for the data, using a Levenberg-Marquardt algorithm to minimize the target function defined in Eq. 5.86, yields a value of the fractional spatial exponent $\alpha = 1.42$ and fractional temporal exponent $\beta = 0.43$ The associated transport exponent is $H = \beta/\alpha = 0.3$, that is strongly subdiffusive

of a similar nature was observed *independently of the origin of the shear flow* by carrying out simulations that replaced the self-consistent, time-evolving shear flow with a *time-independent, externally imposed shear flow with a radial profile given by the time-average of the self-consistently generated one.* The value of H was then found to be very similar in both cases, although the non-Gaussian features were absent in the externally-driven cases [17]. Finally, the reasons for the appearance of non-Gaussian features was explored and discussed in [18].

9.5 Conclusions

In this chapter we have shown that transport across shear flows may become endowed with non-diffusive features if the shearing capability of the flow is sufficiently strong. We have also proved that the use of propagators can be a very useful tool to investigate the non-diffusive nature of transport in complex systems. In fact, propagators have been used extensively in the literature, particularly in the context of numerical simulations. For instance, without leaving the field of magnetic confinement fusion, propagators have also been employed to characterize transport dynamics in near-marginal conditions similar to those discussed in Chap. 6. The codes used in these studies did not use particle techniques and, therefore, required the introduction of tracer particles [27]. Lagrangian techniques, such as those discussed in Sect. 5.4.2, provide an alternate approach that has also been used in this context [8, 28]. The usefulness of all these methods is by no means restricted to fusion plasmas, though, and many of them have also been applied to many other contexts where complex dynamics are at work [29, 30].

References

1. Chandrasekhar, S.: Hydrodynamic and Hydromagnetic Stability. Dover, New York (1961)
2. Townsend, A.A.: The Structure of Turbulent Shear Flow. Cambridge University Press, Cambridge (1980)
3. Terry, P.W.: Suppression of Turbulence and Transport by Shear Flows. Rev. Mod. Phys. 72, 109 (2000)
4. McIntyre, M.C., Palmer, T.N.: The "surf zone" in the Stratosphere. J. Atmos. Terr. Phys. 46, 825 (1984)
5. Biglari, H., Diamond, P.H., Terry, P.W.: Influence of Sheared Poloidal Rotation on Edge Turbulence. Phys. Fluids B 2, 1 (1990)
6. Newman, D.E., Carreras, B.A., Diamond, P.H., Hahm, T.S.: The Dynamics of Marginality and Self-Organized Criticality as a Paradigm for Turbulent Transport. Phys. Plasmas 3, 1858 (1996)
7. Sanchez, R., Newman, D.E.: Self-Organized Criticality and the Dynamics of Near-Marginal Turbulent Transport in Magnetically Confined Fusion Plasmas. Plasma Phys. Controlled Fusion 57, 123002 (2015)
8. Ogata, D., Newman, D.E., Sanchez, R.: Investigation of the Interaction Between Competing Types of Nondiffusive Transport in Drift Wave Turbulence. Phys. Plasmas 24, 052307 (2017)
9. Dif-Pradalier, G., Hornung, G., Garbet, X., Ghendrih, Ph., Grandgirard, V., Latu, G., Sarazin, Y.: The $E \times B$ Staircase of Magnetised Plasmas. Nucl. Fusion 57, 066026 (2017)
10. Wagner, F., Becker, G., Behringer, K., Campbell, D.: Regime of Improved Confinement and High-Beta in Neutral-Beam-Heated Divertor Discharges of the ASDEX Tokamak. Phys. Rev. Lett. 49, 1408 (1982)
11. Shimada, M., et al.: Progress in the ITER Physics Basis - Overview and Summary. Nucl. Fusion 47, S1 (2007)

12. Wagner, F.: A Quarter-Century of H-Mode Studies. Plasma Phys. Controlled Fusion 49, B1 (2007)
13. Hirsch, M., Amadeo, P., Anton, M., Baldzuhn, J., Brakel, R., Bleuel, D.L., Fiedler, S., Geist, T., Grigull, P., Hartfuss, H., Holzhauer, E., Jaenicke, R., Kick, M., Kisslinger, J., Koponen, J., Wagner, F., Weller, A., Wobig, H., Zoletnik, S.: Operational Range and Transport Barrier of the H-Mode in the Stellarator W7-AS. Plasma Phys. Controlled Fusion 40, 631 (1998)
14. Diamond, P.H., Itoh, S.I., Itoh, K., Hahm, T.S.: Zonal Flows in Plasmas - A Review. Plasma Phys. Controlled Fusion 47, R35 (2005)
15. Weynants, R.R., Van Oost, G.: Edge Biasing in Tokamaks. Plasma Phys. Controlled Fusion 35, B177 (1993)
16. Sanchez, R., Newman, D.E., Leboeuf, J.N., Decyk, V.K., Carreras, B.A.: Nature of Transport Across Sheared Zonal Flows in Electrostatic Ion-Temperature-Gradient Ggyrokinetic Plasma Turbulence. Phys. Rev. Lett. 101, 205002 (2008)
17. Sanchez, R., Newman, D.E., Leboeuf, J.N., Carreras, B.A., Decyk, V.K.: On the Nature of Radial Transport Across Sheared Zonal Flows in Electrostatic Ion-Temperature-Gradient Gyrokinetic Tokamak Plasma Turbulence. Phys. Plasmas 16, 055905 (2009)
18. Sanchez, R., Newman, D.E., Leboeuf, J.N., Decyk, V.K.: Nature of Transport Across Sheared Zonal Flows: Insights from Ggyrokinetic Simulations. Phys. Rev. Lett. 53, 074018 (2011)
19. Horton, W.: Drift waves and transport. Rev. Mod. Phys. 71, 735 (1999)
20. Newman, D.E., Terry, P.W., Diamond, P.H., Liang, Y.M.: The Dynamics of Spectral Transfer in a Model of Drift-Wave Turbulence with Two Nonlinearities. Phys. Fluids 5, 1140 (1993)
21. Sydora, R.D., Decyk, V.K., Dawson J.M.: Fluctuation-Induced Heat Transport Results from a Large, Global Toroidal Particle Simulation Model. Plasma Phys. Controlled Fusion 38, A281 (1996)
22. Leboeuf, J.N., Decyk, V.K., Newman, D.E., Sanchez, R.: Implementation of 2D Domain Decomposition in the UCAN Gyrokinetic PIC Code and Resulting Performance of UCAN2. Commun. Comput. Phys. 19, 205 (2016)
23. Brizard, A.J., Hahm, T.S.: Foundations of Nonlinear Gyrokinetic Theory. Rev. Mod. Phys. 79, 421 (2007)
24. Krall, N.A., Trivelpiece A.W.: Principles of Plasma Physics. McGraw Hill, New York (1973)
25. Balescu, R.: Transport Processes in Plasmas. North-Holland, Amsterdam (1988)
26. Hockney, R.W., Eastwood, J.W.: Computer Simulation Using Particles. Adam Hilger Publishers, Bristol (1988)
27. del-Castillo-Negrete, D., Carreras, B.A., Lynch, V.E.: Fractional Diffusion in Plasma Turbulence. Phys. Plasmas 11, 3854 (2004)
28. Mier, J.A., Sanchez, R., Newman, D.E., Garcia, L., Carreras, B.A.: Characterization of Nondiffusive Transport in Plasma Turbulence via a Novel Lagrangian Method. Phys. Rev. Lett. 101, 165001 (2008)
29. Metzler, R., Klafter, J.: The Random Walk's Guide to Anomalous Diffusion: A Fractional Dynamics Approach. Phys. Rep. 339, 1 (2000)
30. Zaslavsky, G.M.: Chaos, Fractional Kinetics, and Anomalous Transport. Phys. Rep. 371, 461 (2002)

Index

© Springer Science+Business Media B.V. 2018
R. Sánchez, D. Newman, *A Primer on Complex Systems,*
Lecture Notes in Physics 943, https://doi.org/10.1007/978-94-024-1229-1